D1368397

PRENTICE HALL
Biophysics and Bioengineering Series

Abraham Noordergraaf, Series Editor

AGNEW AND MCCREERY, EDS. *Neural Prostheses: Fundamental Studies*

ALPEN *Radiation Biophysics*

GANDHI, ED. *Biological Effects and Medical Applications of Electromagnetic Energy*

LLEBOT AND JOU *Introduction to the Thermodynamics of Biological Processes*

FORTHCOMING BOOKS IN THIS SERIES (*tentative titles*)

COLEMAN *Integrative Human Physiology: A Quantitative View of Homeostasis*

DAWSON *Engineering Design of the Cardiovascular System*

FOX *Fundamentals of Medical Imaging*

GRODZINSKY *Fields, Forces, and Flows in Biological Tissues and Membranes*

HUANG *Principles of Biomedical Image Processing*

MAYROVITZ *Analysis of Microcirculation*

RIDEOUT *Mathematical and Computer Modeling of Physiological Systems*

SCHERER *Respiratory Fluid Mechanics*

VAIDHYANATHAN *Regulation and Control in Biological Systems*

WAAG *Theory and Measurement of Ultrasound Scattering in Biological Media*

Prentice Hall Advanced Reference Series

Physical and Life Sciences

Radiation Biophysics

EDWARD L. ALPEN

Professor of Biophysics
University of California, Berkeley
and Professor of Radiology
University of California School of Medicine,
San Francisco

PRENTICE HALL
Englewood Cliffs, New Jersey 07632

Library of Congress Cataloging-in-Publication Data

Alpen, Edward L.
    Radiation biophysics / Edward L. Alpen.
       p.   cm. -- (Prentice Hall biophysics and bioengineering)
    (Prentice Hall Advanced reference series. Physical and life
    sciences)
       Includes bibliographical references.
       ISBN 0-13-750480-2
       1. Radiation--Physiological effect. 2. Biophysics.
    3. Radiobiology.   I. Title.   II. Series.   III. Series: Prentice
    Hall Advanced reference series. Physical and life sciences.
    QP82.2.R3A55   1990
    574.19'15--dc20   90-6712
                                                                 CIP

Editorial/production supervision
 and interior design: Jane Bonnell
Cover design: Wanda Lubelska Design
Manufacturing buyer: Margaret Rizzi

Prentice Hall Advanced Reference Series

Prentice Hall Biophysics and Bioengineering Series

© 1990 by Prentice-Hall, Inc.
A Division of Simon & Schuster
Englewood Cliffs, New Jersey 07632

Printed in the United States of America
10 9 8 7 6 5 4 3 2 1

ISBN   0-13-750480-2

Prentice-Hall International (UK) Limited, *London*
Prentice-Hall of Australia Pty. Limited, *Sydney*
Prentice-Hall Canada Inc., *Toronto*
Prentice-Hall Hispanoamericana, S.A., *Mexico*
Prentice-Hall of India Private Limited, *New Delhi*
Prentice-Hall of Japan, Inc., *Tokyo*
Simon & Schuster Asia Pte. Ltd., *Singapore*
Editora Prentice-Hall do Brasil, Ltda., *Rio de Janeiro*

Contents

v

Preface

The contents of this text are the result of particularly frustrating experiences for me while teaching a senior course in radiation biophysics at the University of California at Berkeley. The students in the course are, in general, majors in Biophysics or Medical Physics who have sophisticated training in the physical sciences and mathematics, as well as more than adequate skills in biology. In choosing a course text it was necessary to propose to them several volumes, only parts of which would ultimately be used in the course. The choices among texts presently offered are those that are best suited for training radiation diagnostic and therapy physicists, which emphasize the physical interactions of radiation with matter with only a polite nod to the biology and chemistry of radiation interactions, or the radiation biology texts, which, with only infrequent exceptions, assume that a biologist has never seen a differential equation.

It is rare today to find undergraduate majors in biology that have not had fairly sophisticated cross-training in mathematics and the physical sciences. It is also true that many undergraduates in the physical sciences would like to increase their knowledge of the interaction of physical agents with living systems without being affronted by the lack of a quantitative approach to the biology. In a sense, this cross-disciplinary curiosity has been the driving force behind the development of the modern discipline of biophysics. It would be naive of me, however, to assume that all potential users would have the necessary physical science background for the material of the text, and wherever I have felt it necessary, fundamental material at the beginner's level is included.

My aim has been to produce a text that would be useful for advanced undergraduates and graduate students in the appropriate sciences, while at the same time filling a need for a desk-top reference for working scientists in the field. The book should prove useful for health physics students and working health physics professionals, as well as those engaged in the radiological sciences in the health professions, and it should be a useful resource for individuals in related fields such as nuclear power generation and federal and state radiation regulatory activities.

The book is structured for a sixteen-week, one-semester course. It progresses from the physical interactions to the radiochemical interactions to the biological sequelae. Finally, it deals with the late effects of ionizing radiation in organized living systems, in particular, in human beings. Those who use this text will find that they can devote as much or as little time to the physical science chapters as they wish. The biology is pretty much intact on its own.

Chapter 2 on the electromagnetic spectrum and the properties of radiation is essential for all readers who have not had adequate training in electromagnetic radiation and its characteristics. Those who have can easily skip this material. Chapter 5 on energy absorption mechanisms is helpful background for all readers, a review for some and new for others.

I have strived for correctness in all parts of the volume, of course, but particularly in the physical science chapters; it will, I hope, be found that the presentation is rigorous. This presentation will not be entirely complete, however. There has been no effort to present complete derivations of some of the more complicated formulations. The Bethe–Bloch stopping power equation, the Klein–Nishina Compton cross-section calculations, and the Compton scatter energy transfer formulations are examples of these. The final formulation is complete, but the reader is referred elsewhere for complete derivations.

The text has, insofar as possible, used classical Newtonian mechanics to explain physical principles. At times, where relativistic corrections or other quantum mechanical treatments are required, they are given without derivation.

The background expected of the student to effectively utilize this text is mathematics through a first course in differential equations, a good lower-division physics course, including introductory quantum mechanics, organic and/or biological chemistry, and a simple knowledge of cell biology.

Finally, I wish to express my heartfelt thanks to all of my many colleagues and friends who have so helpfully reviewed various chapters and sections of the book.

Edward L. Alpen
Berkeley, California

Introduction:
A Historical Perspective

The technological and medical applications of radiation and radioactivity have become so much a part of our everyday life that it is easy to lose sight of the fact that, viewed historically, all radiation science is recent. The year 1995 will mark the centenary of Röntgen's announcement of the discovery of a "new kind of penetrating ray." The history of the next few months after that announcement is so dramatic, including Becquerel's discovery of natural radioactivity, that it is worth recounting, if for no other reason than to reassure us that, at times, science does indeed make giant strides in a brief period. Unfortunately, the knowledge of the biological effects of ionizing radiation and radioactivity lay unexplored for several decades after Röntgen's initial report.

During 1895, Röntgen was carrying out experiments with electrical discharge in evacuated glass tubes, and he noticed a darkening of photographic plates that lay nearby. This is a story that is well known to all, but the further impacts of his announcement are not nearly so well recognized. On December 28, 1895, Röntgen reported before the Würzburg Philosophical Society his observation that he had discovered a penetrating ray that would darken photographic plates (Röntgen, 1895). Shortly after his Würzburg report was presented, he sent copies of the report, along with samples of what he called x-ray photographs, to the leading physicists in Europe.

On January 20, 1896, two French physicians, Oudin and Barthelemy, submitted an x-ray photograph of the bones of the hand before the Academie Francaise. At the same meeting, Henri Poincaré reported on the paper he had received from Röntgen. Henri Becquerel was present, and, since he was actively working on fluorescence, he became interested and was convinced that in some way the fluorescence produced in the glass wall of Röntgen's Crookes' tube was responsible for the darkening of the photographic plate. What then followed is most remarkable for the intensity of effort and the insight shown by Becquerel. He published the first of four papers (Becquerel, 1896a–d) before the Academy on February 24, 1896, remarking on the relationship of "phosphorescence" and the penetrating rays. On May 18, 1896, he presented the fourth in the series, entitled "Emission of New Radiations by Metallic Uranium," in which he lay claim to discovery of what eventually came to be known as natural radioactivity, only 12 weeks from the first concept to the final conclusion (Becquerel, 1896d). One cannot help but observe that in our time it is unlikely that a paper could even be reviewed in 12 weeks, let alone published.

American scientists had not let Röntgen's discovery go unnoticed, and two individuals, Michael Pupin, a physicist at Columbia University, and Thomas Alva Edison were racing to be the first on this continent to report findings with the new "invisible rays." The race was won by Edison with a publication on February 1, 1896, in the *Century Illustrated Magazine*, only two months after the Würzburg announcement. Edison went on to become a major force in the commercialization of x-ray equipment for medical applications, and it is no accident that General Electric Corporation was an early leader in the marketing of medical x-ray equipment. One of Edison's great contributions to medical radiology was the invention of the fluoroscope. With his usual intensity, Edison paid little heed to the potential hazards of x-rays, and his assistants in particular suffered the consequences. The following quotation is from his biography (Dyer, Martin, and Meadowcroft, 1929).

> When the x-ray came up, I made the first fluoroscope, using tungstate of calcium. I also found that this tungstate could be put into a vacuum chamber of glass and fused to the inner walls of the chamber: and if the x-ray electrodes were let into the glass chamber and a proper vacuum was attained, you could get a fluorescent lamp of several candlepower. I started in to make a number of these lamps, but I soon found that the x-ray had affected poisonously my assistant, Mr. Dally, so that his hair came out and his flesh commenced to ulcerate. I then concluded it would not do, and that it would not be a very popular kind of light; so I dropped it.

The applications for x-rays in medicine were so obvious that there was no stopping this new technology. The first medical x-ray was made in 1896, and history was made in military medicine when portable x-ray machines were deployed with Kitchener's Army of the Sudan in 1898 and were used routinely for diagnostic assistance in traumatic injury. In 1898 the *Medical Record* (New York) carried 28 references on radiological diagnoses and procedures.

Becquerel was to be disappointed for some years that his discovery of natural radioactivity received few accolades and even lesser attention. He did, however, demonstrate that the rays from uranium would ionize gases, and there is a story that I must label apocryphal, since I can find no reliable record of it, that he carried a small vial of the newly discovered radium in his waistcoat pocket and used this radioactive sample to demonstrate that the rays would discharge a gold-leaf electroscope. It also has been reported that he later suffered radiation damage to the skin of his abdomen from this practice, but, again, this fact cannot be reliably confirmed. In any case, J. J. Thomson, the distinguished English physicist, formed an association with a young New Zealander, Ernest Rutherford, and together, in 1896, they reported on the ionizing properties of x-rays (Thomson and Rutherford, 1896). This association of Thomson and Rutherford initiated the latter into the mysteries of the newly discovered radioactivity and radiations.

Rutherford, in association with Owen and others at McGill University in Montreal, was to make a series of discoveries that became central to the future of radiation science. Among these discoveries was the demonstration that the radiations from the naturally occurring sources were made up of three different types of rays, a penetrating ray of great mass that was deflected in an electric field (alpha ray), a second penetrating ray of lesser mass (beta ray), which was also deflected by an electric field, and, finally, a ray that was unaffected by electric fields (gamma ray). His other contributions are mostly known to students of the physical sciences, but probably less well known is the fact that he and Owen were the first to demonstrate the existence of gaseous matter arising from thorium that could itself be radioactive (Rutherford, 1900). This gas was thoron.

For all of Rutherford's vision, he was not infallible. To quote from a speech he made in 1933 before the British Association for the Advancement of Science: "The energy produced by the breaking down of the atom is a very poor kind of thing. Anyone who expects a source of power from the transformation of these atoms is talking moonshine."

The key chapter in this early history of radiation science must be the discovery of radium by Pierre and Marie Curie and the discovery of polonium by Marie after the death of her husband, Pierre. It would be a romantic fiction to assume that Pierre Curie fell victim to the dangerous rays of the substance that he discovered in collaboration with his wife, but he was simply the victim of a carriage accident in the streets of Paris. It is, however, indeed true that both Marie and her daughter were victims of their discovery. Both died of leukemia, which must certainly have been caused by their exposures during the isolation of radium and polonium.

The time from Röntgen's announcement of the discovery of x-rays until Rutherford's (Rutherford and Soddy, 1903) classic paper, which described the transitions among the elements of the natural radioactive series, encompasses only the years from 1895 to 1903, less than a decade. This latter paper of Rutherford's was, in a sense, the realization of an old dream of chemists since it represented transmutation of one element to another. The impact of Rutherford's paper on our understanding of the natural radioactive chains is hard to overstate. In this text, Chapter 3 on serial radioactive decay still uses the notation and formulation set out by Rutherford for the description of decay in a radioactive series as described in his 1903 report.

What about the biology of radiation? What was happening to examine the biological effects of these new rays? Next to nothing, except by sad accident. Pierre Curie carried out a few small experiments on the biological effects of the emanations from radium, and, in particular, he reported before the Academy on his studies on the effects of radium emanations on developing tadpoles. He found that these emanations produced severe developmental abnormalities in the growing tadpoles. Little attention was paid to these findings, and even when early radiologists developed skin lesions and lost fingers there was little attempt to systematically examine the effects of radiation on living systems. It was in 1902 that it was first formally reported that radiation of the skin with x-rays could lead to skin cancer, but even then few scientists were motivated to proceed with the systematic biology of the effects of ionizing radiation.

Radiation biophysics as a quantitative science saw the light of day in the 1920s with Dessauer's (1922) efforts at quantitative investigation of the effects of radiation. These early studies were centered around a statistical analysis of dose–response curves in an effort to understand the mechanisms of radiation action. Except for the truly insightful studies of H. J. Müller in 1927 (reviewed by Müller, 1950) on the mutagenic action of ionizing radiation, little else happened before the war years. One of the great constraints in these early times on the biological investigation of radiation effects lay in the limited ability to quantitate dose. In the immediate postwar years, brought on in part by the atom bomb and its impact, radiation biology and biophysics came into its own with the work of pioneers such as Lea and his colleague Catchside, as well as the German biophysicist, Zimmer. It is interesting to observe that most of the early workers in this field were physicists turning to biology, perhaps because their biological colleagues were slow to enter this difficult field.

The event that probably had the most significant impact on modern radiation biophysics was the publication by Puck and Marcus in 1955 of a new method for the quantitative culture of mammalian cells (Puck and Marcus, 1955). Until this time all radiation experiments that involved the analysis of survival data had to be carried out on prokaryotic cells or yeast. Now, for the first time, target theory, statistical killing models, and repair/recovery mechanisms could be evaluated and analyzed on mammalian cell lines. The enormous impact of this new methodology can be measured by examining Chapter 8 of this text. In particular, the proposal of a model for repair of sublethal damage by Elkind and Sutton (1960) revolutionized our thinking about repair processes after ionizing radiation damage.

We are now on the verge of a new cycle of significant discoveries related to the effects of ionizing radiation. The progress in the molecular biology of DNA is providing new tools almost daily for the examination of damage and repair processes in this important biomolecule. If we return to reexamine the field of radiation bioeffects a decade from now, we may find that this decade is in many ways similar to the 1895–1905 period, which was the golden age of discovery for radiation and radioactivity research.

REFERENCES

BECQUEREL, H. (1896a) On the radiation emitted in phosphorescence. *Compt. Rend., Paris* **122,** 420–421 (24 February).

—— (1896b) On the invisible radiations emitted by phosphorescent substances. *Compt. Rend., Paris* **122,** 501–503 (2 March).

—— (1896c) On the invisible radiations emitted by the salts of uranium. *Compt. Rend., Paris* **122,** 689–694 (23 March).

—— (1896d) Emission of new radiations by metallic uranium. *Compt. Rend., Paris* **122,** 1086–1088 (18 May).

DESSAUER, F. (1922) Über einige Wirkungen von Strahlen, I. *Z. Physik.* **12,** 38–44.

DYER, F. L., MARTIN, T. C., and MEADOWCROFT, W. H. (1929) *Edison, His Life and Inventions* (2 vols.). Harper and Bros., New York, pp. 580–583.

ELKIND, M. M., and SUTTON, H. (1960) Radiation response of mammalian cells grown in culture. I. Repair of x-ray damage in surviving Chinese hamster cells. *Radiation Res.* **13,** 556–593.

MÜLLER, H. J. (1950) Radiation damage to the genetic material. I. Effects manifested mainly in the descendants. *Am. Scientist* **38,** 33–40.

PUCK, T. T., and MARCUS, P. I. (1955) A rapid method for viable cell titration and clone production with HeLa cells in tissue culture: the use of x-irradiated cells to supply conditioning factors. *Proc. Natl. Acad. Sci. U.S.* **41,** 432–437.

RÖNTGEN, W. C. (1895) On a new kind of rays. *Sitzgsber. Physik-Med. Ges., Würzburg* **cxxxvii;** also (1898) *Ann. Physik Chem. N.F.* **64,** 1.

RUTHERFORD, E. (1900) A radioactive substance emitted from thorium compounds. *Lond. Edin. Dublin Phil. Mag. J. Sci.* **49,** 1–14.

——, and SODDY, F. (1903) Radioactive change. *Lond. Edin. Dublin Phil. Mag. J. Sci.* **5,** 576–591.

THOMSON, J. J., and RUTHERFORD, E. (1896) On the passage of electricity through gases exposed to Röntgen rays. *Phil. Mag.* **42,** 392–407.

SUGGESTED ADDITIONAL READING

Classics of Science (1964) Volume II. *The Discovery of Radioactivity and Transmutation*, Alfred Romer, ed. Dover Publications, Inc., New York.
GLASSER, O. (1958) *Dr. W. C. Röntgen*, 2nd ed. Charles C Thomas, Springfield, Ill.

1

Radiation Quantities, Units, and Definitions

QUANTITIES AND UNITS

Every specialized field of human endeavor that requires quantitation in its pursuit must establish meaningful descriptions of whatever property or quantity is being measured. Radiation science and radiation biophysics are no exception to this rule. This field is no exception to another general rule: each discipline tends to create a set of its own units, with special definitions. As disciplinary studies tend to expand and overlap with other specialties, conflicts arise among the specially developed units and quantities that each field has invented. The International Committee for Weights and Measures (Comité Internationale des Poids et Mésures) undertook a rationalization of the descriptions of all quantities and units in use in science and attempted to minimize the "special" quantities and units that had proliferated over the years. The new units, described by the abbreviation SI for Système Internationale, have been introduced now in most countries of the world. (See, for example, National Bureau of Standards *Special Publication 330, 1977.*)

Fundamental Units

Every unit for quantitative description of physical quantities can be reduced to measurement of four basic *quantities*. These quantities are then describable by a numerical value and a *unit*. The fundamental quantities and the units chosen to describe them are as follows: *mass*, for which

7

the basic unit is the *kilogram* (kg); *length*, for which the basic unit is the *meter* (m); *time*, for which the basic unit is the *second* (s); and *electric current*, for which the basic unit is the *ampere* (A). The SI system is sometimes referred to as the kms (kilogram–meter–second) system.

Derived Units

Derived units for physical quantities are units that are formed as combinations of any of the fundamental units just described. For example, the quantity, velocity, is the rate of change of distance (length) per unit time, so the derived unit is the ratio of Δl divided by Δt, and the derived unit for the quantity, v, is m s^{-1} (meters per second).

Special Units

The radiation sciences are characterized by a plethora of "special units," that is, units that have a name that is not expressed in the units of fundamental quantities. In 1975 the International Commission on Radiation Units and Measurements (ICRU) recommended the gradual phase-out of these special units and their replacement by SI units to the extent possible. The ICRU proposed two new special units for radiation measurements: the *becquerel*, a unit describing radioactivity, and the *gray*, a unit describing radiation dose. These are defined in detail in the following paragraphs (International Commission on Radiation Units and Measurements, 1980).

Several special units defined in the SI system are not unique to the field of radiation measurements, but are important in understanding the units of the field. Some of these special units of interest to us are the following.

- For the quantity *force*, the unit is the *newton* (N), which is defined as 1 kg m s^{-2}.
- For the quantity *work*, or *energy*, the unit is the *joule* (J) = 1 newton meter, defined in SI units as 1 kg m^2 s^{-2}.
- The *electron volt* (eV) is a special unit of energy that is particularly useful in the radiation sciences; its use is so firmly entrenched in the field that it will not be superseded by any new special SI unit. The electron volt (eV) is the energy acquired by an electron of charge e as it falls through a potential difference of 1 V. The electron charge e is 1.602×10^{-19} C. One electron volt, then, is 1.602×10^{-19} C $\times$ 1 V, which is 1.602×10^{-19} J. The multiple, MeV, is 1.602×10^{-13} J.
- For the quantity *power* (the rate of doing work), the unit is the *watt* (w) = 1 joule per second, defined in SI units as 1 kg m^2 s^{-3}.

- For the quantity *frequency*, the unit is the *hertz* (Hz), for which the SI unit is s^{-1} (oscillations per second).
- For the quantity *charge*, which is the product of current and time, the unit is the *coulomb* (C), for which the SI unit is A s.
- For the quantity *potential difference*, which is the quotient of work divided by charge, the unit is the *volt* (V). The SI unit is A kg m^2 s^{-3}, but this is such a cumbersome description that it is more informative to describe the unit volt as the quotient of joules divided by charge (J/C).
- The quantity *capacity* describes the ability to store charge and has the unit *farad*, which is described as the ratio of charge to potential (C/V).

Before entering into a discussion of radiation quantities and units, we must emphasize the importance of developing these concepts early in the text for future use. Certainly the reader will have some concern as to the significance of the concepts of quantities and units developed here, but each of these concepts will quickly be seen to be relevant. It is suggested that their early consideration is useful, but the reader should return as often as necessary to this chapter to refresh the mind as to the significance and definition of each quantity and unit as it appears in the body of the text.

RADIATION MEASUREMENT

Definitions

Directly ionizing particles.   Directly ionizing particles are charged particles having sufficient kinetic energy to produce ionization by collision. This energy certainly must be greater than the minimum electron binding energy in the medium in which the interaction is taking place.

Indirectly ionizing particles.   Indirectly ionizing particles are uncharged particles that can liberate ionizing particles through kinetic interaction with the medium or that can initiate a nuclear transformation. For example, neutrons can interact with the medium to produce high-kinetic-energy protons or atomic nuclei.

Gamma rays and x-rays.   Gamma rays and x-rays are electromagnetic radiations, that is, photons, of high enough energy to produce ionization. Gamma rays are identical to x-rays in their physical properties, but, by convention, it has become the practice to call ionizing photons produced in "machines" x-rays, while ionizing photons from radioactive sources are called gamma rays.

Quantities and Units

Exposure, X.   The unit of exposure, X, is taken as the quotient of ΔQ by ΔM, where ΔQ is the sum of electrical charges on all the ions of one sign that are produced in air when all the electrons liberated by photons in a volume element of air, the mass of which is ΔM, are completely stopped in air.

$$X = \frac{\Delta Q}{\Delta m} \qquad (1\text{-}1)$$

The special unit of *exposure* that predates the new SI system is the *roentgen* (R). There is no SI unit for exposure, and the old special unit, roentgen, is falling into disuse, but occasionally it is still encountered, particularly in the older literature.

Dose (absorbed dose), D.   Absorbed dose, D, is defined as the quotient of ΔE_D by Δm, where ΔE_D is the energy imparted by ionizing radiation to the mass, Δm, of matter in a volume element.

$$D = \frac{\Delta E_D}{\Delta m} \qquad (1\text{-}2)$$

The special unit of absorbed dose widely used until 1977 was the *rad*: 1 rad = 100 erg/g. The plural unit is also rad, for example 1 rad or 20 rad. The rad is still widely used even though a new special unit has been introduced. This new special unit defined in the SI system is the *gray* (Gy). The gray is defined as the deposition of 1 joule in 1 kilogram: 1 Gy = 1 J kg^{-1}. The gray is slowly supplanting the rad in modern literature. Every attempt has been made in this text to provide both units of dose. The interconversion of the rad and the gray is simple since 1 gray = 100 rad. It was immediately obvious to many radiation scientists that 1 cGy was equal to 1 rad, and, as a result, the centigray has been popularly adopted for scientific communication.

Energy imparted, ΔE_D.   Energy imparted, ΔE_D, is the difference between the sum of the energies of all the directly and indirectly ionizing particles that have entered a volume element (ΔE_E) and the sum of the energies of all those that have left it (ΔE_L), corrected for any changes in rest mass (ΔE_R) that took place in nuclear or elementary particle reactions within the same volume element.

$$\Delta E_D = \Delta E_E - \Delta E_L - \Delta E_R \qquad (1\text{-}3)$$

The unit for energy imparted, ΔE_D, is the gray or the rad.

Dose Equivalent, H.   The dose equivalent, H, is the product of absorbed dose (D), quality factor (Q), dose distribution factor (N), and any other necessary modifying factors.

$$H = D(Q)(N) \tag{1-4}$$

The special unit for H, the dose equivalent, depends on the special unit used for D. For example:

1. If D is given in *rad*, then the special unit for H is *rem*, and the dose in rem is numerically equal to the dose in rad multiplied by the appropriate modifying factors, Q and N. There is no abbreviation for the unit rem.
2. If D is given in *gray*, then the special unit for H is the *sievert*, and the dose in sievert is numerically equal to the dose in gray multiplied by the same appropriate modifying factors. The abbreviation for sievert is Sv.

Relative biological effectiveness, RBE.   The RBE of a particular radiation is the ratio of the absorbed dose of a reference radiation (often taken to be ^{60}Co gamma rays), D_R, to the absorbed dose of the particular radiation being examined, D_X, required to attain the same level of biological effect:

$$RBE_X = \frac{D_R}{D_X} \tag{1-5}$$

The term RBE is generally used in experimental work in radiation biology and related fields. It is related to, but not identical with, the quality factor mentioned in the discussion of dose equivalent. For the definition of RBE, it is necessary to explicitly define the level of biological effect and the nature of the biological end point used in its definition. The quality factor, on the other hand, is widely used in health protection and has no such constraints put upon it.

Particle fluence, Φ.   Particle fluence, Φ, is defined as the quotient of ΔN by Δa, where ΔN is the number of particles that enter a sphere of cross-sectional area a. The definition requires a three-dimensional interpretation, since the number of particles is defined as those entering a sphere of cross section a, and therefore the direction from which the particles come is not a limiting condition.

$$\Phi = \frac{\Delta N}{\Delta a} \tag{1-6}$$

Particle flux density, ϕ.   The particle flux density, ϕ, is the time rate of particle fluence. It is defined as the quotient of $\Delta\phi$ by Δt, where $\Delta\phi$ is the particle fluence for the time interval Δt.

$$\phi = \frac{\Delta\Phi}{\Delta t} \tag{1-7}$$

Particle fluence is dimensionless, since it is a number only. Particle flux density is number per unit time and has the dimension, t^{-1}.

Energy fluence, Ψ.   Energy fluence, Ψ, is related to the particle fluence by the average energy of all the particles included in Φ. The energy fluence is, however, defined as the quotient of ΔE_f by Δa, where ΔE_f is the sum of the energies, exclusive of rest energies, of all the particles that enter a sphere of cross-sectional area Δa. For the same reason as for particle fluence, the three-dimensional definition precludes a constraint as to direction of arrival of the contributors to energy fluence.

$$\Psi = \frac{\Delta E_f}{\Delta a} \tag{1-8}$$

Energy flux density, ψ.   Energy flux density bears the same relationship to particle flux density as does energy fluence to particle fluence. Energy flux density, ψ, is defined as the quotient of $\Delta\Psi$ by Δt, where $\Delta\Psi$ is the energy fluence in the time Δt.

$$\psi = \frac{\Delta\Psi}{\Delta t} \tag{1-9}$$

Kerma, K.   Generally, when charged particles or photons interact with their environment, they give up kinetic energy to the environment. However, not all the kinetic energy transferred may stay in the volume of interest, since there can be radiative losses and kinetic energy losses associated with the secondary particles produced. The term *kerma* has been formulated to account for the energy transferred to the volume through various processes without correcting for any energy losses after interactions. Kerma is defined as the quotient of ΔE_K by Δm, where ΔE_K is the sum of the initial kinetic energies of all the charged particles liberated by indirectly ionizing particles or photons in a volume element of the specified material.

$$K = \frac{\Delta E_K}{\Delta m} \tag{1-10}$$

Kerma has the dimensions of energy and will use the SI unit, gray, or the older special unit, rad.

Linear energy transfer, LET. Linear energy transfer is defined for charged particles in any medium as the quotient of dE_L by dl, where dE_L is the average energy locally imparted to the medium by a charged particle of specific energy traversing a distance dl. As the reader will see in Chapter 14, LET is a restricted version of the "stopping power" from physics. The important difference between stopping power and LET is that the former is defined in terms of the thickness of the attenuating medium, while the latter is defined in terms of the track length described by the particle as it traverses the medium. In general, the track will be more or less random through the medium, and track length for LET will not be synonymous with thickness of absorber. The dimensions of LET will be energy per unit track length.

Charged particle equilibrium, CPE. Charged particle equilibrium, CPE, exists at a point p, centered in a volume, V, if each charged particle carrying a certain energy out of V is replaced by another identical charged particle that carries the same energy into V. If CPE exists at a point, then $D = K$ (dose = kerma) at that point, provided that bremsstrahlung production by secondary charged particles is negligible. Remember that dose is energy absorbed in unit volume of the medium, while kerma is energy transferred from the original particle or photon in the same unit volume.

RADIOACTIVITY MEASUREMENTS

Decay constant, λ. The decay constant of a radioactive nuclide in a particular energy state is the quotient of dP by dt, where dP is the probability of a given nucleus undergoing a spontaneous nuclear transition from that state in a time interval dt.

$$\lambda = \frac{dP}{dt} \tag{1-11}$$

The dimension of the decay constant is t^{-1}.

As will be seen in Chapter 2, the more usually used unit to describe the rate of radioactive disintegration is the *half-time* or *half-life, $T_{1/2}$.* The relationship between the decay constant and the half-life will be developed in Chapter 2; and it is

$$T_{1/2} = \ln\frac{2}{\lambda} \tag{1-12}$$

and the dimension of the half-life is time, t.

Activity, A.   The activity of a radioactive nuclide at a given time is the quotient of dN by dt, where dN is the expectation value (most likely) number of spontaneous nuclear transitions that will take place in the time interval dt.

$$A = \frac{dN}{dt} = -\lambda N \qquad (1\text{-}13)$$

Since activity is the number of disintegrations per unit time, the dimension of activity is t^{-1}.

The unit for the activity of a radionuclide, as with the rad or gray, is in transition as the result of the introduction of the new special unit in the SI system. The new SI special unit for the activity of a radionuclide is the *becquerel* (Bq). One becquerel = one disintegration per second. It has the dimension of t^{-1}.

The older special unit is the *curie* (Ci): 1 Ci = 3.7×10^{10} s^{-1}. The curie was originally derived from the measured activity of a sample of ^{226}Ra, so the value of the curie fluctuated as the purity of the element was improved. Several decades ago the curie was arbitrarily established as the disintegration rate just given, 3.7×10^{10} s^{-1}. The initial definition of the curie was given in terms of "that amount of any radionuclide such that it gives rise to 3.7×10^{10} disintegrations per second." However, that definition would imply a mass-related dimension. That is not to be taken literally.

REFERENCES

INTERNATIONAL COMMISSION ON RADIATION UNITS AND MEASUREMENTS (1980) Report 33, *Radiation Quantities and Units*. ICRU, Washington, D.C.

NATIONAL BUREAU OF STANDARDS (1977) *Special Publication 330*. U.S. Government Printing Office, Washington, D.C.

2

Electromagnetic Radiation: Its Nature and Properties

INTRODUCTION

The perplexing question for the physicists of the nineteenth century was the nature of light. How could energy be transmitted through space, even through a vacuum? The mechanical model for light had already been developed as a wave propagating through space, with all the properties that the mechanical model could assign: frequency, ν, and wavelength, λ, and the relation of these parameters of the wave, $\nu\lambda = C$, where C was a constant and not yet known to be equal to the speed of light. This function formed a perfectly satisfactory basis for the measurement and prediction of the observed properties of light, such as refraction and diffraction, but the mechanical model left unanswered the question of the propagation of the waves through apparently empty space. Many unsatisfactory hypotheses were proposed, among which the "ether" concept was dominant for some years. This concept proposed that an unknown substance existed, even in a vacuum, that supported the propagation of wave motion. This substance was called "ether" for lack of a better term. The medium, according to physicists of that day, had to possess certain elastic properties to support the propagation of wave motion, and the medium interacted with the wave in some way, since the speed of the wave was affected by the properties of the medium through which it was passing. During the latter part of the nineteenth century, an entirely new line of reasoning developed, derived in great part from Maxwell's prediction of the velocity of an electromagnetic wave from first principles and the development of his famous equation, which stated the speed of an electromagnetic wave in a vacuum.

$$C = \frac{1}{\sqrt{\mu_0 \epsilon_0}}$$ (2-1)

where μ_0 is the permeability constant and ϵ_0 is the permittivity constant. The important insight from this equation is that the permeability constant derives from Ampere's law relating electric current and magnetic field strength, and the permittivity constant derives from Coulomb's law relating charge, distance, and force. In a qualitative way, we deduce from these relationships that an electromagnetic wave might be a wave propagated through space that is characterized by a time-varying electric field and a time-varying magnetic field, which are related by the well-known laws of electromagnetism.

A generalization can be made that electromagnetic waves are known to be energy propagated through space, and that one of the simplest models for this generation is an oscillating electric charge. For example, an electron oscillating in a linear path will produce time-varying electric and magnetic fields that are propagated through space and that constitute an electromagnetic wave front. The new theories also suggested that there was no limit on the frequencies that could be generated, and, indeed, the theories predicted the existence of radio waves before they were produced and measured.

Electromagnetic waves exist in a continuous spectrum that extends over many decades of frequency and that includes radiations from the lowest frequencies (radio waves) to the highest, which are of principal interest for this text. These are the x- and gamma rays, which have sufficient energy to ionize atoms with which they interact; they are therefore called *ionizing radiation*. Figure 2.1 outlines the principal regions of the electromagnetic spectrum and highlights the region of ionizing radiation frequencies.

From Figure 2.1, the regular relationship between the frequency ν and the wavelength λ is easily observable. The product of the frequency and the wavelength is a constant equal to the speed of light.

$$C = \nu\lambda$$ (2-2)

The speed of light in a vacuum (and in air) is $2.998 \times 10^8 \, \mathrm{ms}^{-1}$ (for simplicity, $3 \times 10^8 \, \mathrm{ms}^{-1}$). The relationship between frequency or wavelength and the energy of the radiation is not at all as obvious as the simple relationship between frequency and wavelength, but in the early 1900s Planck (1901), following on his further developments of Wien's formula for the spectral radiance of cavity radiators, suggested a most radical departure for the physics of his time. A theoretical explanation for the cavity radiation problem had remained an unsolved problem for many

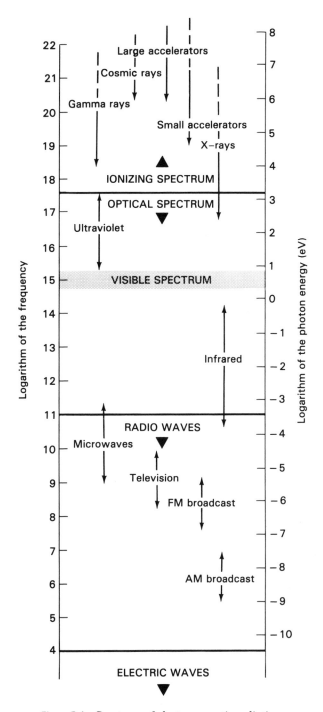

Figure 2.1 Spectrum of electromagnetic radiations.

years. Wien's formulation contained arbitrary constants and was only a moderately good fit to the actual data. In his 1901 paper, Planck proposed a simple modification of Wien's formulation that was a better fit to the data but was, after all, still empirical and required fitting arbitrary constants. He went on to seek a more satisfactory theorem, based on atomic theory as it was then known. He assumed the atoms constituting the radiating surface of the cavity to be electromagnetic oscillators with characteristic frequencies. It was further assumed that these oscillators would absorb and emit electromagnetic radiation from and to the cavity. He also proposed that these electromagnetic oscillators were constrained to certain energy levels. That is, the oscillators could not have *any* energy, but were constrained to energies that were defined by his new formula. In effect he proposed a *quantum theory* of electromagnetic radiation, in which the following assumptions were assumed to be true.

1. An oscillator can have only those energies given by equation 2-3a.

$$E = nh\nu \tag{2-3a}$$

2. The oscillator does not emit radiation continuously, but rather in "packets" or *quanta*, the energy of which will vary as the irradiator moves from one *quantum* level to another. This quantity, ΔE, can then be described as follows:

$$\Delta E = \Delta n(h\nu), \quad \text{and when } \Delta n \text{ is } 1, \quad \Delta E = h\nu \tag{2-3b}$$

QUANTUM THEORY OF ELECTROMAGNETIC RADIATION

As suggested, Maxwell's classical theory of electromagnetism proposed that energy is emitted in the form of electromagnetic waves when an electric charge is accelerated. Also, a charge undergoing simple harmonic motion will generate electromagnetic waves with the same frequency as the moving charge. Planck's formulation, for its time a radical and antiintuitive proposal, suggests that the energy emitted by an oscillator is determined by its frequency, ν, a constant, h, and an integer number, n. It says that the oscillator can have only a number of *quantized* energy steps determined by the value of n. The oscillators do not radiate energy in a continuum, but only in quantized steps determined by the quantum number n. Much later developments have shown a minor discrepancy in Planck's formula, which is now known to be as follows:

$$E = (n + \tfrac{1}{2})h\nu \tag{2-4}$$

Planck's constant is 6.63×10^{-34} J s (4.14×10^{-15} eV s). The concept of the quantized nature of an oscillator offended many of Planck's colleagues. They could indeed observe that the energy of, for example, an oscillating spring system or an electrical RCL circuit could have any value in a continuum. However, readers can prove for themselves that there is indeed no conflict with the observed behavior of oscillators in macro systems. The quantum numbers of such oscillators will be enormous, and the "jumps" in energy will not be apparent (see Problem 1 at the end of this chapter). Planck presented his concepts to the Berlin Physical Society on December 14, 1900. At this time his concept of the radiation emitted was that of an electromagnetic wave. It remained for Einstein to define the concept of the photon, a packet of energy that has both the wave properties of electromagnetic radiation necessary for the understanding of diffraction and refraction and the properties of a "bundle of energy," the energy of which is defined by Planck's law.

Einstein's formulation of the concept of the photon (Einstein, 1906) developed from his study of the photoelectric effect. If monochromatic light is allowed to fall on a metal plate and an electrical field exists, of which the metal plate is the cathode, an electric current may be detected, which is the result of the emission of electrons from the surface of the metal plate. This photostimulated emission of electrons, the photoelectric effect, has certain properties not explainable by classical physical principles. If the potential across the electrodes is reversed at the time of electron emission, a potential, V_0, can be found that is just adequate to stop the current flow. This potential is the *stopping potential*. At the stopping potential the photoelectric current is reduced to zero even though the light intensity remains unchanged. This potential, multiplied by the electron charge, e, is the maximum kinetic energy of the emitted photoelectrons.

$$K_{max} = eV_0 \qquad (2\text{-}5)$$

The value of K_{max} is not influenced by the intensity of the radiation, but depends on the frequency of the radiation. Several observed properties of the photoelectric effect are not consistent with the wave theory of light.

1. Wave theory would predict that K_{max} depends on the intensity of the light source.
2. Wave theory predicts the occurrence of the effect for any frequency of incident light, but experiment shows that there is a lower frequency limit, ν_0, below which no electron emission occurs.
3. There should be a time delay before the emission of the photoelectron while it gathers kinetic energy from its surroundings. Such a delay is not detectable.

Einstein provided the solution for this dilemma by proposing that light travels through space in packets, which he named photons. He proposed that the energy of each photon was the product of its frequency, ν, and Planck's constant, h. Einstein proposed that the photoelectric effect could be described by the following equation:

$$h\nu = E_0 + K_{max} \tag{2-6}$$

where E_0 is the minimum energy required for the electron to escape from its position in the metal surface. The term E_0 is usually called the *work function* for a particular surface and is characteristic of the material. The work function explains the existence of a cutoff frequency for the photoelectric effect. Equation 2-6 explains why the stopping potential is a function of the frequency of the radiation, not the intensity. The nonexistence of a time lag is explained by the delivery of the energy in a bundle, $h\nu$, which is delivered to a single electron in a packet sufficient for the escape of the electron if $h\nu$ exceeds E_0. The kinetic energy of the escaping electron cannot exceed K_{max}, and to the extent that the escaping electron transfers energy to the surrounding medium before its escape from the surface, the electron energy will be reduced.

The fundamental test of the Einstein hypothesis for the existence of the photon is in the independent determination of Planck's constant (Millikan, 1914). This determination is in complete agreement with the determination of Planck's constant using Planck's radiation equation. Einstein received the Nobel prize in physics in 1921 for his studies of the photoelectric effect, not, as is generally believed, for his theory of relativity.

The modern theories of the nature of radiant electromagnetic energy are that it has both the wave properties suggested by Planck and the particulate nature proposed by Einstein.

The quantum nature of electromagnetic radiation is central to the mode of action of this radiation in producing ionization or excitation of molecules in biological systems. If we conceive of photons as delivering packets of energy in small volumes through scattering events, which will be described later, these packets will be in a state where their energy is concentrated around a target molecule. For example, the first ionization potential of water is 12.6 eV. Any event that can deposit this much or more energy in close proximity to the water molecule could produce ionization of that molecule.

MASS–ENERGY EQUIVALENCE: EINSTEIN'S FORMULATIONS

Until the advent of Einstein's theory of special relativity, it was one of the fundamental dogmas of physics that matter could neither be created nor destroyed. His simple equation, which has had such a fundamental effect on physics and certainly everyday life, states, in contradiction to the principle of conservation of mass, that

$$E = mc^2 \qquad (2\text{-}7)$$

This equation states that relativistic mass and energy are equivalent and that they form a single invariant. It furthermore states that units that describe one can be used to describe both. For example, it is common practice to describe the rest mass of atomic and subatomic particles in terms of energy units. For example the rest mass of an electron can be given as 0.51 MeV, where MeV is an energy unit. It is equally possible to assign both mass and momentum unit descriptors for a massless quantity such as a photon.

Does this new concept of energy–mass equivalence violate the fundamental physical laws of conservation of mass and conservation of energy? If we conceive of mass and energy, in the Einstein $E = mc^2$ context, as interchangeable and related by c, the speed of light, then we accept that the sum of mass and energy must be constant, and nothing can be destroyed from this sum. The classical laws can be viewed as special considerations where no interconversion is taking place. It will be seen that for nuclear reactions in particular this interconversion of mass and energy is essential to understanding the processes of nuclear conversion.

RELATIVISTIC CONSIDERATIONS OF MASS AND VELOCITY

In the previous section two terms were used without explanation: rest mass and relativistic mass. In this section the roles of mass and velocity (and, implicitly, energy) are considered. Again, from Einstein's theories it was predicted that as the speed of particles increased their mass would increase. These theories have now been shown experimentally to be correct. When a nuclear particle moves at higher and higher speeds, the mass of the particle increases, and it can be shown that it increases in a simple way. Einstein formulated an expression to state the relationship of mass and velocity.

$$m = \frac{m_0}{\sqrt{1 - (v^2/c^2)}} \qquad (2\text{-}8)$$

The numerator in equation 2-8 is the rest mass of the particle in question, and it is also the mass of the particle when its average velocity is zero. With this consideration, the energy of a particle at rest is m_0c^2; this is sometimes called the *rest energy*. It is obvious from equation 2-8, since the velocity of the particle squared is divided by the speed of light squared, that this fraction will be extremely small until the velocity of the particle is a significant fraction of the speed of light. We can calculate from equation 2-8 that if the particle has a speed equal to one-tenth the speed of light its mass will increase by only 5%. If it has a speed equal to 98% of the speed of light, its mass will increase by a factor of 5. Certainly for everyday values of velocity the relativistic correction is trivial.

For nuclear reactions, radiation scattering, and the world of high-energy physics, where particles are accelerated to very high velocities, an important consideration is the mass correction for kinetic energy associated with the high velocity. The normal expression for kinetic energy is inadequate, and corrections for high-velocity effects on mass must be considered (often referred to in this text and elsewhere as relativistic corrections).

ELECTRIC OSCILLATOR AND ATOMIC STRUCTURE

The discovery of the electron by J. J. Thomson in 1897 introduced a new dimension into explanations of the structure of the atom. Thomson's original model proposed that the positive nuclear charge of the atom was "spread out" more or less uniformly throughout the structure of the atom, with the electrons scattered throughout this charge space. Rutherford reinterpreted the experiments of Geiger (1909) and of Geiger and Marsden (1913) in which they placed very thin metal foils between a source of alpha rays and a series of detectors for determining the trajectory of the particles passing through the gold foil. The alpha particles, which are helium nuclei, were only rarely affected by their passage through the gold foil, and in the few cases where they were scattered from their original trajectory, the deflection was very large. Rutherford (1911) interpreted these experiments such that the structure of the gold foil was one in which the nuclei of the gold atoms were quite far apart, and he further proposed that the internuclear space contained the electrons of the gold atom. From these findings he proposed the planetary model for the atom. In Rutherford's model the nuclear charge is confined to an extremely small spherical volume at the center of the orbital electrons.

This planetary model of Rutherford had a significant shortcoming, which was corrected by an alternate model proposed by Bohr. Since, according to Maxwell's classical laws of electromagnetism, an accelerated charge emits electromagnetic radiation, the orbiting electrons of Rutherford's model would emit radiation continuously, and, since they were losing energy and no energy is being supplied by an external source, they would spiral inward with continuously decreasing orbital radius.

THE BOHR ATOM

Bohr (1913) introduced two novel ideas to elaborate on the simple Rutherford atom. These were nonclassical by the laws of physics of his time. The first assumption was that there were discrete, quantized orbitals for the motion of the electron about the nucleus and that, while in these orbitals, the oscillating electron failed to radiate. This proposal suggests that, rather than a continuously variable emitted energy associated with a slowing down electron, energy, when emitted, would be quantized. The second assumption was that the electron would radiate or absorb energy when it moved from one orbital to another. The energy of the photon absorbed or emitted with the orbital shift would be exactly equal to the energy differential of the two orbital positions. Bohr first explicitly defined these rules for the hydrogen atom, but it was quickly generalized to all atomic structures. The result of these postulates is that any atomic structure of a nucleus and orbiting electrons would have a characteristic, multilevel spectrum for the absorption and emission of radiation that is determined by the energy levels of the orbital electrons. When an atom makes a transition from one energy state, E_1, to a state with a lower energy, E_2, then radiation would be emitted with a quantized energy equal to the difference of the two energy levels.

$$h\nu = E_1 - E_2 \qquad (2\text{-}9)$$

The quantized state of the energy levels of the electron of the hydrogen atom were defined by Bohr as follows: The *angular momentum* of the electron around the nucleus is

$$L = \frac{nh}{2\pi} \qquad (2\text{-}10)$$

where n is an integer ($n = 1, 2, 3, \ldots$).

The energy levels associated with the various quantum numbers n are inversely proportional to the square of the quantum number. To demonstrate this for the general case, assume an electron of mass m, charge $-e$, and a constant uniform orbital speed of v, at a radius of r from a nucleus with charge Ze. Z is the atomic number. Substituting in equation 2-10, then, the angular momentum is

$$m\nu r = \frac{nh}{2\pi} \qquad (2\text{-}11)$$

Since at equilibrium orbit the centripetal force, mv^2/r, on the electron must equal the coulombic attraction, kZe^2/r^2, where k is the constant arising from Coulomb's law and has the value 8.9875×10^9 newton meter^2 per coulomb^2, the orbital radius is

$$r = \frac{kZe^2}{mv^2} \qquad (2\text{-}12)$$

Eliminating v between equations 2-11 and 2-12,

$$r = \frac{n^2 h^2}{4\pi^2 k Z e^2 m} \qquad (2\text{-}13)$$

Simplifying equation 2-13,

$$r = 0.529 \times 10^{-10} n^2/Z \quad \text{meter}$$

For the hydrogen atom the radius of the lowest orbit, $n = 1$ and $Z = 1$, is 0.529×10^{-10} m.

The orbital velocity v can be obtained by eliminating r between equations 2-11 and 2-12.

$$v = \frac{2\pi k Z e^2}{nh} = 2.19 \times 10^6 \frac{Z}{n} \, \text{m s}^{-1} \qquad (2\text{-}14)$$

When $Z = 1$ and $n = 1$, as for the hydrogen atom, the speed of the electron in the lowest energy state or *ground state* is 1/137 of the speed of light. The energy associated with this lowest orbital position is -13.6 eV. The energy required to ionize hydrogen is, then, 13.6 eV.

The permitted values for the energy states of the electrons can be calculated, and from these we can calculate the energy associated with a photon emitted or absorbed by a transition from one level to another. These permitted values are determined as the sum of the kinetic energy (KE) and potential energy (PE) of the electron for any value of n:

$$\text{KE} = \frac{1}{2} m v^2 = \frac{2\pi^2 k^2 Z^2 e^4 m}{n^2 h^2} \qquad (2\text{-}15)$$

and

$$\text{PE} = -\frac{k Z e^2}{r} = -\frac{4\pi^2 k^2 Z^2 e^4 m}{n^2 h^2} \qquad (2\text{-}16)$$

and the sum of these is

$$E_n = -\frac{2\pi^2 k^2 Z^2 e^4 m}{n^2 h^2} \qquad (2\text{-}17)$$

Equation 2-17 simplifies to $E_n = 2.18 \times 10^{-18} \, Z^2/n^2$ J, or

$$E_n = -13.6 \, Z^2/n^2 \quad \text{eV}$$

The wavelength of photons emitted during quantum level transitions can be calculated from Bohr's theory, using the preceding equations. For example, for a transition from $n = 3$ to $n = 2$, the energies will be $-13.6/(3)^2 = -1.51$ eV and $-13.6/(2)^2 = -3.40$ eV, and the transition energy will be $(-1.51) - (-3.40) = 1.89$ eV. The positive sign indicates the radiative release of energy of 1.89 eV.

Bohr's model of the hydrogen atom was an astounding advance and a departure from classical physics. It was in itself, however, inadequate to explain the behavior of more complex atoms. The discovery of the general theories of quantum mechanics in the late 1920s allowed a satisfactory extension to more complex structures.

DE BROGLIE WAVE THEORY

The earlier theories suggested that electromagnetic radiation could possess the properties of waves as well as the properties of particles in the form of the massless photon. It was proposed that the apparently massless photon can also have associated with it the property of momentum, and this momentum, P, is expressed as follows. (It will be particularly important to recall this expression when discussions of scattering interactions of photons with matter arise.)

$$P = \frac{E}{c} = \frac{h\nu}{c} = \frac{h}{\lambda} \qquad (2\text{-}18)$$

In 1921, de Broglie (1921) proposed the reverse aspect of the particulate nature of the photon. He suggested that all particles have the characteristics of an electromagnetic wave in addition to their well-understood particulate nature. These particles in motion could be assigned a wavelength, and, indeed, particles in motion could have wave properties. His suggestion was that the same relationship as equation 2-18 held for particles in motion. That is, the momentum and wavelength are related by Planck's constant. Since the momentum of a particle in motion is determined by its mass and velocity, $P = mv$, the wavelength associated with that particle is

$$\lambda = \frac{h}{mv} \qquad (2\text{-}19)$$

This quantity is the de Broglie wavelength of the particle. In nonrelativistic terms, when the velocity of the particle is small relative to the speed of light, the kinetic energy is given as usual as $E = \frac{1}{2} mv^2$, and we may express the velocity in terms of the kinetic energy thus:

$$\lambda = \frac{h}{\sqrt{2Em}} \qquad (2\text{-}20)$$

The de Broglie wavelength for an electron over a wide energy range approximates atomic dimensions, and therefore these wave descriptions are essential to an understanding of atomic structure.

Bohr's angular momentum quantization formula (2-9) can be rearranged as follows: equation 2-11, $mvr = nh/2\pi$, rearranges to $2\pi r = nh/mv$, and equation 2-19 allows us to substitute λ for h/mv, giving

$$2\pi r = n\lambda \qquad (2\text{-}21)$$

From equation 2-21 it is clear that the circumstances $(2\pi r)$ of the quantized orbits in hydrogen are integral multiples of the de Broglie wavelength. This suggests that the actual orbital picture of the electron circling the nucleus of the hydrogen atom, rather than being a moving point charge, is described best as a standing wave of known wavelength, and that the electronic charge distribution is reflected by this wave function. Because of its wave nature, the position of the electron is not specified within the dimensions of the wave envelope.

REFERENCES

BOHR, N. (1913) On the constitution of atoms and molecules. *Phil. Mag. (Lond.)* **26**, 1–25.

DE BROGLIE, L. (1921) Corpuscular spectra of the elements. *Compt. Rend., Paris* **172**, 527–529.

EINSTEIN, A. (1906) Zur Theorie der Licht Erzeugung und Lichtabsorption. *Annalen d. Physik* **20**, 199–206.

GEIGER, H. (1909) The scattering of alpha particles by matter. *Proc. Roy. Soc. (Lond.)* **lxxxii**, 492–504.

———, and MARSDEN, E. (1913) The laws of deflexion of a particle through large angles. *Phil. Mag. (Lond.)* **25**, 604–621.

MILLIKAN, R. A. (1914) A direct determination of "h." *Phys. Rev. Ser. 2* **4**, 73–75.

PLANCK, M. (1901) Über das Gesetz der Energieverteilung im normalspectrum. *Annalen d. Physik* **4**, 553–563.

RUTHERFORD, E. (1911) The scattering of α and β particles by matter and the structure of the atom. *Phil. Mag. (Lond.)* **21**, 669–688.

SUGGESTED ADDITIONAL READING

JEANS, J. H. (1924) *Report on Radiation and the Quantum Theory*. Physical Society of London. Fleetway Press Ltd., London.

PROBLEMS

1. Assume an oscillating spring that has a spring constant, k, of 20 N m^{-1}, a mass of 1 kg, and an amplitude of 1.0 cm. If Planck's radiation formula describes the behavior of this system, what is the quantum number, n? What is ΔE if n changes by 1? The frequency of a simple oscillator is given by

$$v = \frac{1}{2\pi} \sqrt{\frac{k}{m}}$$

2. An atom is shown to absorb light at a wavelength of 375 nm. It emits light at 580 nm. What was the energy absorbed by the atom from one incoming photon?

3. The rest mass of an electron is equal to 9.11×10^{-31} atomic mass units. If this rest mass is converted into electromagnetic radiation and the electron disappears, what is the wavelength and momentum of the photon that takes its place?

4. A proposed surface to be used in a photoelectric light detector has a work function of 2.0×10^{-19} J. What is the minimum frequency radiation that it will detect? What will be the maximum kinetic energy of electrons ejected from the surface when it is irradiated with light at 3550 Å?

5. In Problem 4, what is the de Broglie wavelength of the maximum kinetic energy electron emitted from the surface? What is its momentum?

6. Using the Bohr equations for the hydrogen atom, assume that the hydrogen atom undergoes a transition from $n = 3$ to $n = 1$. What is the energy, momentum, and frequency of the emitted photon? What is the de Broglie wavelength of the electron after this transition?

3

Radioactivity

INTRODUCTION

Radionuclides are unstable nuclides that achieve greater stability by undergoing nuclear transformations. The nuclide may emit alpha or beta particles when the ratio of neutrons to protons in the nucleus is unfavorable for the state of stability. If, after the emission of the particle, the nuclide is still in an energetically unstable state, then it may emit a gamma ray for transition from the excited nuclear level to the stable ground level with no further change in the neutron to proton ratio.

The nuclear emissions possess high kinetic energy, which is in the range of a few thousand electron volts (keV) to several millions of electron volts (MeV).

UNIT OF RADIOACTIVITY

The intensity of a radioactive source (*activity*) is determined by the rate of nuclear transformations per unit time. The older unit of radioactivity, the *curie* (Ci), was originally defined as the activity associated with 1 g of ^{226}Ra, but that changed as the purity of the nuclide improved. It was finally adjusted to an absolute value of 3.7000×10^{10} disintegrations per second. Submultiples of the curie are the millicurie (1 mCi = 0.001 Ci) and the microcurie (1 μCi = 0.001 mCi).

Because of the change to SI units in 1974, there was need to change the unit for *activity*. For simplicity the special unit was established as the *becquerel*, the value of which was set at 1 disintegration per second.

The units of radioactivity measure only the rate of nuclear transformations and do not deal with the kinetic energy released in the process. The reader is cautioned to observe that disintegration rate is the number of radioactive nuclei undergoing decay per unit time. The number and kind of emissions are not specified. Whether one emission accompanies a disintegration or whether four emissions result from a disintegration, it is counted as a single event for the determination of activity.

LAW OF RADIOACTIVE DECAY

The probability with which a radionuclide decays is, with one trivial exception, a characteristic and immutable constant associated with a particular radionuclide, and this probability cannot be influenced by ambient conditions or their variation. The trivial exception is that tritium gas under very extreme pressures is shown to decrease slightly its decay rate.

The *decay constant* (or transformation constant) λ is the probability of decay of a single radioactive atom per unit of time and is related to the rate of disintegration and the number of radioactive atoms as follows:

$$-\frac{dN}{dt} = \lambda N \qquad (3\text{-}1)$$

N is the number of atoms of the radionuclide in the source, and t is the time (expressed in whatever unit is compatible with the decay constant).

Integration of equation 3-1 yields

$$N = N_0 e^{-\lambda t} \qquad (3\text{-}2)$$

N_0 is the initial number of radioactive atoms in the source, e is the natural base of logarithms, and N is the number of radioactive atoms present at any time t. If we let $N = N_0/2$ and $t = T$ (where T is taken to be the time required for a source to decrease its activity to $N_0/2$), substitution into equation 3-2 gives, on rearrangement,

$$\lambda = \frac{\ln 2}{T} = \frac{0.693}{T} \qquad (3\text{-}3)$$

T is known as the *half-time* or the *half-life* and is conveniently measured from a semilogarithmic graphic plot of the number of radioactive atoms remaining as a function of time.

Definition of Activity

The quantity $-dN/dt$ defines activity as the rate of radioactive decay expressed in disintegrations per second. The activity can be expressed in the following way:

$$A \text{ (in becquerel)} = \lambda N \tag{3-4}$$

$$A \text{ (in curies)} = \frac{\lambda N}{3.700 \times 10^{10}} \tag{3-5}$$

We frequently do not have easy access to the actual activity of a radioactive sample, particularly in the laboratory environment. Under these conditions the term *intensity* is often used, and it is taken to mean the observed rate of measured radiation events.

Mean Life, τ

The actual lifetime of any particular radioactive atom is indeterminate and can have any value from zero to infinity. The average lifetime of a radionuclide atom is an important and definable quantity, predictable on a statistical basis.

If, starting at t_0, we were to measure the individual lifetimes for all the individual atoms to decay, t_1, t_2, t_3, . . . , t_n, sum these times, and divide by the number of atoms present at time t_0, the result would be the *average lifetime* for all the atoms. Expressing this in integral form, the equation for mean life is

$$\tau = -\frac{1}{N_0} \int_{N_0}^{0} t \, dN \tag{3-6}$$

Substituting $-dN = \lambda N \, dt$ and $N = N_0 e^{-\lambda t}$ from the earlier equations and integrating yields

$$\tau = \frac{T}{0.693} = 1.44\,T \tag{3-7}$$

Remember from equation 3-3 that $\lambda = 0.693/T$, so τ is also $= 1/\lambda$. The mean life is one of the quantities useful in calculating radiation dose from a radionuclide.

RADIOACTIVE DECAY OF MIXTURES

The total activity of a source at time t of a mixture of nuclides is the summation of the activities of the individual components.

$$A_{\text{total}} = A_1 + A_2 + A_3 + \cdots \tag{3-8a}$$

$$A_{\text{total}} = \lambda_1 N_1 + \lambda_2 N_2 + \lambda_3 N_3 + \cdots \tag{3-8b}$$

Chain Decay

Chain decay is the process of nuclear transformation in which a radionuclide decays consecutively through a series of radionuclides to a stable nuclide. A simple case of chain decay is the decay of a radionuclide to a second radionuclide, which then decays to a stable element:

$$A \xrightarrow{\lambda_A} B \xrightarrow{\lambda_B} C \text{ (stable)}$$

A and B are usually referred to as *parent* and *daughter*, respectively. The decay of the parent radionuclide and the growth of the daughter radionuclide are related kinetically and can be treated in the following manner.

At any time t_0, the activity of the parent is $\lambda_p N_p$, and the activity of the daughter is $\lambda_d N_d$ if N_p and N_d are the number of atoms of each at time t_0. Then

$$\frac{dN_p}{dt} = -\lambda_p N_p \tag{3-9}$$

and

$$\frac{dN_d}{dt} = -\lambda_d N_d + \lambda_p N_p \tag{3-10a}$$

Equation 3-10a represents that the activity found for the daughter at any time will be equal to the initial amount of daughter corrected for its decay $(-\lambda_d N_d)$, plus the new daughter activity produced by the decay of the parent $(+\lambda_p N_p)$.

Equation 3-10a may be simply rearranged to

$$\frac{dN_d}{dt} = \lambda_p N_p - \lambda_d N_d \tag{3-10b}$$

Integrating equation 3-10b yields a solution for the number of atoms of daughter (N_d) at any time t.

$$N_d(t) = N_p(t_0) \frac{\lambda_p}{\lambda_d - \lambda_p} (e^{-\lambda_p t} - e^{-\lambda_d t}) + N_d(t_0) e^{-\lambda_d t} \tag{3-11a}$$

Equation 3-11a may be reformulated in terms of activity rather than numbers of atoms by the use of the two equations expressing the activity of daughter and parent in terms of N_d and N_p: $A_d = \lambda_d N_d$ and $A_p = \lambda_p N_p$. Such a substitution leads to the following form for the activity of the daughter at any time t:

$$A_d(t) = A_p(t_0) \frac{\lambda_d}{\lambda_d - \lambda_p} (e^{-\lambda_p t} - e^{-\lambda_d t}) + A_d(t_0) e^{-\lambda_d t} \tag{3-11b}$$

Great care must be taken with these two expressions (3-11a, b) since they are quite alike except for the decay constant ratio. In applying the equations we must be sure whether we are computing activity, A, or numbers of radioactive atoms, N. The special case for the application of equations 3-11a and b is that in which the starting activity of the daughter nuclide is zero; that is, there has just been a chemical removal of the daughter from the parent. In that special case the last term of either 3-11a or b disappears.

General Cases for Chain Decay

There are three general cases for formulating chain decay:

1. *Secular equilibrium*, which is established when the half-life of the parent is much greater than the half-life of the daughter; that is, $\lambda_p \ll \lambda_d$, $T_p \gg T_d$. In this case the half-life of the parent is so long that there is negligible decrease of the activity of the parent for the real time of observation.
2. *Transient equilibrium*, which is established when the half-life of the parent is not much larger than the half-life of the daughter; that is, $\lambda_p < \lambda_d$, $T_p > T_d$.
3. *Nonequilibrium*, which exists when the half-life of the parent is smaller than that of the daughter; that is, $\lambda_p > \lambda_d$, $T_p < T_d$.

Let us examine the formulations associated with each of these special cases.

Secular Equilibrium

For the conditions given, and assuming that the activity for the daughter at t_0 is zero, equation 3-11b can be simplified by assuming that $\lambda_d - \lambda_p = \lambda_d$ (that is, $\lambda_p = 0$) and that $e^{-\lambda_p t} = 1$.

$$A_d(t) = A_p(t_0)(1 - e^{-\lambda_d t})   \qquad (3\text{-}12)$$

And since, after several half-lives, $e^{-\lambda_d t}$ will approximate zero,

$$A_d(t) = A_p(t_0)   \qquad (3\text{-}13)$$

At this time the two nuclides are said to be in secular equilibrium and their activities are equal, and this activity is the activity of the parent nuclide.

What these equations demonstrate is that the activity of the daughter is determined solely by the decay constant of the parent. After equilibrium is achieved, the daughter will appear to decay with the very long half-time of the parent. The activities of the two nuclides remain constant because of the very long half-life of the parent. Note that in equation 3-13 the activity of the parent is expressed as the activity of the parent at t_0. Since it is given that the parent has a very long lifetime, no significant change in its initial radioactivity will occur during the period of observation.

Transient Equilibrium

If the half-life of the parent is longer than that of the daughter, but decay of the parent is detectable and significant in the time frame of the observer (starting, remember, with $A_d = 0$), the activity of the daughter will increase until it exceeds the activity of the parent, and then the daughter will decay with a half-time that is the same as the half-time of the parent.

For small values of t, equation 3-11b must be used to compute the activity of the daughter, A_d, but after a sufficiently long time, usually 10 half-lives or more, $e^{-\lambda_d t}$ will be negligible compared to $e^{-\lambda_p t}$. Still assuming A_d to be zero at time zero, equation 3-11b may be rewritten as follows:

$$A_d(t) = A_p(t_0) \frac{\lambda_d}{\lambda_d - \lambda_p} e^{-\lambda_p t} \qquad (3\text{-}14)$$

Since $A_p(t_0) e^{-\lambda_p t}$ is the activity of the parent at any time t, the activity of the daughter is a constant derived from the decay constants of the parent and daughter, times the activity of the parent (after sufficient half-lives have passed).

$$A_d(t) = A_p(t) \frac{\lambda_d}{\lambda_d - \lambda_p} \qquad (3\text{-}15)$$

When the conditions of equation 3-15 are achieved, the two nuclides are said to be in transient equilibrium. After equilibrium is achieved, the activities of parent and daughter are in a constant proportion, which is determined by their respective half-lives. It must be remembered, however, that the activities of both parent and daughter are both decreasing during the period of observation. This transient equilibrium condition is of particular importance to nuclear medicine, since it is possible to prepare a form of the parent attached in a nonremovable state to some substrate. The daughter is produced by decay of the parent, and the daughter can be removed separately from the parent by the use of a suitable eluting agent. One of the most widely used of these is a generator for the production of ^{99m}Tc from its parent, ^{99}Mo.

The ratio of daughter activity to that of the parent at equilibrium is easily calculated from equation 3-15.

$$\frac{A_d}{A_p} = \frac{\lambda_d}{\lambda_d - \lambda_p} \tag{3-16}$$

If this equation is rewritten in terms of half-lives rather than decay constants, the following expression is developed. (The student may wish to do the simple algebra to show that equation 3-17 is correct.)

$$\frac{A_d}{A_p} = \frac{T_p}{T_p - T_d} \tag{3-17}$$

The derivation will not be shown (see Suggested Additional Reading), but the time at which the maximum daughter activity is reached is given by the following expression if the activity of the daughter, $A_d(0)$, is taken to be zero at time zero.

$$t_{max} = \frac{1.44 T_p T_d}{T_p - T_d} \ln \frac{T_p}{T_d} \tag{3-18}$$

Nonequilibrium

If the half-life of the parent is smaller than the half-life of the daughter, the activity curves for the parent and daughter will show an increasing and falling activity for the daughter and an ever decreasing activity for the parent. There are no simplifying assumptions, and equations 3-11a or 3-11b, as appropriate, must be used for all calculations.

Figures 3.1, 3.2, and 3.3 show examples of decay by each of these chain processes.

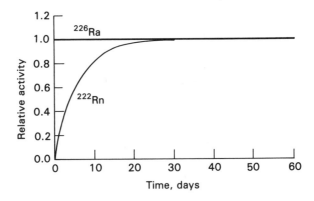

Figure 3.1 *Secular equilibrium*: the growth of radon 222 into a sample of radium 226. The half-life of radon 222 is 3.8 days. The half-life of radium 226 is 1600 years.

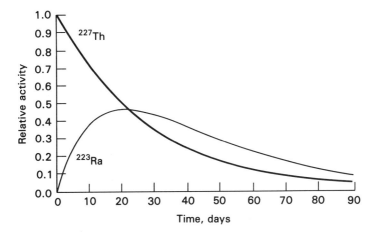

Figure 3.2 *Transient equilibrium*: the growth of radium 223, half-life of 11.43 days, from thorium 227, half-life of 18.5 days. Equations in the text predict a maximum activity for radium 223 at 20.7 days and an equilibrium ratio of daughter to parent of 2.45. The observed peak is at 20 days, and at 90 days the ratio of daughter to parent is 2.20, so transient equilibrium is not yet completely established.

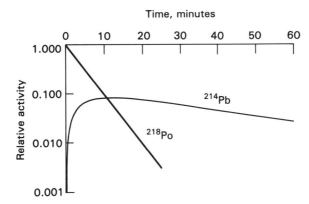

Figure 3.3 *Nonequilibrium*: the decay of polonium 218, half-life of 3.05 min, to lead 214, half-life of 26.8 min. The activity of the lead 214 is relatively low at its peak; therefore, the data are plotted on a logarithmic scale. Note the log-linear decay of the parent and the apparent log-linear decay of the daughter after the parent is exhausted. The half-life for the decay of the daughter at that time is that of pure lead 214.

BRANCHING DECAY PROCESSES

Another possible pathway of complex radioactive decay is that which we refer to as *branching decay*. In this circumstance we postulate that the

parent radionuclide is capable of undergoing *either* of two possible transformations:

$$A \xleftarrow{\quad \lambda_{(C\ to\ A)} \quad} C \xrightarrow{\quad \lambda_{(C\ to\ B)} \quad} B$$

The transformation constant for the decay of C to A is $\lambda_{(C\ to\ A)}$, and the
transformation constant for the decay of C to B is $\lambda_{(C\ to\ B)}$. The total
transformation constant for the radionuclide C is the sum of the individual transformation constants:

$$\lambda_C = \lambda_{CA} + \lambda_{CB} \qquad (3\text{-}19)$$

since, by definition, the half-life of a substance is related to its rate of
disappearance regardless of the mechanism involved. An example of
branched decay is the radionuclide ^{40}K, which undergoes β^- decay (89%)
to ^{40}Ca and transformation by electron capture to ^{40}Ar (11%).

NUCLEAR NOMENCLATURE OF RADIOACTIVE DECAY

- *Atomic number*: The number of protons in the nucleus (Z). This is
  also the net positive charge on the nucleus.
- *Atomic mass number*: The total number of nucleons in the nucleus
  (A). This is not quite the same as the actual mass of the atom of an
  element. Why?
- *Neutron number*: The number of neutrons in the nucleus (N).

The usual notation in describing a nuclide is

$$^A_Z\text{chemical symbol}$$

For example, $^{23}_{11}Na$.

The chemical nature of the element is completely specified by its
atomic number, Z. Therefore, in usual practice the designation of Z and
the elemental symbol is redundant. More usually the following incomplete
notation is seen.

$$^A\text{chemical symbol} \qquad (^{23}Na)$$

Any particular species, that is, for a unique A and Z, is called a *nuclide*,
and if it is unstable, it is referred to as a *radionuclide*.

Isotopes are defined as nuclides having the same Z but different A's.
Obviously they must also have different N's. For example,

$$^{37}_{17}Cl \qquad ^{35}_{17}Cl$$

Isobars are nuclides having the same A's but different Z's. Again
they must obviously have different N's. For example,

$$^{64}_{28}Ni \qquad ^{64}_{30}Zn$$

Isotones are nuclides having the same N's, different Z's, and hence different A's. (Remember that $N = A - Z$.) For example,

$$^{40}_{18}\text{Ar} \qquad ^{41}_{19}\text{K} \qquad ^{42}_{20}\text{Ca}$$

CHARTING OF DECAY SCHEMES

Information regarding mode of decay, decay energy, energy levels of the daughter nuclide, associated gamma ray emissions, and decay constants or half-times is usually embodied in a diagram called the *decay scheme*. In the decay diagram the ground state of the parent is shown as a horizontal line. Since a transformation represents energy release, the product is in a lower energy state, and this fact is represented by arrows in the downward direction.

If the transformation *increases* the atomic number, the arrow slants downward to the *right*. This process is β^- decay. If the transformation *decreases* the atomic number, the arrow slants downward to the *left*. These processes are α decay, β^+ decay, and decay by electron capture (EC). The β^+ decay is differentiated from EC by allowing the arrow to be displaced some distance downward before slanting to the left to signify the mass or energy requirement of the β^+ decay.

With any nuclear transformation, if the newly formed daughter nuclide is in an excited state, γ rays will be emitted. The γ ray emission is indicated by a vertical descending arrow connecting the initial and final nuclear energy levels. If γ rays are cascading between two nuclear levels with some intermediate level, they are said to be in coincidence. Some sample decay schemes are shown in Figures 3.4, 3.5, and 3.6. If several

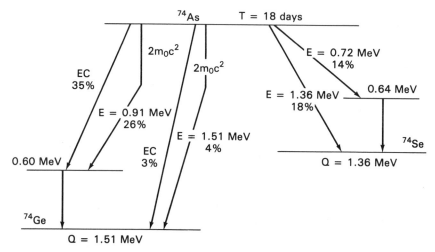

Figure 3.4 Decay scheme for arsenic 74, which has a branching alternative to either germanium 74 or selenium 74.

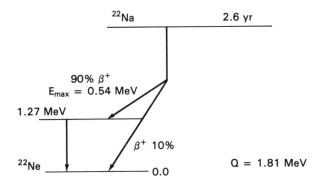

Figure 3.5 Decay scheme for sodium 22, which decays by positron decay to neon 22. Note that the 0.54 MeV shown is the kinetic energy of the positron and does not represent the mass–energy equivalence of two electrons, 0.511 MeV.

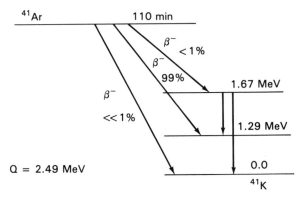

Figure 3.6 Decay scheme for argon 41, which decays by one of three possible beta transformations to potassium 41.

alternative decay processes are competing for the transformation, the fractional contribution of each mode is shown at the side of the downward pointing arrows. These transformation processes will be described in detail in the next section.

NUCLEAR STABILITY

What makes a specific nuclide radioactive? The general rule is that the most stable nuclides are those that have the same number of neutrons and protons in the nucleus. These are rare except at low atomic number. As the atomic number increases, the N/Z ratio generally increases. All elements whose $A > 230$ are unstable, that is, radioactive. For these heavier nuclides the ratio N/Z approaches 1.5. As a general rule, decay processes will tend to lower the N/Z ratio, although individual steps of the decay process may produce an increase in this ratio. The shell model of the nucleus has given rise to a set of *stability rules*, or *even/odd rules*, which tend to be quite predictive. These are as follows:

1. Stable nuclei of even Z are more numerous than those of odd Z.
2. Stable nuclei of even N are more numerous than those of odd N.
3. Stable nuclei of even A are more numerous than those of odd A.
4. In general, stable nuclei of even A have even Z. Some exceptions exist, however, such as ^{2}H, ^{6}Li, ^{10}B, and ^{14}N.

Only one stable structure is known for which Z is greater than N: $^3_2$He.

Here is an example of what happens to the N/Z ratio in a typical alpha decay.

$$^{226}\text{Ra} \Rightarrow {}^{222}\text{Rn} + \alpha + \gamma? + Q$$

$$
\begin{array}{rcc}
Z = & 88 & 86 \\
(A - Z) = N = & 138 & 136 \\
N/Z = & 1.5682 & 1.5814
\end{array}
$$

NUCLEAR MASS AND BINDING ENERGY

Mass Defect

It is a general principle that the most stable state of any system is that with the lowest energy content, which, from the Einstein mass–energy equivalence, would be the state with the least mass and the lowest kinetic energy. As an example consider the nucleus $^4_2$He composed of two protons and two neutrons and known to be stable. According to the stability principle, the mass of the assembled nucleus must be somewhat less than the sum of the masses of its component nucleons. A loss of mass, accompanied by the emission of energy, will usually be found when we assemble nucleons into nuclides. The measure of this loss is called the *mass defect*, Δ, defined as the difference between the atomic mass (the measured mass), sometimes called the isotopic mass, M, and the mass number, A.

$$\Delta = M - A \qquad (3\text{-}20)$$

where Δ will be positive for light nuclei and for those with $A > 180$, and negative for intermediate mass numbers. The *packing fraction*, f, is simply the mass defect per nucleon, multiplied by 10^4 to bring the small numbers to a more convenient size.

Mass Decrement

The *mass decrement*, δ, which is more useful for calculation of transformation energetics, is defined as the difference between the calculated sum of the masses of the constituent nucleons and electrons, W, and the isotopic mass, M.

$$\delta = W - M \tag{3-21}$$

Example

Calculate the mass decrement δ and the mass defect Δ for ⁴He; $Z = N = 2$.

$$W = 2 \times 1.007593 + 2 \times 1.008982 + 2 \times 0.0005487 = 4.034248 \text{ amu}$$

The measured isotopic mass of ⁴He is 4.003873 amu. The mass decrement δ is $4.034248 - 4.003873 = 0.030376$ amu. The mass defect Δ is $4.003873 - 4.000000 = 0.003873$ amu.

The energy equivalent of δ is $0.030376 \times 931.145 = 28.28$ MeV, where the value 931.145 is the energy equivalent in MeV of one atomic mass unit. The mass defect has no meaningful energy equivalent.

This energy, 28.28 MeV, represents the energy that would be released in the synthesis of one atom of helium 4 from two protons and two neutrons. The helium 4 nucleus (an alpha particle) is then stable against disintegration by that amount of energy. The binding energy is an indicator of stability against nuclear change, but binding energy alone does not predict stability against radioactive decay. Beryllium 8 has a greater binding energy per nucleon than beryllium 9, but the former undergoes spontaneous fission to two alpha particles.

The mass defect Δ is generally not very useful in calculations of the energetics of nuclear transitions. The atomic mass unit is taken historically and arbitrarily as one-sixteenth of the mass of one atom of the most abundant isotope of oxygen, ¹⁶O. Since that value does not account for the mass decrement in the combining of the 16 nucleons of oxygen, the mass defect only represents changes relative to oxygen. For example the mass defect of ¹⁶O is, by definition, zero. To further confuse the issue, chemists use one-twelfth of the mass of ¹²C as their standard, rather than the physicists' standard of ¹⁶O. Using the ¹²C standard, the mass of ¹⁶O is 15.9994. For these reasons, the mass defect, or the packing fraction, which represents the mass defect per nucleon, is of little interest.

The mass decrement δ of oxygen can be calculated as follows.

mass of eight electrons, $8 \times 0.000548756 = 0.004390$ amu
mass of eight neutrons, $8 \times 1.008982 \quad = 8.071856$
mass of eight protons, $8 \times 1.007593 \quad = \underline{8.060744}$
$$\text{Total} = 16.136990 \text{ amu}$$

The mass decrement δ is $16.136990 - 16.000000 = 0.136990$ amu.

RADIOACTIVE DECAY BY ALPHA PARTICLE EMISSION

The class reaction for radioactive decay by alpha particle emission is described by the following general equation:

$$_Z^A X \rightarrow {}_{Z-2}^{A-4}Y + {}_2^4 He + \gamma? + Q$$

where X represents the chemical symbol of the parent nuclide and Y represents the daughter nuclide.

This equation is the state and energy expression for the nucleus only. Since Z is decreased by 2, two orbital electrons are also lost, and the overall energy balance would have to account for the rest mass equivalent of these two electrons. The gamma ray emission may or may not be a necessary product of the reaction, depending on the energy state of the product nucleus.

Remember from the example given earlier in this chapter of the decay of ^{226}Ra, it is apparent that the N/Z ratio usually increases for alpha decay. This will generally mean that further transformations will still be necessary, and this accounts in part for the fact that alpha decays are usually part of a long chain of nuclear transformations.

The Q term in the preceding equation represents the total energy released by the transition. If the exact masses of all the components and the energy of any gamma ray emitted are known, it is possible to compute the kinetic energy of the alpha particle. Remember that only the system within the nucleus is in consideration. No energetics correction is made for the two lost electrons.

Properties of Alpha Decay

1. The energy of the alpha particle emitted is closely related to the half-life of the parent. In general, the shorter the half-life is the higher the energy of the emitted alpha particle.

2. Alpha decay is restricted to the very heavy nuclides, Z greater than 83, except for two important exceptions, ^{8}Be and ^{8}B.

3. All the alpha particles emitted for a particular nuclide decay pathway have identical energies. There may be several different energy alphas, however, since alternative decay schemes are common.

4. The range of the alpha particle is unique, since its energy is unique.

Because of the high ionization density in the track of alpha particles (high LET) and the general chemical nature of the alpha emitters, they have little value in radiodiagnostics or therapy. They are generally very radiotoxic if absorbed.

NEGATIVE ELECTRON EMISSION DECAY (β^- DECAY)

When a nuclide is neutron rich, that is, above the stability line, it may decay by the emission of a negative electron. Ignoring the other subnucleon particles, the transition that is occurring within the nucleus can be pictured as

$$\text{neutron} \Rightarrow \text{proton} + \text{negative electron} + \text{antineutrino}$$

This class of decay is fundamentally different from that which occurs in alpha decay. The alpha particle can have a real and finite lifetime within the parent nucleus before transition takes place, and while in this state, the alpha particle may escape the binding energy of the nucleus. In the simplest sense, alpha decay is the fragmentation of a heavy nucleus.

In both negative electron (beta) and positive electron (positron) decay, a reaction occurs within the nucleus that releases as its external sign particles that did not exist within the nucleus. In the process, one nucleon, the neutron, is converted to another, a proton.

The class reaction for negative electron decay may be written:

$$^A_Z X \rightarrow ^A_{Z+1} Y + \beta^- + \bar{\nu} + \gamma? + Q$$

Alpha particles are essentially monoenergetic. Beta particles, on the other hand, show a continuous distribution of energies downward from the maximum value. This would appear to violate the law of conservation of energy if it were not for the handy neutrino (antineutrino) invented by Pauli (1933) to explain the discrepancy. This particle was finally demonstrated to exist by Reines and Cowan (1953). The neutrino (antineutrino) is required to have a rest mass very nearly zero, a zero charge state, and it is assigned quantum numbers for spin, momentum and energy and very, very weak interaction with matter. The interaction cross section for neutrinos is about 10^{-23} cm^2, suggesting that there would only be a "few" interactions in a light-year thickness of lead.

At the time of beta decay there is a distribution of energy between the beta particle and the antineutrino, which is probabilistically determined. The result is a predictable and continuous energy spectrum for the beta particle with a maximum energy value equal to the Q for the transition. In the latter case, all or nearly all of the transition energy is in the electron and nearly none in the antineutrino. The case where all the kinetic energy is in the electron and none in the antineutrino is not possible (discussed later).

In making the energy calculations for the typical β^- transition, it is convenient to neglect the partition of energy between the electron and the antineutrino. If the assumption is made that the maximum energy is possessed by each ejected negative electron, then energy conservation is maintained, even though fictional.

Example

In the β^- decay of ^{41}Ar to ^{41}K, 1 mass unit has an energy equivalent of 9.3148×10^8 eV, or 931.48 MeV. One MeV has a mass equivalent of 1.074×10^{-3} mass units.

Product, ^{41}K	isotopic mass is	40.97847 amu
Beta energy	1.20 MeV $\times 1.074 \times 10^{-3}$	0.00129
Associated gamma	1.29 MeV $\times 1.074 \times 10^{-3}$	0.00139
	Total	40.98115 amu

Thus, for the *parent*, ^{41}Ar, the isotopic mass is 40.98115 amu.

POSITIVE ELECTRON EMISSION DECAY (POSITRON DECAY)

Negative electron decay was an example of the radioactive transformation of a nuclide with a neutron excess. This species will lie to the left of the stability line and will tend toward stability by conversion of a neutron to a proton. Other nuclides may undergo positron decay even though they have a neutron excess.

The internal nuclear transformation is

$$\text{proton} \Rightarrow \text{neutron} + \text{positive electron} + \text{neutrino}$$

The prototypical reaction is

$$^A_Z\text{X} \rightarrow {}^{A}_{Z-1}\text{Y} + \beta^+ + \nu + \gamma? + Q$$

There is one important difference between negatron and positron transitions that does not appear explicitly, since it relates to orbital electrons. Remember, for beta decay there was no allowance for the loss of the electron mass from the system and none was required. With beta decay, when the electron leaves the nucleus another must take up an orbital position to balance the new charge (Z increased by 1), leading to a balance of mass change as far as electrons are concerned; therefore, no allowance must be made in the energy/mass calculations for electron shifts. In positron decay, when a positron is emitted there is a net loss of one positive charge on the nucleus (Z decreased by 1). Another electron must leave an orbital position to balance the new nuclear charge, leading to the need for accounting for *two* electron masses in the energy/mass calculations.

Compare the mass/energy calculation for the following positron transition.

$$^{22}_{11}\text{Na} \rightarrow {}^{22}_{10}\text{Ne} + \beta^+(0.54 \text{ MeV}) + \gamma(1.27 \text{ MeV})$$

Product nuclide ^{22}Ne isotopic mass	21.99138 amu
Mass of two electrons, 0.0005487×2	0.00110
Energy of the positron, $0.54 \times 1.074 \times 10^{-3}$	0.00058
Gamma ray, $1.27 \text{ MeV} \times 1.074 \times 10^{-3}$	0.00136
Total	21.99420 amu

The isotopic mass of the parent, ^{22}Na, is 21.99420 amu.

Annihilation Reaction

The other important characteristic of positron decay is the associated annihilation reaction. Since the positive electron is an antimatter particle, if it and a nearby negative electron are near rest energy, they will interact to cause the disappearance of both the positron and the electron and the transformation of the mass of both into electromagnetic energy. This radiation will appear as two photons of equal energy emitted at nearly 180° to each other. Each will have an energy of 0.511 MeV. The sum of the two annihilation gamma ray energies is 1.022 MeV. This energy is equivalent to the rest mass of the two particles, the negative and positive electrons, which have disappeared. There must also be conservation of momentum. The 180° opposed directions of the two photons accomplish this conservation of momentum. To the extent that the positron is not completely thermalized at the time of annihilation, the angular separation of the two photons may be slightly different from 180°.

Finally, since positron decay is associated with the appearance of a neutrino, the energy and range relationships for the emitted positron will not be significantly different from those for the negative electron. In energy/mass calculations for positron decay, it is a convention, as in beta decay, to assign all the energy of the positron–neutrino pair to the positron.

DECAY BY ELECTRON CAPTURE

One mechanism has just been considered whereby neutron-deficient species can adjust their nuclear composition toward stability. It has also been noted that in the course of the process of positron emission there is the ultimate disappearance of two electrons or their energy equivalent. If the transition energy to the product nucleus of the transformation cannot be made to fit this pattern of energy loss (that is, $Q > 1.022$ MeV), then an alternative decay pathway is available.

The alternative mechanism is *electron capture*. Electron capture as a mechanism does not suffer the same constraint, that the transformation energy must exceed 1.022 MeV to be allowable. Electron capture allows the same reduction of Z by 1 with constant A that positron emission effects. Electron capture does not preclude the possibility of positron emission if the latter is energetically possible, so for all radionuclides that meet the 1.022-MeV energy constraint, the two processes will be competing pathways for transformation. In some of these cases the electron capture pathway is nonexistent or nearly so.

To understand how electron capture takes place, we must recollect the wave nature of the electron and the fact that the wave mechanical description of the motion of the electron asserts that the electron can come very close to the nucleus and can, indeed, at times be within the nucleus.

The description of the particle process is as follows:

$$p + e^- \Rightarrow n + \nu$$

Since the emission of a negative electron in the reaction

$$n \rightarrow p + e^- + \bar{\nu}$$

causes the emission of an antineutrino, the capture of an electron in the reverse reaction must lead to the emission of a neutrino.

$$p + e^- \rightarrow n + \nu$$

The prototype reaction for electron capture is

$$e^- + {}_Z^A X \rightarrow {}_{Z-1}^A Y + \nu + \gamma? + Q$$

k-Capture Fluorescent Radiation

Following electron capture there must be an adjustment of the orbital electrons to fill the vacancy. Associated with these transitions are the characteristic electromagnetic radiations resulting from the orbital transitions. For example, if a free electron fills a k orbital, energy equivalent to the binding energy of a k orbital electron is released in the form of a photon. This photon is usually called a k-capture x-ray.

What radiations are seen externally in the EC process? If there is no ground state adjustment by emission of a gamma ray (often none appear in the EC process), then the only externally perceived radiation is the k-capture x-ray. The neutrino is very difficult to detect.

INTERNAL CONVERSION

When the nucleus is in the excited state after a transformation, it tends to relieve its excitation by the emission of a gamma ray. In some cases, however, the gamma ray is not emitted, and the energy is transferred to one of the orbital electrons, which, in turn, is ejected with high kinetic energy from the nucleus. It is as if the electron captured the transition gamma ray internally, within the atomic structure, by the equivalent of the photoelectric process. The gamma ray disappears, and an electron appears with an energy equivalent to that of the gamma ray less the orbital binding energy of the ejected electron.

There is an empty orbital position to be filled. When a free electron fills this position, there is the emission of a k-capture x-ray with an energy equal to the binding energy of the orbital position filled. Ultimately, all the original excited state energy carried off by the original gamma ray is accounted for in the kinetic energy of the internal conversion electron and the orbital capture gamma rays.

In general, internal conversion competes with gamma ray emission, so a mixture of the two processes may be seen. There must be enough energy in the original gamma transition to exceed the binding energy of the orbital electron displaced. The k-capture process is straightforward, with the energy of the fluorescent photon being equal to the binding energy of the empty orbital.

The Auger Electron

There is, however, a competing process that is less well understood. The process is the emission of *Auger electrons*. The Auger process is diagrammed in Figure 3.7. Auger electron emission is predominant for

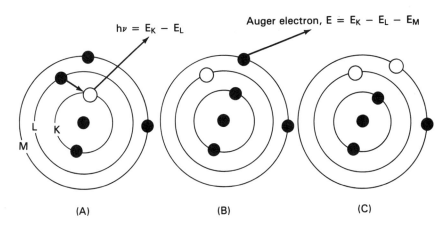

Figure 3.7   (A) The usual emission of a k characteristic x-ray, $h\nu$, energy equal to $E_K - E_L$, the difference in binding energy for the two orbital electrons, K and L. (B) $h\nu$ has been absorbed and a monoenergetic Auger electron is emitted, in the example shown, from the M shell, the energy of which is $E_K - E_L - E_M$. (C) In its final state the atom has vacancies in the L and M orbitals.

nuclides of low Z and is always in competition with the normal k-capture photon emission.

REFERENCES

PAULI, W., JR. (1933) *Rapports Septiems Conseil Physique*, Solvay, Bruxelles. Gautier-Villars, Paris.
REINES, F., and COWAN, G. L. (1953) Detection of the free neutrino. *Phys. Rev.* **92**, 830–831.

SUGGESTED ADDITIONAL READING

EVANS, R. D. (1955) *The Atomic Nucleus*. McGraw-Hill Book Co., New York.
LAPP, R. E., and ANDREWS, H. L. (1972) *Nuclear Radiation Physics*, 4th ed. Prentice-Hall, Inc., Englewood Cliffs, N.J.

PROBLEMS

For the following problems use the following constants:

$$9.31145 \times 10^2 \text{ MeV} = 1 \text{ amu}$$
$$\text{mass of the electron} = 0.000548756 \text{ amu}$$
$$\text{mass of the proton} = 1.007593 \text{ amu}$$
$$\text{mass of the neutron} = 1.0008982 \text{ amu}$$

1. The mass of $^{64}_{29}$Cu is 63.929757 amu. It undergoes positron decay with a half-life of 12.9 h. The product of this decay is $^{64}_{28}$Ni. The mass of this product is 63.927956. What is the total energy of the positron and the neutrino resulting from the decay? Is the product likely to be stable or radioactive? Why?

2. The radionuclide $^{90}_{37}$Rb has a mass of 89.914871 and a half-life of 2.9 min. A product of the decay is $^{90}_{40}$Zr, the mass of which is 89.904710. If the total energy produced by the decay of 1 mCi of the ^{90}Rb were measured, it would be found to be what? Can you propose a decay scheme for this transformation? The identity of any intermediate nuclides need not be provided in the answer.

3. A *source* of $^{99m}_{43}$Tc arrives at the laboratory for use at 10 A.M. on Monday morning, at which time this daughter product is eluted for diagnostic use. The parent, $^{99}_{42}$Mo, has a decay constant of 0.01039 h^{-1}. If, after the separation of the daughter, the parent was found to have an activity of 5.0 $\times$ 10^9 Bq, what is the activity of the parent and the daughter following Thursday at 10 A.M.?

4. The nuclide $^{131}_{53}$I has a half-life of 8.06 days. Find the mean life and the decay constant for this radionuclide. A source of this isotope has an activity of 2.500 mCi. Find the activity remaining after the elapse of 12 days from the measurement just given. Express the result in millicuries and becquerels.

5. The radionuclide $^{231}_{91}$Pa, has a half-life of 3.25 $\times$ 10^4 yr. The equilibrium *atomic ratio* for this isotope with its parent, ^{235}U, is 2.89 $\times$ 10^6. What is the half-life of the uranium parent and what is the activity of the daughter at equilibrium?

4

Interaction of Radiation
with Matter

INTRODUCTION

When a radiation beam passes through tissue or other absorbing media, energy is lost from the incident beam. Some of this energy is imparted to the medium in which the interacting events take place, while some of it leaves the volume of interaction. These energies are respectively called *energy imparted*, ΔE_{ab}, sometimes called *energy absorbed*, and *energy lost*, ΔE_L. The total energy leaving the incoming beam is called *energy transferred*, ΔE_{tr}.

$$\Delta E_{ab} = \Delta E_{tr} - \Delta E_L \qquad (4\text{-}1)$$

It must be recognized that, particularly with gamma rays or other high-energy radiations, there is often no transfer of energy from some photons, which may proceed through the material unaffected. Their energy is not included in the term *energy transferred*.

For the purposes of this chapter, there will be no consideration of possible changes in rest mass and its equivalent energy conversion. That subject will be left for further examination in Chapter 5.

The important processes by which energy is transferred in tissue or tissue equivalent systems are the following:

1. Photoelectric process
2. Compton scattering process
3. Pair production

In addition, there is a less important scattering process in matter that has come to be known as Rayleigh scatter. Rayleigh scatter is a *coherent* scattering process; that is, it is a cooperative phenomenon involving the interaction of all the electrons of the atom. Photons are scattered by bound electrons in a process in which the scattering atom is neither excited nor ionized. Since coherent scatter is only important at high values of Z and for energies of a few keV, it is not an important process for tissue equivalent materials; and since little or no energy is imparted to the tissue, it has no biological importance.

After the absorption of energy associated with the preceding processes, there is another chain of events that ultimately leads to tissue damage, both reversible and irreversible. These processes are schematized in Figure 4.1. Before discussing the processes of energy transfer, the primary interaction event of Figure 4.1, it is first necessary to describe the attenuation process itself, the rate of energy transfer per unit path length. It will be seen that, depending on the nature of the primary interaction event, various properties of the absorber will be of importance in determining the rate of energy transfer.

LINEAR ATTENUATION COEFFICIENT

Assume that a slab of homogeneous material of thickness ΔX is placed in the path of a very narrow incident photon beam, and that the number of incident photons for the time of measurement is N (see Figure 4.2). Then, since the chance of a single scattering interaction that will remove a photon from the incident beam depends on N (the incident photon fluence) and the properties of the attenuator (its linear attenuation coefficient), the equation for the number of photons removed $(N - n)$ can be written as

$$\Delta N = \mu \, \Delta X N \tag{4-2}$$

where ΔN is the fraction of transmitted photons scattered, and μ is a constant of proportionality. In other words, the fraction of photons removed depends on the thickness of the absorber and a constant of proportionality, μ, which is determined by the properties of the homogeneous absorber and which will be different for different radiations and different materials. This constant, μ, is known as the *linear attenuation coefficient*:

$$\mu = \frac{\Delta N}{\Delta X} \tag{4-3}$$

"Good" and "Bad" Geometry

In Figure 4.2, detector P is positioned to measure only photons that have not undergone a scattering event. This arrangement of source and detector is known as a "good" geometry measure of attenuation, since all

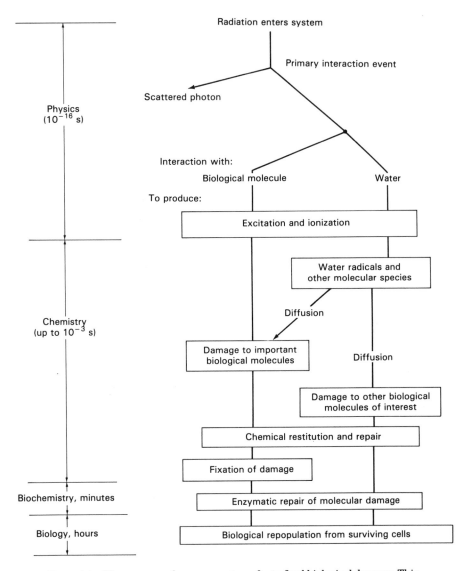

Figure 4.1 The processes from energy transfer to final biological damage. This chapter deals with the primary interaction event.

events that have undergone any energy transition will suffer a momentum and energy change and on their new scattered path will not be seen at this location. The other requirement for good geometry is that the beam, N, be very small in its physical dimensions. If a broad beam is used, there is the possibility that photons not included in the definition of N will be scattered in from other areas of the attenuator. Therefore, good geometry

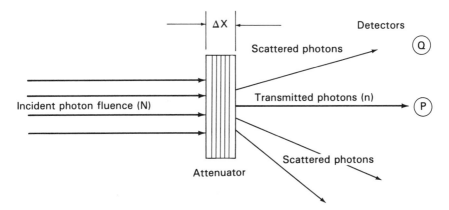

Figure 4.2 Transmission of photons through a plane scattering medium of thickness, ΔX. A detector placed at P will measure the transmission under "good" geometry in which it will see only photons that are transmitted without scatter (change in energy and momentum). At position Q the detector will measure scattered photons that have interacted in the attenuator. This latter is classed as "bad" geometry or broad-beam attenuation for purposes of evaluating the attenuation coefficient.

for attenuation measurements is described as measurements made with a small, highly collimated pencil beam of incident photons and a measurement system that "sees" only unscattered photons. "Bad" geometry is sometimes described as broad-beam geometry, since that nonpejorative term describes a condition that will provide unwanted scatter-in photons.

There is no adequate analytical expression for measurement of attenuation in broad-beam conditions. The narrow beam, or good geometry expression for attenuation from an incident photon beam can be obtained by integration of equation 4–3.

$$N = N_0 e^{-\mu x} \tag{4-4}$$

N_0 is the incident fluence. N is the fluence remaining after a thickness, x, of specified absorber, and μ is the linear attenuation coefficient.

Mass, Electronic, and Atomic Attenuation Coefficients

Further examination of the interaction processes will show that these interactions of radiation with matter may depend on either the density of the electrons with which the photons interact or the density of the atoms with which, under certain other circumstances, the photon will interact. The former will be shown to be the most important for tissue equivalent systems; but for the general case, attenuation coefficients are described in terms of the linear dimension, the unit mass dimension, which is simply related to the linear coefficient by the density, the probability of interaction per electron, and the probability of interaction per atom. These various attenuation coefficients are respectively, the linear, mass, electronic, and atomic attenuation coefficients.

The linear attenuation coefficient is μ. It is a measure of the probability of interaction of a photon per unit of linear path length in the absorber. It is often more convenient, as has just been suggested, to consider the attenuation coefficient in some other terms than the linear thickness of absorber. The alternatives are the *mass attenuation coefficient*, the *electronic attenuation coefficient*, and the *atomic attenuation coefficient*. There are simple expressions that interrelate the various coefficients and allow us to make the necessary conversions.

N_a = Avogadro's number, 6.02205×10^{23} atoms per mole.

N_a/A = number of atoms per gram of material, where A is the atomic weight or molar weight of the element.

N_e = number of electrons per gram of material = $(N_a \times Z)/A$ and $N_e \times 10^3$ = the number of electrons per kilogram of material.

ρ = density of the absorber (kg/m^3).

In Table 4.1 the relationships between the various attenuation coefficients are shown, along with their symbols and units.

It is interesting that the value of N_e does not vary a great deal among the various elements, with the exception of hydrogen. For most elements the ratio Z/A is approximately 0.5, so N_e should approximate one-half of Avogadro's number. For example, N_e for carbon is 3.01×10^{23}, and N_e for lead is 2.38×10^{23}. Hydrogen has a Z/A of 1.0, so its value for N_e is close to Avogadro's number. N_e for hydrogen is 5.97×10^{23}.

TABLE 4.1 Relationships among Attenuation Coefficients

Coefficients	Symbol	Units
Linear	μ	m^{-1}
Mass	μ/ρ	m^2/kg
Electronic	$_e\mu = \mu/\rho \, (1/1000N_e)$	m^2/el
Atomic	$_a\mu = \mu/\rho \, (Z/1000N_e)$	m^2/at[a]

[a] Note that the conversion for atomic attenuation coefficient uses N_e, the number of electrons per gram, for the calculation.

Definition of Cross Section

The dimensions for all except the linear attenuation coefficient are such that a term representing "length squared" (area) is in the numerator. This has led the physics community to the convention of referring to these coefficients in terms of *cross section*. Taking, for example, the electronic cross section, we can visualize the attenuation coefficient in terms of the effective cross-sectional area of an electron in its interactions with an incoming photon. If the electron of an absorber is thought of as having a certain area or cross section for interaction with a photon, then this area multiplied by the number of electrons in a cubic centimeter of the material would be the total effective area of the electrons in the cubic centimeter, and this is equal to the electronic attenuation coefficient.

It is better not to think of the cross section in a literal geometric sense, but simply to remember that it is the probability constant for interaction of the photon per electron in the material. The electronic cross section, for example, simply describes a "larger and larger" electron as the probability of interaction increases, while, of course, the electron is not changing its physical characteristics.

ENERGY TRANSFER AND ENERGY ABSORPTION

In general, when interaction of a photon takes place in an absorber, part of the photon energy is then in a form that is basically short range in nature (fast electrons, for example), and part of the energy reappears as the energy of photons of different (lower) energy. These "scattered" photons may or may not leave the system that is under consideration without further energy transfer (scattering events). Higher-energy electrons may leave the system, or they may radiate part of their energy as bremsstrahlung, which may or may not leave the system. The processes vary widely and are strongly dependent on the initial energy of the photons. See Table 4.2 for examples.

In a single scattering event it is not possible to predict the partition between energy that is transferred to the absorber and energy that is absorbed. It will depend on the volume of the absorbing medium, the mean free path of the scattered particle and its range, and the number of radiative events (bremsstrahlung) that take place. The values given in Table 4.2 are average values over many events. Consider the partition of energy in a typical case. Take the interaction of 10-MeV photons with carbon, using the coefficients provided in the table. The energy transferred into kinetic energy of electrons is 7.30 MeV. Of that, 7.04 MeV is ultimately absorbed in the medium of interest. The balance, 7.30 − 7.04 MeV is reradiated as bremsstrahlung and is lost to the medium. The balance of

TABLE 4.2   Energy Transferred and Energy Absorbed for
Incident Photons of Various Energy (for Carbon)

Photon Energy (MeV) E_{Tot}	Average Energy Transferred (MeV) E_{tr}	Average Energy Absorbed (MeV) E_{ab}
0.01	0.00865	0.00865
0.10	0.0141	0.0141
1.0	0.440	0.440
10.0	7.30	7.04
100.0	95.63	71.90

the initial photon energy, 10.0 − 7.30 MeV is scattered and also lost to the medium of interest. In this example, 70.4% of the energy of the incident photon is deposited in the medium by the slowing down of high-kinetic-energy electrons, but 96% of the energy transferred is deposited in the medium. It is also clear from the table that for low-atomic-number absorbers such as carbon, the energy lost as bremsstrahlung is very small, and to all intents the energy transferred is equal to the energy absorbed.

For the highest photon energy in the table, 100 MeV, conditions are significantly different. Transferred energy is 95.63% for photons of this energy scattered in carbon, but of this energy only 71.9% of the energy is finally absorbed in the medium. Much of the energy reappears from the pair production process, which will be described shortly.

It is often useful to state cross sections or attenuation coefficients in terms of energy transferred or energy absorbed. These values are the *energy transferred attenuation coefficient* and the *energy absorbed attenuation coefficient*, and they can be calculated from the appropriate attenuation coefficients and the average energy transferred and absorbed as tabulated in Table 4.2.

From equation 4-2, let n' be the number of scattering events occurring in the linear element ΔX, and n' will be $(N - n)$; then

$$n' = \mu N \, \Delta X \tag{4-5}$$

If the *average* energy transferred is E_{tr}, then the increment of energy transferred in ΔX is

$$\Delta E_{\text{tr}} = E_{\text{tr}} \mu N \, \Delta X \tag{4-6}$$

Multiplying the numerator and denominator on the right by $h\nu$, the energy per photon, to dimensionally adjust the expression to the dimensions of μ,

$$\Delta E_{tr} = \left(\mu \frac{E_{tr}}{h\nu} \right) Nh\nu \, \Delta X \qquad (4\text{-}7)$$

The quantity $h\nu$ inserted in the numerator and denominator converts E_{tr}, which is expressed in energy units, into fluence, the dimensions of μ. The quantity in parentheses is the transfer coefficient, μ_{tr}. The expression for the transfer coefficient is

$$\mu_{tr} = \mu \left(\frac{E_{tr}}{h\nu} \right) \qquad (4\text{-}8)$$

The equation for energy transferred in ΔX is

$$\Delta E_{tr} = \mu_{tr} Nh\nu \, \Delta X \qquad (4\text{-}9)$$

We can use the same approach for the energy absorption coefficient:

$$\mu_{ab} = \mu \left(\frac{E_{ab}}{h\nu} \right) \qquad (4\text{-}10)$$

and

$$\Delta E_{ab} = \mu_{ab} Nh\nu \, \Delta X \qquad (4\text{-}11)$$

The importance of the mass–energy absorption coefficient, in particular, μ_{tr}/ρ, has been emphasized by Hubbell (1977) in an extensive report on the energy transfer and energy absorption coefficients for tissue elements. The imparted charged particle kinetic energy, as symbolized by ΔE_{ab}, is a reasonable approximation, depending on the dimensions of the absorber, the photon energy, and the atomic number of the absorber, of the amount of energy made available in the medium for chemical and biological effects. The mass–energy absorption coefficient is central in the determination of dose.

MECHANISMS OF ENERGY TRANSFER FROM GAMMA RAYS

Photoelectric Scattering Process

The photoelectric absorption process is one in which the incoming photon interacts with an orbital electron of the scattering atom (see Figure 4.3). The incoming photon disappears and a photoelectron is ejected from its orbital position in the K, L, or M shells of the atom. The orbital position vacated by the photoelectron is filled by an electron from an outer orbital. The result of this orbital shift is the emission of a photon, the energy of which is the difference in the orbital binding energy of the photoelectron

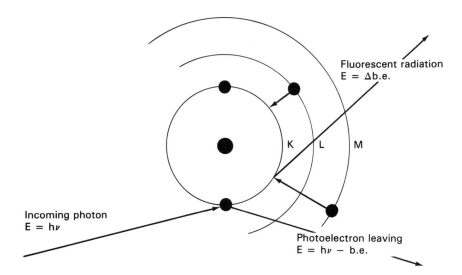

Figure 4.3 The events in the photoelectric scattering process. The diagram indicates the incoming photon ejecting the photoelectron, followed by the movement of an outer orbital electron into this vacancy. Shown in the diagram is the movement of an M orbital electron into the K orbital position with the emission of a fluorescent photon. Also shown is the possible alternative, which is the movement of an L orbital electron into the vacancy.

and the orbital binding energy of the electron filling the vacated orbital. The energy of the incoming photon is $h\nu$. If this photon is to interact with an orbital electron, the energy of the photon must be equal to or exceed the binding energy (b.e.) of the electron that is displaced. The kinetic energy of the photoelectron that appears is equal to the energy of the incoming photon less the binding energy of the electron displaced.

$$\text{photoelectron KE} = h\nu - \text{b.e.} \qquad (4\text{-}12)$$

where b.e. is the binding energy for the appropriate orbital electron that has appeared as a photoelectron.

The energy relationships for the photoelectric process are complex and beyond the scope of this text (see, for example, Evans, cited in Suggested Additional Reading). The photoelectric interaction is most likely to occur if the energy of the incident photon is just greater than the binding energy of the electron with which it interacts. As a result, the plot of the attenuation coefficient as a function of the photon energy is a complicated relationship, with sharp peaks at the binding energies of the various orbital shells and with strong dependence on the atomic number of the atom. Theoretical calculations of the photoelectric cross sections have been done by several workers, and measured cross sections are in very good agreement with theory. Scofield (1973) has reported the theoretical cross sections from 1 to 1500 keV, and Hubbell (1977) has compared experimental data with theory for nitrogen and found them to be in very good agreement.

The ordinate shown for the graph in Figure 4.4 is a logarithmic scale of the photoelectric mass partial attenuation coefficient. The partial cross section, as it is sometimes called, is indicated by the symbol τ/ρ. The appropriate subscripts would be used on τ for the electronic and atomic partial cross sections and for the energy transferred and energy absorbed mass attenuation coefficients.

For all the interaction processes that will be discussed in this chapter, the most important factors for complete understanding of dose are the variation of the cross section with atomic number of the medium and with the energy of the photon for each process. For the photoelectric process, the rules are as follows:

1. The photoelectric process involves *only bound electrons.*

2. The partial cross section for the photoelectric process, τ/ρ, varies as $(h\nu)^{-3}$ as an approximation. The exponent will depend somewhat on

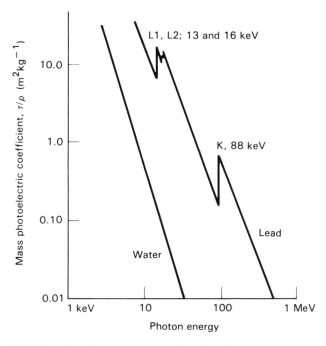

Figure 4.4 Photoelectric cross section, τ/ρ, plotted as a function of photon energy for water and for lead. The K and L orbital edges for lead are visible. Since the K and L binding energies for water are so low, they are not visible in this plot.

the atomic number of the medium. For example, for H, C, and N it is 3.0. For Pb it is 2.96.

3. Near the absorption edge (the energy that just equals the orbital binding energy for the electron in question) there is a rapid change in cross section. The probability of interaction is maximum if the photon energy just exceeds the binding energy.

4. The electronic attenuation coefficient per gram varies with atomic number, approximately as Z^3 for high Z elements and as about $Z^{3.8}$ for low Z elements.

5. The photoelectric cross section is nearly 80% accounted for by K orbital interactions.

6. Since each atom contains Z electrons, the atomic cross section, $_a\tau$, will vary as $Z^{4.8}$ for low Z materials:

$$_a\tau = \frac{\tau/\rho}{Z/1000N_e}, \qquad _e\tau = \frac{\tau/\rho}{1000N_e}$$

Then

$$_a\tau = \frac{_e\tau}{Z}$$

7. In tissue, $E_{tr} = E_{ab}$ and both are very nearly equal to $h\nu$.

Energy Transferred–Energy Absorbed Relationships; Photoelectric Effect

If an interaction takes place, the kinetic energy of the photoelectron produced is $h\nu - E_{b.e.}$, where $E_{b.e.}$ is the binding energy for the shell in consideration. In filling the vacant orbital position the energy, $\Delta E_{b.e.}$, the difference in the binding energy of the two orbitals involved, is emitted as the fluorescent photon. For tissue elements (those of low Z), the values of the binding energies are so small (500 eV or less) that essentially all the incident photon energy will be converted to kinetic energy. The energy of the photoelectron will be deposited locally. The fluorescent photon will be of such low energy, that is, equal to the binding energy difference of the two shells, that all of its energy will also be deposited locally. So, to all intents, energy transferred is energy absorbed in a system of low Z such as tissue.

Spatial Distribution of the Emission of Photoelectrons

Photoelectrons are emitted generally in a forward directed fashion from the point of interaction, taking the path of the incoming scattered photon as forward. The distribution is lobular, with a maximum at both sides of the forward direction. The angular relationship of these scattering angles is dependent on the energy of the scattered photon. If the energy of the scattered photon is low, the photoelectron is emitted almost normally to the incoming photon track. As the scattered photon energy increases, this pattern tends to be in a more forward directed but still bilobular pattern.

COMPTON SCATTERING PROCESS: INCOHERENT SCATTERING

The process just described, photoelectric scattering, is a phenomenon that involves the atom and its associated electrons. It cannot occur with unbound electrons. Incident photons can also interact by two mechanisms with free electrons, either by *coherent* scattering processes or by *incoherent* scattering. Coherent scattering is a process in which no energy of the incident photon is converted into kinetic energy in the medium. The electromagnetic wave of the incident photon interacts with the electrons of the medium, causing them to oscillate with the same frequency, $h\nu$, and they, in turn, as oscillating charges, emit electromagnetic radiation of the same frequency as that of the incident photon. The emitted electromagnetic radiations from each of the electrons combine and/or interfere with each other to result in a wave of frequency $h\nu$, and with direction different from that of the incoming photon. Since the scattering is a cooperative phenomenon, involving all the electrons of the medium, it is called coherent scattering. The net result of coherent scattering (Rayleigh scatter) is to increase somewhat the angular dispersion of the incoming wave without depositing energy in the medium. Since there is no transfer of kinetic energy into the medium, and since the cross section for coherent scatter is very small for low-atomic-number materials at energies greater than 100 keV or so, the process is of no interest for biological or chemical events in the medium. Rayleigh scatter occurs mostly at low photon energies in high Z materials where the electron binding energies interfere with the Compton scattering process. Hubbell's compilations (1977) include coherent cross sections for those who might have an interest.

Under circumstances where electrons can scatter independently, the process is called incoherent scattering or Compton scattering. It will be seen that this is one of the most important interaction mechanisms for tissue equivalent materials because of the energy cross-section considerations that will be discussed.

Figure 4.5 shows a schematic of the Compton collision of a photon of energy $h\nu$ and momentum $p = h\nu/c$, with a *free electron*. When an incident photon is scattered by a loosely bound (or virtually free) electron, the phenomenon is called *Compton scatter*. As a result of the photon interaction with the electron, the electron is set in motion with a kinetic

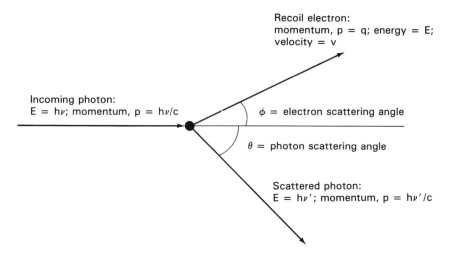

Figure 4.5 The events in the Compton (incoherent) scattering process.

energy and momentum as described in Figure 4.5. Energy and momentum are conserved in the scattering event.

Since, in the Compton process, the electron binding energies are very small (free or nearly free electron), we can generally use the calculations of classical mechanics with conservation of energy and momentum. It is important, however, to understand the underlying quantum mechanical nature of the process, and in particular the transfer of energy implied in equation 4-13 and the assignment of momentum to a photon. This classical mechanical simplification is, to a first order, adequate, but for very low energy photons, or high Z materials, or circumstances where the electron velocity is relativistic, appropriate corrections are required.

From Figure 4.5, the energy of the recoil electron (E_e) equals the incident photon energy $(h\nu)$ minus the recoil photon energy $(h\nu')$:

$$E_e = h\nu - h\nu' \tag{4-13}$$

The scattered photon will have a lower energy than the incident photon and therefore its wavelength will be longer. Since momentum must be conserved, the recoil angle of the electron, ϕ, is uniquely determined if the angle of photon scatter, θ, is known. Since the velocity of the electron is relativistic, it is necessary to represent the energy of the electron in relativistic terms.

$$h\nu - h\nu' = E_e = m_0 c^2 \left[\frac{1}{\sqrt{1 - (v^2/c^2)}} - 1 \right] \tag{4-14}$$

Since linear momentum is also conserved, the sum of the momentum vectors, q and $h\nu'/c$, must be equal to the incoming momentum, $h\nu/c$. Thus, if $h\nu/c = p$ and $h\nu'/c = p'$, then

$$q^2 = p^2 + (p')^2 - 2pp' \cos \theta \qquad (4\text{-}15)$$

Conservation of linear momentum in the forward direction may be written as

$$\frac{h\nu}{c} = \frac{h\nu'}{c} \cos \theta + \frac{m_0 v}{\sqrt{1 - (v^2/c^2)}} \cos \phi \qquad (4\text{-}16)$$

and conservation of momentum in the orthogonal direction may be written as

$$\frac{h\nu'}{c} \sin \theta = \frac{m_0 v}{\sqrt{1 - (v^2/c^2)}} \sin \phi \qquad (4\text{-}17)$$

The reader may wish to carry out the derivations of equations 4-16 and 4-17 from equations 4-13 and 4-14 and may also wish to show that, by eliminating v and ϕ from equations 4-16 and 4-17, the final expressions for the energy of the scattered photon and the scattered electron are as follows:

$$E_e = h\nu \frac{a(1 - \cos \theta)}{1 + a(1 - \cos \theta)} \qquad (4\text{-}18)$$

and

$$h\nu' = h\nu \frac{1}{1 + a(1 - \cos \theta)} \qquad (4\text{-}19a)$$

In equations 4-18 and 4-19a, a is substituted for the ratio shown in 4-19b, where $h\nu$ is expressed in MeV and the usual value of 0.511 MeV is assigned to $m_0 c^2$. The term a is widely used as a shorthand term in scattering formulations, and it should be remembered for future use:

$$a = \frac{h\nu}{m_0 c^2} \qquad (4\text{-}19b)$$

Limits of Energy Transferred

If the photon interacts with the electron in such a fashion that the angle of recoil is forward at 0°, then the scattered photon will be scattered straight back; that is, $\theta = 180°$ and $\phi = 0°$. This is the interaction that imparts *maximum energy* to the recoil electron and leaves the *minimum energy* with the scattered photon, with $\theta = 180°$, $\cos \theta = -1$. For that case, expressions 4-14 and 4-15 simplify to equations 4-20 and 4-21.

$$E_{e(\text{max})} = h\nu \frac{2a}{1 + 2a} \qquad (4\text{-}20)$$

and

$$hv'_{min} = hv \frac{1}{1 + 2\alpha} \qquad (4\text{-}21)$$

For *minimum energy* transfer, $\theta = 0$, $\cos \theta = 1$, and expressions 4-18 and 4-19a give:

$$E_e = 0 \qquad (4\text{-}22)$$

and

$$hv = hv' \qquad (4\text{-}23)$$

With the help of equations 4-20 through 4-23, it is now possible to examine the effect of the energy of the incident photon on the fraction of its energy that is transferred to kinetic energy of the scattered electron. In principle, after a Compton scattering event it is possible to assume that all the energy in the scattered photon will leave the absorbing system. The energy transferred will be the kinetic energy that is imparted to the electron. It is not so conveniently easy in the case of Compton scatter to assess the fraction of the energy transferred that will become energy absorbed. To the extent that the scattered electron loses energy through radiative mechanisms (bremsstrahlung), that energy will be lost to the system. Addressing only the question of the fraction of the energy of the incident photon that is transferred to kinetic energy of the electron, we may calculate $E_{e(max)}$ as a function of the incident photon energy. These calculations have been carried out in Table 4.3.

The calculations in Table 4.3 demonstrate the importance of the incident photon energy on the fraction of energy that is transferred in

---

TABLE 4.3   Effect of Incident Photon Energy on the Fraction of Energy Transferred in the Compton Process

Photon Energy, 5.11 keV	Photon Energy, 5.11 MeV
$\alpha = \dfrac{5.11 \text{ keV}}{0.511 \text{ MeV}} = 0.010$	$\alpha = \dfrac{5.11 \text{ MeV}}{0.511 \text{ MeV}} = 10$
$E_{e(max)} = 5.11 \text{ keV} \times \left(2 \times \dfrac{0.01}{1.02} \right)$	$E_{e(max)} = 5.11 \text{ MeV} \times \left(2 \times \dfrac{10}{21} \right)$
$= 0.10 \text{ keV}$	$= 4.87 \text{ MeV}$
$hv' \text{ (min)} = 5.11 \text{ keV} \times \dfrac{1}{1.02}$	$hv' \text{ (min)} = 5.11 \text{ MeV} \times \dfrac{1}{21}$
$= 5.01 \text{ keV}$	$= 0.24 \text{ MeV}$
Energy transferred: 2%	Energy transferred: 95%

the Compton process, and they show clearly that the effectiveness of Compton scatter for transfer of energy in biological tissues is not great until the incident photon energy is in excess of 100 keV or so. Note that this consideration does not yet include the influence of the cross section for the Compton process, which is high at low energy. The effect of cross section and photon energy will be examined in more detail in the next section, where details of the energy transfer and energy absorbed process are considered.

For low-energy photons, when the scattering interaction takes place, little energy is transferred, regardless of the probability of such an interaction. As the energy increases, the fractional transfer increases, approaching 1.0 for photons at energies above 10 to 20 MeV. Thus we must convolute the independently variable cross section with the calculation of the energy transferred to know the total energy transferred by a complex spectrum of gamma rays.

Energy Absorbed

For the photoelectric process, there is very efficient absorption of transferred energy in tissuelike systems where the photoelectron can be assumed to deposit most of its energy locally. Because of cross-sectional considerations, the photoelectric process makes only a modest contribution to energy absorbed for photons in the energy range that is of interest for biological damage. Reconsideration of Figure 4.4 will demonstrate that for water, for example, the photoelectric cross section is low for photon energies of a few tens of keV.

In the Compton process the absorbed energy will generally vary as the inverse of the energy transferred, since for high-kinetic-energy electrons a fraction of the kinetic energy will partly reappear as bremsstrahlung radiation.

Summary of the Compton Process

1. The process involves interaction of a photon and an electron (weakly bound), but the original photon reappears as a scattered photon of lower energy.
2. The process is almost independent of atomic number.
3. Compton scattering cross sections decrease as energy increases.
4. On average, the fraction of energy transferred to kinetic energy per interaction increases with photon energy.
5. The Compton process is most important for energy absorption for soft tissues for energies in the range from 100 keV to 10 MeV.

Scattering Coefficients

Klein and Nishina (1929) made one of the earliest applications of Dirac quantum dynamics in computing theoretically the differential cross section for the Compton process. It speaks well for the quality of their computations that these data are still used today as the standard of comparison for new developments. The derivation is too extensive to quote in detail here, but the final format is worthy of careful consideration. They found that the differential cross section for the number of photons scattered into unit solid angle per electron (electronic differential cross section) is given by

$$\frac{d_e\sigma_t}{d\Omega} = \frac{e^4}{2m_0^2c^4} \left[\frac{1}{1 + \alpha(1 - \cos\theta)} \right]^2$$
$$\left[1 + \cos^2\theta + \frac{\alpha^2(1 - \cos\theta)^2}{1 + \alpha(1 - \cos\theta)} \right] \qquad (4\text{-}24)$$

When the photon scattering angle is zero, there is no transfer of kinetic energy to the electron, and equation 4-24 simplifies to the classical Thomson scattering value of $e^4/m_0^2c^4$. If the photon energy is small, α approaches zero and equation 4-24 reduces to the classical, prequantum mechanical Thomson scattering expression:

$$\frac{d_e\sigma_t}{d\Omega} = \frac{e^4}{2m_0^2c^4}(1 + \cos^2\theta) \qquad (4\text{-}25)$$

The term $e^4/2m_0^2c^4$ can be substituted by $r_0^2/2$, where r_0 is the classical electron radius, ke^2/m_0c^2. The coefficient k is the constant from Coulomb's law.

$$\frac{d_e\sigma_t}{d\Omega} = \frac{r_0^2}{2}(1 + \cos^2\theta) \qquad (4\text{-}26)$$

For the total cross section over all scattering angles for the classical equation (4-26), it is necessary to integrate equation 4-26 over all scattering angles from zero to 2π. The solid angle between θ and $d\theta$ is $2\pi \sin\theta\, d\theta$. With this expression, equation 4-26 can be integrated to give, for the special classical Thomson scattering case,

$$_e\sigma_0 = \frac{8}{3}\pi r_0^2 \qquad (4\text{-}27)$$

This Thomson classical scattering coefficient for a free electron is for the case where no energy is imparted to the electron. As the photon energy increases, the amount of kinetic energy in the scattered electron increases.

The Klein–Nishina formulation (4-24) provides correct calculations for the general case. It is demonstrable that the Thomson classical formulation is part of the Klein–Nishina solution. The K–N formulation may be rewritten as the Thomson formula and a "K–N factor," generally notated as F_{KN}. Rewriting the Klein–Nishina formula (4-24), we obtain

$$\frac{d\sigma}{d\Omega} = \frac{d\sigma}{d\Omega} F_{KN} = \frac{r_0^2}{2}(1 + \cos^2 \theta)F_{KN} \qquad (4\text{-}28)$$

and F_{KN} is the following expression:

$$F_{KN} = \left[\frac{1}{1 + \alpha(1 - \cos\theta)}\right]^2 \left[1 + \frac{\alpha^2(1 - \cos^2\theta)^2}{[1 + \alpha(1 - \cos\theta)](1 + \cos^2\theta)}\right] \qquad (4\text{-}29)$$

The factor F_{KN} is always less than 1.0, reaching the value of 1.0 only when the scattering angle θ is zero and scattering is described in the classical way. Also, as α approaches zero, the factor also approaches 1.0 and the scattering is classical. The integrations of equation 4-28 over all scattering angles produces the following expression, which states the total Compton scattering coefficient:

$$_e\sigma = \frac{3}{4}{_e\sigma_0}\left\{\frac{1 + \alpha}{\alpha^2}\left[\frac{2(1 + 2\alpha)}{1 + 2\alpha} - \frac{\ln(1 + 2\alpha)}{\alpha}\right]\right. \\ \left. + \frac{\ln(1 + 2\alpha)}{2\alpha} - \frac{1 + 3\alpha}{(1 + 2\alpha)^2}\right\} \qquad (4\text{-}30)$$

In this expression $_e\sigma_0$ is the classical zero angle Thomson scattering term of equation 4-27.

The data for the energy transfer coefficient in Figure 4.6 are obtained by multiplying the differential cross section expression of equation 4-28 by the ratio $E_e/h\nu$, which is determined from equation 4-20. The energy transfer differential cross section per unit solid angle is given in equation 4-31.

$$\frac{d\,_e\sigma_{tr}}{d\Omega} = \frac{d\,_e\sigma}{d\Omega}\frac{\alpha(1 - \cos\theta)}{1 + \alpha(1 - \cos\theta)} \qquad (4\text{-}31)$$

The integration of this expression over all angles yields the value of the energy transfer coefficient, $_e\sigma_{tr}$. The integrated expression is dauntingly complicated and of no particular pedagogical interest. Johns and Laughlin report the results of this effort in Hine and Brownell, *Radiation Dosimetry* (see Suggested Additional Reading), beginning on their page 49. The results are shown in Figure 4.6. The coefficient for energy scattered, $_e\sigma_s$, is simply the difference between the total coefficient and the energy transfer coefficient as calculated by equation 4-27.

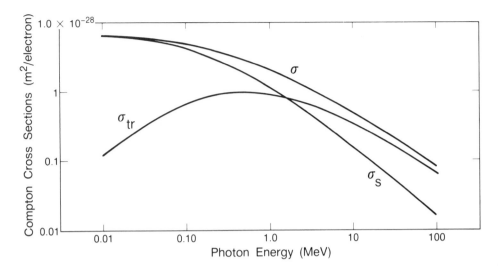

Figure 4.6   The total Compton coefficient, the scatter coefficient, and the energy transfer coefficient in water. These are calculated from the Klein–Nishina formulation as summarized by Johns and Cunningham in their Appendix A-2a (see Suggested Additional Reading).

Energy of Compton Electrons

It is possible to derive from the original Klein–Nishina formulation, equation 4-26, by a suitable manipulation, the spectrum of Compton electron energies. Readers are referred once more to Hine and Brownell's *Radiation Dosimetry* for the details. In general, the electrons may have any energy up to a maximum that is somewhat less than the energy of the photon, $h\nu$. For example, for an incident photon energy of 0.5 MeV the maximum electron energy is 0.331 MeV, and for an incident photon of 1.0 MeV the maximum electron energy is 0.796 MeV. The energy distributions go through a minimum, so the likelihood is greater for both low- and high-energy electrons than for the intermediate values.

Electron Binding Energy Effects

The implicit, and at times explicit, assumption of all the foregoing calculations of Compton scattering is that the electrons interacting with the incident photon are free. That is, their binding energy is zero or nearly so. The total cross section, assuming free electrons, as shown in Figure 4.6 indicates that the coefficient decreases downward in a monotonic fashion with increasing energy, and the cross section is at a maximum at lowest energies. Hubbell (1977) has recalculated the Klein–Nishina coefficients taking into account the electron binding energies. There is a sharp fall-off of the total coefficient at low energies (below 50 keV or so); but since the energy transferred coefficients are insignificant at these low energies, the impact on the energy transferred is minimal. Hubbell's calculation for carbon as the absorber shows that at 100 keV photon energy there is no detectable effect of binding energy of the electron.

PAIR PRODUCTION

If an energetic photon enters matter and if it has an energy in excess of 1.02 MeV, it may interact by a process called *pair production* (see Figure 4.7). In this mechanism of energy transfer, the photon, passing near the nucleus of an atom, is subjected to strong field effects from the nucleus and may disappear as a photon and reappear as a positive and a negative electron pair. The mechanism is one in which the scattering nucleus plays a more or less passive role. The state of the scattering nucleus before and after the event is the same, except for a very small change in its kinetic energy and momentum. There is no other excitation of the scattering nucleus. A photon has simply been converted into a positive and a negative electron. The kinetic energy transferred to the nucleus is generally insignificant; but since there is some nuclear recoil, energy and momentum are not conserved within the system of the disappearing photon and the two emerging electrons. It should be made clear that the two electrons produced, $e-$ and $e+$, are not scattered orbital electrons, but are created, *de novo*, in the energy/mass conversion of the disappearing photon. Because of the uncertainties about the energy and momentum gained by

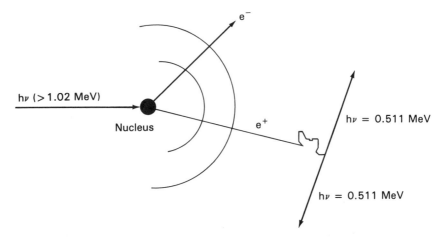

Figure 4.7 The events in the pair production scattering process.

the recoiling nucleus, it is not possible to undertake such elegant and complete analyses of the scattering process as with the Compton effect.

The *atomic cross section* for pair production in the neighborhood of a nucleus is proportional to Z^2. However, for photon energies above 20 MeV, we must correct to an *effective Z* in order to allow for the screening effect of the true nuclear charge by atomic electrons. This is, in detail, a complicated correction, but for first-order purposes the following equation describes the relationship of the atomic cross section for pair production:

$$_a\kappa \text{ (approx.)} = \ln h\nu \qquad\qquad (4\text{-}32)$$

where $_a\kappa$ is the atomic cross section due to pair production.

What will the kinetic energy distribution between the two electrons be? Since the energy equivalent of two electron masses is 2×0.511, or 1.022 MeV, the kinetic energy of the electrons produced by the process will be the difference between the energy of the incoming photon and 1.022 MeV.

$$h\nu - 1.022 = E_{\text{positron}} + E_{\text{electron}} \qquad\qquad (4\text{-}33)$$

The partition between the two electrons of the energy predicted by equation 4-33 is a complex function of the energy of the incoming photon and the atomic number of the scattering nucleus, but, to a very good first approximation, all partitions are equally probable, except for the extreme case of one electron having zero kinetic energy. The probability for this special case approaches zero. The annihilation radiation process for the positron is identical to that which has been considered before in the discussions of positron decay of a radioactive nuclide.

Annihilation Reaction

As the positron loses energy, finally reaching the Fermi energy, it will transfer its kinetic energy to the medium in the same way that such transfers occur with the negative electron; however, as it slows to near rest energy it will interact with a free electron to produce two annihilation photons of 0.511 MeV, the direction vectors of which are opposed. The release of a total of 1.022 MeV of energy accounts for the energy equivalence of the mass of the two electrons that have disappeared. If the momentum of the two-body system, electron and positron, is zero in the laboratory system, the two quanta of electromagnetic radiation will appear with exactly opposed direction vectors. This condition is rarely achieved, and, because of thermal agitation of both particles, the average momentum of the center of mass can be shown to be about $0.009 \, m_0 c^2$. As a result of this small momentum residue, there is a spread of about 0.5° in the angular distribution about the mean of 180°. If one photon is plane polarized, the other photon will have a plane of polarization at right angles to the first.

In addition to the usual annihilation reaction of the electron and the positron at near zero momentum, occasionally a high-kinetic-energy positron will be annihilated. This occurs in about 2% of the interactions. The momentum at the time of annihilation is not zero, and the resulting quanta will have energies in excess of 0.511 MeV and appropriate direction vectors for momentum balance.

Energy Transferred and Energy Absorbed

The energy transferred in the pair production process is uniquely determined by the incident photon energy less the mass equivalent of two electrons. The transferred energy is written as

$$E_{tr} = h\nu - 1.022 \tag{4-34}$$

The energy absorbed is much more difficult to determine, except experimentally, since the two electrons may lose energy to their environment by bremsstrahlung, as well as by energy transfer processes that result in energy absorption. There will always be the appearance of 1.022 MeV of energy as annihilation photons, which, in general, can be considered as lost to the system. Thus, at energies around 1.022 MeV or so, the threshold for pair production, little kinetic energy will be transferred by this process, regardless of the cross section for its occurrence. Annihilation photons have the same role as the scattered photon in the Compton process and the fluorescent photon in the photoelectric process in that they are considered as energy lost to the system (that is, not transferred).

Summary of the Pair Production Process

1. The threshold for the process is an incident photon energy of 1.022 MeV.
2. The process has a rapidly increasing cross section above this threshold energy.
3. The stopping power per atom varies approximately as Z^2.
4. The energy transferred is $(h\nu - 1.022)$ MeV.
5. The two annihilation photons must generally be considered to be energy that is not absorbed. They may undergo further scattering processes before they leave the system.

Triplet Production

Another version of the pair production process is one in which the incident photon interacts with an orbital electron, rather than with an atomic nucleus. The scattering electron, since it has small mass, will take on significant kinetic energy and will itself appear as a product of the interaction. In addition, an electron and a positron appear as the result of the disappearance of the incident photon. This process has a threshold energy of 2.04 MeV; but since its cross section for low Z matter is very low, it is of little significance for the interaction of radiation with biological materials. The process is called triplet production because two electrons and one positron are ejected from the site of the scattering event.

Variation of Cross Section with Energy

The variation of cross section of the pair production process for different materials as a function of energy is shown in Figure 4.8. As can be seen, there is only minimal contribution from this process at energies

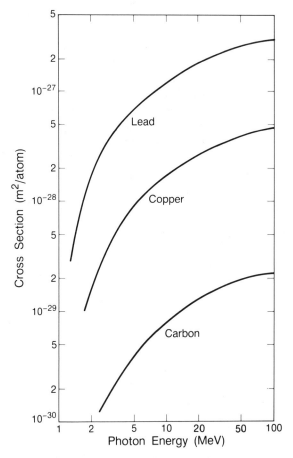

Figure 4.8 Atomic cross section for the pair production scattering process as a function of energy for lead, copper, and carbon. Note the strong dependence on Z of the absorber. The data are from Plechaty, Cullen, and Howerton (1975).

above the 1.022-MeV threshold and below 10 MeV for tissue equivalent elements such as oxygen and carbon.

BREMSSTRAHLUNG: RADIATIVE ENERGY LOSS

We have made several references to energy loss from high-kinetic-energy electrons by a radiative process. There are a number of mechanisms by which high-kinetic-energy particles can transfer energy to their surrounding medium. The more likely event for fast electrons is scattering interactions that transfer energy in the form of excitation or ionization. Another important process is the radiative transfer of energy when an electron is accelerated. As a high-speed electron (or any other particle) passes a nucleus, there will be a coulombic force that will deflect the path of the electron. Since in the case of the electron the mass of the nucleus is so much larger, there will be little disturbance of the nucleus. As a particle of mass M_p and charge ze passes near a nucleus of mass M_n and charge Ze, the particle will experience an electrostatic force resulting in an acceleration of the particle. The heavy nucleus is assumed not to be significantly accelerated, since it is taken that $M_n \gg M_p$. Because a charge undergoing acceleration radiates energy at a rate that is proportional to the square of the acceleration, the accelerated (decelerated) particle will emit electromagnetic radiation. Since by coulombic reasoning the acceleration can be shown to be proportional to $(zZe^2/M)^2$, the radiative energy loss will depend strongly on Z and the inverse of the mass M, since both are squared terms. Indeed, high-atomic-number absorbers are much more important for radiative energy loss, and electrons are manyfold more important contributors than even a proton.

From quantum mechanical considerations, Heitler (1954) devised a solution for the rate of radiative energy loss for electrons of energy less than 100 MeV.

$$S_{\text{rad}} = \frac{1}{\rho} \frac{dE}{dx} = 4r_0^2 \frac{N_e Z}{137} \left[\ln \frac{2(E + \mu_0)}{\mu_0} - \frac{1}{3} \right] \tag{4-35}$$

This process has been called *bremsstrahlung* from the German word for "braking radiation," in reference to the deceleration of the electron.

REFERENCES

HEITLER, W. (1954) *The Quantum Theory of Radiation*. Oxford University Press, New York.

HUBBELL, J. H. (1977) Photon mass attenuation and mass energy-absorption coefficients for H, C, N, O, Ar, and seven mixtures from 0.1 keV to 20 MeV. *Radiation Res.* **70**, 58–81.

KLEIN, O., and NISHINA, Y. (1929) Uber die Streuung von Strahlung durch freie Elektronen nach der neuen relativistischen Quantumdynamik von Dirac. *Z. Phys.* **52,** 853–868.

PLECHATY, E. F., CULLEN, D. E., and HOWERTON, R. J. (1975) *Tables and Graphs of Photon Interaction Cross-sections from 1.0 keV to 100 MeV Derived from LLL Evaluated Nuclear Data Library.* Report No. UCRL-50400, vol. 6, 1st revision. University of California Lawrence Livermore National Laboratory, National Technical Information Service, Springfield, Va.

SCOFIELD, J. H. (1973) *Theoretical Photoionization Cross Sections from 1 to 1500 keV.* Report No. UCRL-51326. National Technical Information Center, National Bureau of Standards, U.S. Department of Commerce, Springfield, Va.

SUGGESTED ADDITIONAL READING

EVANS, R. D. (1955) *The Atomic Nucleus.* McGraw-Hill Book Co., New York.

HINE, G. J., and BROWNELL, G. L. (1956) *Radiation Dosimetry.* Academic Press, Inc., New York.

JOHNS, H. E., and CUNNINGHAM, J. R. (1983) *The Physics of Radiology,* 4th ed. Charles C Thomas, Springfield, Ill.

PROBLEMS

1. It is given that the photoelectric effect cannot occur with a free electron and an incident gamma ray. It is also given that this fact has as its basis the conservation of momentum and energy. Show through equations for momentum and energy balance that photoelectric interaction with a free electron is not possible.

2. For each of the following initial photon energies, compute the maximum kinetic energy that the Compton scattered electron may have. Plot the results on appropriate coordinate paper.

$$\text{Photon energies (in keV): } 1, 5, 50, 100, 500$$
$$\text{(in MeV): } 1, 10, 50, 100$$

What can you conclude about the relationship of the photon energy to the maximum scattered electron energy?

3. Given the following data, compute the total mass attenuation coefficient of lead for 10-MeV photons, knowing that the total attenuation coefficient is the sum of those for the individual processes. Lead has a density of $11,350 \text{ kg m}^{-3}$.

$$\tau \text{ (photo)} = 0.168 \times 10^{-28} \text{ m}^2/\text{atom}$$
$$\sigma \text{ (coherent)} = 0.0093 \times 10^{-28} \text{ m}^2/\text{atom}$$
$$\sigma \text{ (Compton)} = 4.193 \times 10^{-28} \text{ m}^2/\text{atom}$$
$$\kappa \text{ (pair)} = 12.400 \times 10^{-28} \text{ m}^2/\text{atom}$$
Triplet coefficient is insignificant

4. For 10-MeV photons, the mass attenuation coefficient for air is 0.204. The average energy transferred (E_{tr}) is 7.37 MeV. The average energy absorbed (E_{ab}) is 7.10 MeV. Assuming a fluence of 10^4 photons traverse a volume of air of "thickness" 1.5 g/cm^2, compute the number of scattering events, the energy appearing as kinetic energy of electrons in the medium, the energy removed from the incident beam, and the amount of the original incident energy that reappears as bremsstrahlung.

5. Using the differential cross-section formulas made available for Compton scattering, compute the energy of a scattered photon appearing at a scattering angle of 45°, 90°, and 180° from the incident axis.

6. A photon of wavelength 0.20 nm is incident on a slab of carbon, and its wavelength is shifted by 0.0100%. What is the photon scattering angle and what is the maximum energy that the scattered electron could gain?

5

Energy Transfer
Processes

INTRODUCTION

Now that examination of the most important scattering processes for the interaction of high-energy photons with matter, with emphasis on tissue equivalent systems, is complete, it is necessary to examine the relative importance of each of these processes for tissue equivalent systems in particular, and to emphasize the next step in the transfer of energy ultimately from the photon to biologically important molecules in the living cell. It will be seen that the high-kinetic-energy electrons produced in the primary scattering process are the key element in the transfer of energy from the scattered photon to the molecular systems of the living cell.

It was shown in Chapter 4 that partial stopping powers (attenuation coefficients) can be assigned for each of the scattering mechanisms. In this way the relative importance of each of the processes can be examined separately at whatever energy is of interest. It was also shown in Chapter 4 that it is possible to partition the total stopping power for each of the scattering processes into the part that accounts for energy transferred and the part that accounts for energy absorbed.

The separate cross sections or attenuation coefficients can be constructed for the portion of energy transferred as well as the portion finally absorbed. The symbolism generally used is the following:

$\dfrac{\mu}{\rho}$ = total mass attenuation coefficient

$\dfrac{\mu_{tr}}{\rho}$ = total energy transferred mass attenuation coefficient

$\dfrac{\mu_{ab}}{\rho}$ = total energy absorbed mass attenuation coefficient

$\dfrac{\tau}{\rho} = \begin{array}{l}\text{photoelectric mass attenuation coefficient}\\ \text{(The subscripts } tr \text{ and } ab \text{ have the same meaning as before)}\end{array}$

$\dfrac{\sigma}{\rho} = \begin{array}{l}\text{Compton mass attenuation coefficient}\\ \text{(The subscripts have the same meaning, but often one sees}\\ \sigma_s/\rho,\ \text{referring to energy either scattered or, } \sigma_{tr}/\rho,\ \text{transferred)}\end{array}$

$\dfrac{\kappa}{\rho} = \begin{array}{l}\text{pair production mass attenuation coefficient}\\ \text{(Again the subscripts have the same significance)}\end{array}$

IMPORTANCE OF THE COMPTON PROCESS IN TISSUE SYSTEMS

It was suggested earlier that, for most of the photon radiations that might contribute to the ultimate dose in tissue equivalent material, the Compton process would be the most significant contributor to absorbed energy. The cross sections for the Compton process were discussed at length in Chapter 4, with attention to the Klein–Nishina formulations of these cross sections. The Klein–Nishina calculations are those made for a free electron, and they do not make allowance for the effects of electron binding energies on the total cross section. The total Compton cross section for a free electron is a continuously decreasing function, with the cross section essentially unchanged from its maximum value until the photon energy exceeds 50 keV. Depending on the binding energy of the orbital electron involved in the Compton scattering event, there can be a significant deviation of the actual cross section from that calculated from the Klein–Nishina formulation. There is very little difference between the cross sections for low-atomic-number atoms and for the free electron when the incident photon energies are above 100 keV. At lower energies, and particularly for higher Z absorbing media, there will be significant differences between the K–N calculations and those that take into account the electronic binding energies. More recent calculations have provided such cross sections, in which allowance has been made for the binding energy. The detailed cross sections for several elements were calculated by Hubbell (1977). These calculations are not reexamined here, but the results are shown in Figure 5.1, which gives the total electronic Compton coefficients for a free electron, as well as the corrected electronic Compton coefficients for carbon ($Z = 6$). Also shown are the electronic Compton coefficients for energy transferred, σ_{ab}. Note again that the curves plotted are for *electronic* cross sections.

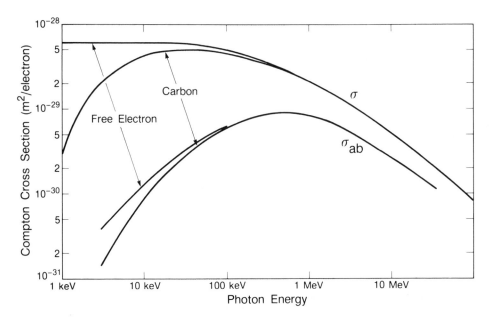

Figure 5.1 Electronic Compton cross sections for the free electron, and carbon as a function of photon energy. σ_{ab} is the cross section for energy absorbed, and σ is the total Compton electronic cross section. Also shown is the total cross section for the free electron.

Total Attenuation Coefficient

In addition to the Compton process, which is of overriding importance for photon energies of 10 keV to 10 MeV, both the photoelectric process and, to a smaller extent, pair production contribute to the total deposition of energy. The contributions of each of the three processes are shown in Figure 5.2A and B. It should be clear that only one type of interaction can occur with a photon in any one event. The likelihood of each event is defined by the attenuation coefficient or the cross section for each event. The likelihood of *any* single event occurring is the sum of the likelihoods for all classes of events. Since the attenuation coefficient defines the probabilities for all possible scattering events, it is possible to state the total probability of an event as follows:

$$\mu_{total} = \tau_{photoelectric} + \sigma_{coherent} + \sigma_{Compton} + \kappa_{pair} \qquad (5\text{-}1)$$

It was earlier shown that the coherent cross section was not responsible for the deposition of significant energy, and indeed the cross section is only of significance at energies below 10 keV when we are considering low-atomic-number materials. For this reason, the coherent cross sections are not usually tabulated or included in total cross-section calculations for tissue equivalent material.

The cross sections in equation 5-1 can be made applicable as desired to mass, electronic, and atomic cross sections by the use of the appropriate subscripts to indicate the intention of the user. The appropriate cross sections are tabulated as a function of photon energy in many standard reference works. The reader interested in the details is referred to Hubbell (1977).

From the tabular data of Hubbell, Figure 5.2 has been prepared to show the total mass attenuation coefficients for lead and water. Also shown are the energy absorbed coefficients (mass, μ_a/ρ) and energy transferred coefficients (mass, μ_s/ρ).

From the data presented in Figure 5.2, it is possible to see the relative significance of each of the scattering processes and to observe how these relative contributions are affected by Z. The presence of the K edge for lead in the total attenuation coefficient at just under 100 keV identifies the large contribution of the photoelectric process for this Z (lead) and this energy. For water, which to all intents can be considered to be tissue equivalent when corrected for density, there is a significant contribution from the photoelectric process only at the lower energies, that is, below a few tens of keV. The key curve to examine in Figure 5.2B is the total mass energy absorption coefficient for water, labeled μ_a/ρ. This curve shows very well that the significant contribution to absorbed energy in this tissue equivalent material is from the Compton process, which contributes the hump in the cross section from 100 keV to 10 MeV. The principal contributor to dose in all soft tissue systems will always be that which arises from the Compton scattering process for all the energies that might present significant exposure hazards for biological systems. High Z materials, such as copper and particularly lead, have large contributions from the photoelectric process, even for photon energies well in excess of 1 MeV.

Total Attenuation Coefficients for Mixtures

The total attenuation coefficients (and any partial coefficient, for that matter) can be calculated by using the simple proportionate contribution of each element in a mixture by attributing the correct attenuation coefficient or cross section to the appropriate *mole-fraction* represented by each element in the mixture.

Example

What is the total Compton mass attenuation coefficient for water for photons of 1.25 MeV?

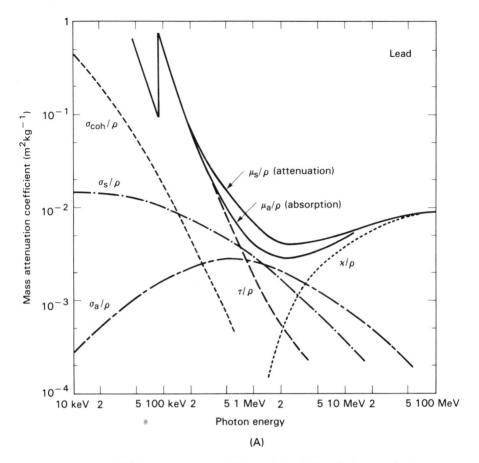

Figure 5.2 (A) Mass attenuation coefficients for lead. The relative contributions of the three scattering processes as a function of energy are shown. Also shown are the energy absorbed cross sections. (B) Mass attenuation coefficients for water. Same notation as for part A.

$$\frac{\sigma}{\rho} \text{ for hydrogen at 1.25 MeV} = 0.1129 \text{ cm}^2/\text{g}$$

$$\frac{\sigma}{\rho} \text{ for oxygen at 1.25 MeV} = 0.0570 \text{ cm}^2/\text{g}$$

Since 1 mole of water contains 16 g of oxygen and 2 g of hydrogen, the mole fraction of each is $\frac{2}{18}$ for hydrogen and $\frac{16}{18}$ for oxygen. The total cross section can then be computed from the tabulated elemental total cross sections.

$$\left(\frac{\sigma}{\rho_{\text{water}}}\right) = \frac{2}{18}(0.1129) + \frac{16}{18}(0.0570)$$
$$= 0.0632 \text{ cm}^2/\text{g}$$

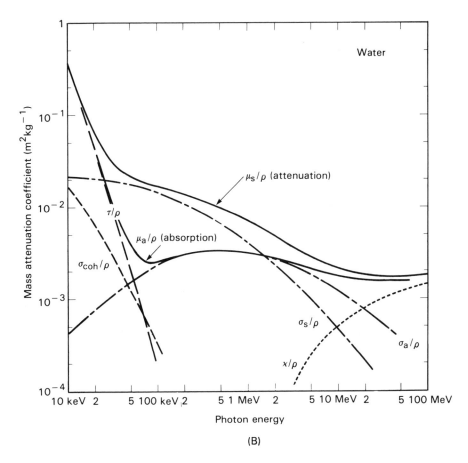

Figure 5.2 *Continued*

INTERACTION OF CHARGED PARTICLES WITH MATTER

In the processes that have been described in the previous section there is one characteristic of all the scattering processes by which photons transfer their energy to their environment. In every case of a photon scattering process, an energetic electron is one of the products. A truism that will appear repeatedly in this text is that the principal mechanism by which energy is finally deposited in the chemical systems of a biological environment is through the interactions of these high-kinetic-energy electrons with the absorbing medium. To make a useful example, suppose a 1-MeV photon undergoes a scattering event, presumably a Compton event. Suppose that the scattered electron leaves the event with a kinetic energy equal to 200 keV. The scattering of the photon has produced a single electron with a large kinetic energy. This energy must be deposited in the environment as the electron passes. In the example the 200-keV electron could give rise to approximately 15,000 ionizations of water molecules. It is clear that in this process lies the most important transfer of energy from a photon to tissue component molecules.

All energetic charged particles, whatever their mass, will lose kinetic energy to their environment chiefly through coulombic interactions of the charge on the moving particle with the charges on the electrons and nuclei of the matter through which they are passing. Since electrons are significantly more plentiful then charged nuclei in any medium, the preponderance of the interactions will be with the electrons of the medium.

The general formulation for this interaction is diagrammed in Figure 5.3. For this formulation, the mass of the electron of the medium in which the interaction is taking place is m_0. The distance b in the figure is the distance of closest approach of the two particles. It is sometimes called the *impact parameter*. Consider, for the moment, that the mass of the particle moving with velocity v is large, of mass M. As M moves along its trajectory, e^- will experience a force along the line passing through the positions of M and e^-. This force will have a component normal to the trajectory line, shown as F_y in Figure 5.3. There is a force component in the horizontal direction shown as F_x. As M passes the point normal to the position of the electron, there will be a reversal of the direction of the vector, F_x, and the net force on the electron in this direction is zero. The magnitude of F along the line MQ is

$$F = \frac{kze^2}{r^2} \tag{5-2}$$

where the constant k is $8.9875 \times 10^9 \, \text{N m}^2/\text{C}^2$. This constant is the constant of proportionality in Coulomb's law.

Since M is taken to be large compared to the electron mass, it will be deflected very little from its trajectory, but some energy ΔE will be

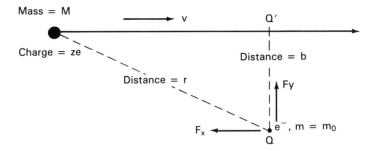

Figure 5.3 Interaction of a moving heavy particle of mass M passing an electron of the medium.

taken from the moving particle. The momentum imparted to the electron, Δp, is given by

$$\Delta p = \int_{-\infty}^{+\infty} F_y \, dt = \int_{-\infty}^{+\infty} \frac{zke^2}{r^2} \cos \theta \, dt \tag{5-3}$$

From classical coherent scattering theory, an expression has been developed in terms of what has come to be called the *classical electron radius*, r_0. This expression is

$$r_0 = \frac{ke^2}{m_0 c^2} = 2.81794 \times 10^{-15} \quad m \tag{5-4}$$

If equation 5-3 is written with all variables in terms of θ and the definition of r_0 is taken from equation 5-4, the following simplification can be written:

$$\Delta p = \frac{zke^2}{bv} \int_{-\pi/2}^{+\pi/2} \cos \theta d\theta = \frac{2zke^2}{vb} = \frac{2zr_0 m_0 c^2}{vb} \tag{5-5}$$

Δp is the momentum transferred to the resting electron. The energy taken from the heavy particle is

$$\Delta E(b) = \frac{(\Delta p^2)}{2m_0} = \frac{2z^2 r_0^2 m_0 c^4}{v^2 b^2} = \frac{z^2 r_0^2 m_0 c^4 M}{b^2 E} \tag{5-6}$$

where E, the kinetic energy of the moving particle, is $E = \frac{1}{2}Mv^2$.

The parameter b, the distance of closest approach of M and e, is called the *impact parameter. A very important point must be made.* The transfer of energy is inversely proportional to the kinetic energy of the moving particle. It is also, then, inversely proportional to the square of the velocity of the moving particle. The energy transferred in the interaction depends on the charge on the particle, Z, and its velocity. The result is that the rate of energy transfer per unit path length rises dramatically as the moving particle approaches the end of its path.

The derivations given in equations 5-2 to 5-6 are not complete. The reader is referred to any textbook of modern physics for a complete derivation. (See, for example, Attix, Roesch, and Tochilin in Suggested Additional Reading.)

The special case given in Figure 5.3 is for a moving heavy particle and a Fermi energy electron of the medium. The special case has general applicability. There are several possibilities.

1. M is large compared to m_0, that is, M is a charged nucleus or nucleon.
2. M is the same as m_0, that is, M is an electron.
3. Either M or m_0 is at rest.

Case 1.   This case is assumed in the preceding model. The details of the expressions in equations 5-2 to 5-6 are not important, nor is its derivation, for the applications of radiation biophysics. The important consideration is to examine the denominator of the expression. To restate a point made previously, the impact parameter b will vary stochastically, but the important determinant of the rate of energy loss is the squared velocity term, which appears in the denominator. *Most of the moving particle's energy will be given up as it begins to approach the rest state.*

Case 2.   In the case of two electron masses interacting, we can no longer assume that the trajectory of either is unaffected. The calculation is classically the same but complicated. The important point is that the path of both the projectile electron and the target electron are affected. Energy loss is still inversely proportional to the velocity of the projectile electron squared, but the track of the projectile electron will be unpredictable and stochastically determined.

Case 3.   This particular case has no practical relevance.

FINAL STEPS IN ENERGY ABSORPTION

It is certainly clear at this point that the high-energy photon that enters an absorbing system will not cause ionization damage in important biomolecules in a single step. In every case, the photon undergoes scattering events that give rise to high-kinetic-energy electrons, which will ultimately deposit energy in molecular species. These electrons will generally have kinetic energies in the kilo- to megaelectron-volt range, and to be effective in producing ionization damage to important biomolecules, these electrons must undergo energy transfer events that are on the order of tens of electron volts. For example, water makes up most of the milieu of the living cell. The first ionization potential for water is 12.6 eV. How do these transfer processes in the few eV range happen?

The four principal processes by which energy is transferred from a high-kinetic-energy electron are discussed next. These processes are of varying significance, depending on the energy of the electron and the Z of the absorber. Figure 5.4 shows the life history of an entering fast electron in a low Z absorber.

Multiple Collision Energy Transfer

In this process the electron undergoes multiple small ionization losses by coulombic interaction with the bound electrons of the medium. Since the momentum transfer per collision is small for individual events, there will be only small changes in the direction vector of the incoming electron per event, but overall there will be significant change in path

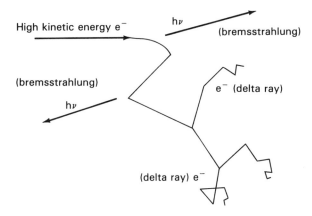

Figure 5.4 Life history of a fast electron.

after many collisions. The multiple collision process is the process of predominant importance in the deposition of energy in the molecular species in the target. Occasionally, the energy transfer in one of these interactions will be large enough that the target electron will have sufficient kinetic energy that it, in turn, engages in multiple collision energy transfers. These electrons are shown in Figure 5.4 as delta rays.

Delta ray (δ ray) is shorthand nomenclature for an electron from the absorbing medium to which has been imparted enough energy that this secondary electron will have a track of its own, and multiple collision energy transfers can occur from this new electron. When the primary charged particle is an electron, the δ ray may have a kinetic energy as high as one-half of the kinetic energy of the primary electron. If the δ ray has a high enough energy, it, in turn, can produce a δ ray.

Photoelectric Process

The photoelectric process with an electron can occur under appropriate conditions of kinetic energy of the electron and the Z of the absorbing medium. As in the photoelectric process with a photon, the primary electron is absorbed, ejecting usually a K or L orbital electron. The empty orbital position is then filled by an outer orbital electron, and a photon is emitted with an energy equal to the differential in the binding energies of the two orbitals involved. Little energy is transmitted to the absorbing medium, and the regenerated photon must undergo the normal scattering processes for its energy to be transferred. The process follows the usual rules for the photoelectric process as for photons. It is generally only important at higher Z.

Bremsstrahlung Generation

This process is the emission of photon radiation as the result of deceleration of the charged particle through coulombic nuclear interactions. This process is important when the electron is still at high kinetic energy. The process can occur even for low-energy particles, but the contribution of the process to energy loss at low energies can generally be disregarded. For electrons the loss due to bremsstrahlung becomes important for energies above 10 MeV in lead and above 100 MeV in water. In general, bremsstrahlung radiation becomes important only when the kinetic energy of the particle is considerably greater than the rest mass energy of the particle. For a detailed discussion of the production of bremsstrahlung, Merzbacher and Lewis (1958) is a recommended source.

Direct Collisions

Collisions in which the electron is stopped and gives up all its energy as bremsstrahlung are very rare but do occur occasionally.

LIFE HISTORY OF A FAST ELECTRON

The electron enters at high energy and gives up part of its energy in bremsstrahlung interactions. These interactions may or may not occur before ionization-type collisions take place, which, in any case, predominate. While the electron still has a relatively high kinetic energy, it can impart enough of this energy to another electron. This second electron may have significant velocity and thus may produce further ionizations. These secondary electrons are called *delta rays* and are themselves very important in producing ionization damage in important molecules.

Remember that as the initial electron undergoes multiple collisions it is losing energy and therefore velocity. As this happens, the amount of energy transferred per interaction is increasing according to the inverse-velocity-squared principle mentioned earlier.

Finally, the initial electron comes to rest. The energy imparted in the many collisions will be present as both ionization (separation of an electron from its nuclear charge, leaving a positively charged residue and a negatively charged electron, an ion pair) and excitation (the raising of an orbital electron from its usual ground state in the orbital to an excited state).

ABSORBED DOSE AND KERMA

To reiterate a point made several times, of the energy transferred in scattering processes, only the energy absorbed in the local medium is of any importance in establishing the final level of damage in the medium.

In Chapter 1, two quantities were discussed, *kerma*, which is a measure of energy transferred, and *dose*, which is a measure of energy absorbed. To restate,

$$\text{absorbed dose } D = \frac{dE_{ab}}{dm} \tag{5-7}$$

or dose is the absorbed energy per unit mass. We can restate dose in terms of the incident fluence as follows:

$$D = \Phi \frac{\mu}{\rho} (_{av}E_{ab}) \tag{5-8}$$

where $_{av}E_{ab}$ is the average energy absorbed per interaction event.

Kerma is defined in terms of fluence in a similar way for energy transferred.

$$K = \Phi \frac{\mu}{\rho} (_{av}E_{tr}) \tag{5-9}$$

where $_{av}E_{tr}$ is the average energy transferred per interaction event.

As the photon fluence enters into an absorber from another medium, the initial kerma will be a maximum at the entrance interface and then will decrease according to the laws of attenuation discussed earlier. Dose, on the other hand, will have a value less than kerma at the entrance surface; but after penetrating the medium to some depth, the dose will exceed the kerma. The explanation for this is simple. First, Figure 5.5 is a diagram of dose and kerma as a function of depth for a homogeneous medium. The reason for the difference between kerma and dose is the presence or absence of a condition called *charged particle equilibrium* (CPE). The definition of charged particle equilibrium given in Chapter 1

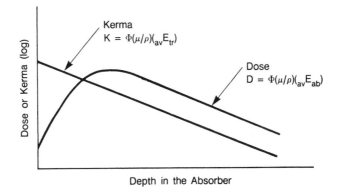

Figure 5.5 Logarithm of dose or kerma as a function of depth in the absorber. The scales are relative and not meant to be exact. The differences between kerma and dose are exaggerated to make the issue clear.

may be paraphrased as follows: for a given volume element, for every particle carrying energy V out of the volume, there is another particle carrying energy V into the volume. If it is assumed in the example of Figure 5.5 that the medium just before the zero depth interface is of a much lower density than that of the medium in which dose determination is taking place (in most cases for biological dosimetry this lower density is air and the absorber is tissue), the kerma just before the interface will be determined by the mass attenuation coefficient for air. Just after the interface the kerma will be determined by the mass attenuation coefficient for the medium. Since the density of air is so much lower, the cross-sectional frequency for production of high-kinetic-energy electrons just before the interface will be very much lower. These electrons, generally directed in the forward direction, will contribute to the low *dose* at the surface of the new medium. After the interface, there will be a slow buildup of fast electrons from interaction in the medium. These will contribute to dose at a finite distance in the forward direction. At some point, charged particle equilibrium is reached, where the rate of forming new fast electrons is just equal to the rate of their stopping. Since fluence is falling, following the attenuation law, kerma is falling. Dose will reach a maximum near a point not far beyond the point where charged particle equilibrium was established. Electrons contributing to dose at any position x arose at some position $x - dx$, which is at a lesser depth in the medium; therefore, the position of maximum dose will be slightly beyond the CPE point. After CPE is established, the dose will decrease following the decrease in the particle fluence.

NEUTRON INTERACTIONS IN TISSUE

Neutrons are uncharged particles that are incapable of the typical coulombic interactions attributed to charged particles. Neutrons are, however, often major contributors to tissue dose, and because of their high LET (*vide infra*) their biological effectiveness is often significant. Neutrons are generated over a very wide range of energies, depending on the processes from which they arose. Thermal neutrons, having kinetic energies of a small fraction of an electron volt, and which owe their name to the fact that their energy is approximately kT, are produced in nuclear reactors, along with neutrons with energies up to several MeV, from the fission process. Neutrons with energies of 14 MeV or so are the product of the fusion process, and neutrons of several hundreds of MeV are produced by reactions in space and in and around large physics accelerators. The processes of energy transfer to tissue are different at all these energies.

The processes of energy transfer from neutrons are the following:

1. Elastic scatter
2. Inelastic scatter
3. Nonelastic scatter
4. Neutron capture
5. Spallation

The significance of each will depend on the energy of the neutron. Any combination of these processes is possible for a sufficiently energetic neutron. Capture is, however, a terminal process, leading to the disappearance of the neutron. For tissue systems the spallation reaction is important only at high energies, well above 20 MeV per neutron. The details of each process will be discussed next.

Elastic Scatter

Elastic scatter is the kinetic interaction of an energetic neutron with a nucleus of the absorbing medium in which classical kinematics describes the energy transfer. The elastic scatter process is important for neutrons with energies up to 14 MeV or so. The principal scattering nucleus in the absorbing medium for tissue equivalent systems is the proton. Additional scattering also occurs with oxygen, carbon, and nitrogen, in that order of importance. For the dosimetry of neutron absorption, it is important that elastic scattering, particularly for hydrogen, can be considered to be nearly isotropic for energies below 14 MeV.

When a neutron is scattered elastically by an atom, the transferred energy E_t to the target nucleus is given by

$$E_t = E_n \frac{4M_a M_n}{(M_a + M_n)^2} \cos^2\theta \qquad (5\text{-}10)$$

The initial neutron energy is E_n, and θ is the recoil angle in the laboratory system of coordinates. M_a is the mass of the target nucleus, and M_n is the mass of the neutron. The average energy imparted to the target nucleus is described by

$$\overline{E}_t = \frac{2E_n M_a M_n}{(M_a + M_n)^2} \qquad (5\text{-}11)$$

The elastic cross sections are highest at the lowest energy, and except for the resonance peaks that are found for oxygen, carbon, and nitrogen, the cross section decreases smoothly to a plateau level at about 14 MeV. For the higher energies, elastic scatter with hydrogen accounts for about 85% of that part of the dose that is from elastic processes, while for neutron energies below 250 keV, hydrogen elastic scattering accounts for 95% of the dose.

The average energy transferred to the hydrogen recoil nucleus can be computed from equation 5-11 to be about one-half of the neutron energy. This energy is then converted into ionization and excitation in the absorbing medium by the processes described in the preceding section for moving charged particles in general. In this case the high-kinetic-energy particle is a proton, which interacts with the electrons of the medium.

Inelastic Scatter

A specialized definition of inelastic scatter is applied for purposes of neutron dosimetry. A neutron may undergo a capture process with a nucleus of the medium to form a transitory compound nucleus that reemits a neutron of different energy. The product nucleus usually is in an excited state, which leads to the emission of a gamma ray. Since a neutron is absorbed and a neutron is reemitted, the product nucleus is identical with the starting nucleus. A typical inelastic scatter event can be the following:

$$^{14}\text{N} \, (\text{n, n}') \, ^{14}\text{N}: \quad E_\gamma = \text{approx. 10 MeV}$$

A number of these reactions are possible with nitrogen, carbon, and oxygen in tissue equivalent material. The reactions are only kinematically possible if the energy of the neutron is greater than the threshold energy necessary for conservation of energy and momentum. The energy transfer to the medium in inelastic scatter is by way of the interactions of the de-excitation photon. Because of the generally high energy of the de-excitation photons, the amount of energy transferred to tissue equivalent systems from them is small.

Nonelastic Scatter

Nonelastic scatter differs from the inelastic scattering in that a secondary particle, not a neutron, is emitted after capture of the initial neutron. A reaction typical of nonelastic scatter is

$$^{12}\text{C} \, (\text{n}, \alpha) \, ^{9}\text{Be}: \quad E = 1.75 \text{ MeV}$$

With this type of reaction, there is energy transfer to tissue from the charged particle produced (an alpha particle in the example) and in many cases from a de-excitation gamma ray.

Nonelastic scattering reactions are generally important for higher-energy incident neutrons. There are very few reactions for which the cross sections are significant below 5 MeV, and these reactions are typically predominant for the very high energy neutrons.

Neutron Capture

Neutron capture differs from nonelastic scatter only through convention. The capture processes occur at low energies. Capture processes account for a very significant fraction of the tissue dose for very low energy neutrons, and capture is almost the entire source of tissue dose for thermal and near thermal neutrons. An example is the capture of a thermal neutron (E_n = 0.025 eV) by nitrogen in the following reaction:

$$^{14}\text{N (n, p) } ^{14}\text{C:} \quad \text{proton energy} = 0.58 \text{ MeV}$$

Another very important reaction for thermal neutrons is the capture by hydrogen:

$$^{1}\text{H (n, }\gamma\text{) } ^{2}\text{H:} \quad \text{gamma energy} = 2.2 \text{ MeV}$$

The latter reaction predominates for contribution to dose from thermal neutrons. Since the hydrogen capture gamma ray has an energy of 2.2 MeV, its contribution to dose will vary significantly with the size of the absorbing target. The larger the target volume is the more important this reaction becomes.

Spallation

Spallation is a process in which, after the neutron is captured, the nucleus fragments into several parts. The process is important only at energies in excess of 100 MeV for the incident neutron, and the cross sections increase to energies as high as 400 to 500 MeV. The carriers of energy for contribution to dose are the several neutrons and the de-excitation gamma rays that are emitted in the spallation reaction.

Kerma and Dose from Neutrons

The computation or measurement of kerma or dose for neutrons is an extremely complicated matter, depending not only on the energy spectrum of the neutrons, but also on the elemental composition of the absorber medium and the volume of the absorber. The dose will be derived from energy transfer from short-range recoil protons, from other particles such as alphas and protons resulting from capture and inelastic reactions, and from gamma rays ranging in energy from less than 1 MeV to upward of tens of MeV. The details are beyond the scope of the text, and the reader is referred to Attix, Roesch, and Tochilin (see Suggested Additional Reading).

Some generalizations are worth recording. The important neutron energies for health protection purposes are from thermal up to a few MeV. For these energies some processes are predominant. For the thermal energy domain, hydrogen capture gamma rays will be the significant contributor to dose. For higher energies up to a few MeV, the hydrogen recoil (elastic scatter) reaction will be the predominant dose contributor.

REFERENCES

HUBBELL, J. H. (1977) Photon mass attenuation and mass-energy absorption coefficients for H, C, N, O, Ar, and seven mixtures: from 0.1 keV to 20 MeV. *Radiation Res.* **70,** 58–81.
MERZBACHER, E., and LEWIS, H. W. (1958) X-ray production by heavy charged particles. In *Handbuch der Physik*, S. Flugge, ed., **34,** 169–192. Springer, Berlin.

SUGGESTED ADDITIONAL READING

ATTIX, F. H., ROESCH, W. C., and TOCHILIN, E. (1968) *Radiation Dosimetry*, Volume I, *Fundamentals*, 2nd ed. Academic Press, Inc., New York.
EVANS, R. D. (1955) *The Atomic Nucleus*. McGraw-Hill Book Co., New York.

PROBLEMS

1. A 3.3-MeV neutron undergoes an elastic scattering event with a proton. The emerging scattered photon is seen to have a displacement of 120° from the trajectory of the incoming neutron. What is the kinetic energy of the scattered photon?

2. A thermal neutron, with an energy of 5 eV, scatters off a resting state hydrogen nucleus. Show that the interaction must be elastic and that energy and momentum are conserved.

3. Polystyrene, $(C_8H_8)_x$, has an unknown total mass attenuation coefficient at 1.25 MeV. Its density is 1.040. The total mass attenuation coefficient for water at this energy is 0.00632 cm²/g. What is the total mass attenuation coefficient of polystyrene?

4. Establish that the dimensions given in equation 5-4 for the classical electron radius are correct, and show that the value of r_0 is 2.817×10^{-15} m.

5. A proton with an energy of 200 MeV in passing through a tissue medium gives rise to a high-kinetic-energy electron with an energy of 1.2 MeV. Evaluate the impact parameter, b, in appropriate units. For the same value of b, what would be the kinetic energy of the tissue electron if the projectile had been an electron?

6

Radiation
Chemistry

INTRODUCTION

The development in the previous chapters on scattering theory and energy transfer from high-kinetic-energy electrons made it clear that the interaction of radiation with molecules in the absorbing medium is not by way of the initial photon, but rather through the transfer of energy from the high-kinetic-energy charged particles created by the scattering processes. Except in unusual cases, such as the interaction of neutrons with tissue systems, these secondary charged particles will be high-kinetic-energy electrons. The interaction of these particles results in the transfer of part of the electron's kinetic energy to molecules of the absorbing medium. The typical energy transfer per event for 20-keV electrons is shown in Figure 6.1.

STOCHASTIC NATURE OF ENERGY TRANSFER

The stochastic quality of the electron track as it deposits energy is not adequately described by the usual mathematical formulations. The Bethe–Bloch stopping power equation (which is discussed in Chapter 13) describes the process of energy transfer from a slowing down electron as a continuous process. This has come to be known as the *continuous slowing down approximation*. This approximation, which is suitable for general physics questions, proves to be inadequate for describing the

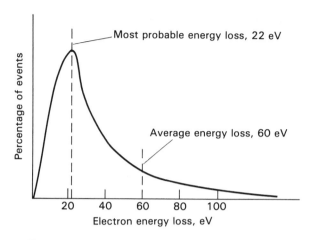

Figure 6.1 Histogram for first collision energy loss for 20-keV electrons. The modal or most probable energy loss is shown as 22 eV. The average loss per event is about 60 eV. We might compare the modal and average energy losses with the first ionization potential of water, which is 12.6 eV. The energy deposited in a modal event, 22 eV, might be barely sufficient to ionize two molecules of water to the first ionization state.

events that take place that ultimately lead to molecular alterations in the medium. For example, if we examine the tracks of two electrons of equal starting kinetic energy and identical starting location, the distribution of events along the track will be different, and so will be the physical locations of the tracks.

Spurs, Blobs, and Tracks

Figure 6.1 indicates that energy transfer is generally the result of the deposition of energy in discrete events, varying over a wide range of energies. Mozumder and Magee (1966) developed a shorthand terminology for these events. They classified them as follows:

> Spur: from 6 to 100 eV
> Blob: from 100 to 500 eV
> Short track: from 500 to 5000 eV

The need for such colorfully descriptive terms arises as we attempt descriptions of the events following the energy transfer from the secondary electron. The absorbed energy in the molecule of the medium can lead to the ejection of one or more electrons from the molecule, resulting in an ion pair or *ionization* of the target molecule. The absorbed energy may also lead to the raising of electrons from a ground state to a higher energy state, a process called *excitation*. If the energy utilized in raising the orbital electron to some new excited state is greater than the energy needed for ionization, but the electron is still associated with its parent nucleus, the process is often called *superexcitation*.

Molecules in all the forms described (ionized, excited, or superexcited) are very unstable. Electronic configurations are rearranged and interactions with other molecules will occur.

As a result of some of the secondary electron scattering interactions, *delta rays* are also formed, electrons that themselves have enough kinetic energy as the result of the interaction with the secondary electron to create a track of their own. The same classification is possible along this new track. According to Mozumder and Magee (1966), a 1-MeV electron deposits 65% of its energy in isolated spurs, 15% in blobs, and 20% in short tracks. The latter are not delta rays, but rather are scattered electrons that have a relatively small amount of kinetic energy as the result of the scattering interaction with the secondary electron. All this energy, in the case of the short track, is deposited in a more or less continuous fashion. The stochastic nature of these events (spurs, blobs, and short tracks) is important in understanding the direct action of radiation on, for example, water. The energy deposited as the result of a secondary electron interaction will be concentrated in the small volume of the spur, blob, or short track, and, depending on the amount of energy deposited, several or many ionizations or excitations may result. Within this small interaction volume, the products may interact with each other, but the individual events will be far enough apart that interactions of the products of the event between energy deposition sites are unlikely. An example will illustrate the point. Suppose a high-kinetic-energy secondary electron deposits energy in individual spurs that are, on average, about 4000 Å apart. If the medium is water, the H radical formed has a diffusion constant of 8×10^{-5} cm^2/s. In 1 μs, the average lifetime of the radical, it will diffuse about 1800 Å. The important point is that interaction of spurs and other small events will be minimal except for high LET radiations.

To summarize, direct action of the secondary electrons on the medium will deposit energy in randomly located, small interaction volumes in which one to a few reaction products will be formed by the deposition of energy from the secondary electron. Interactions may occur within the small interaction volumes, but interactions will be very unlikely for reaction products between separate interaction sites such as spurs, blobs, and short tracks. The average spur is estimated to be about 10 Å in radius and, in the case of water as an energy-absorbing medium, will contain an average of six or so radical products.

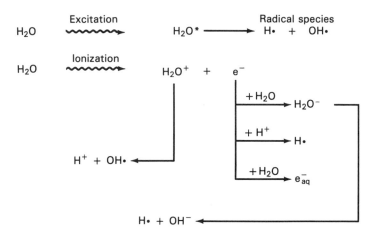

Figure 6.2 Products formed as the result of excitation and ionization of water. After the initial physical event, the excited or ionized water molecules dissociate to form the primary products shown. These, in turn, react with water or hydrogen ion, as shown.

RADIATION CHEMISTRY OF WATER

Primary Products of Radiolysis

Water makes up a large part of the milieu of living systems, and, not surprisingly, for living systems a good deal of the energy transfer of the initial events of ionization and excitation takes place with this molecule. The direct effect of excitations and ionizations by the secondary electron is the predominant reaction in living systems for low LET radiation. The initial reactions, which occur on a time scale of 10^{-16} to 10^{-12} s, are shown in Figure 6.2. The end result of this initial radiation chemistry of excitation and ionization in a water medium is to produce the following products: The *primary products* are H_2O^* and its dissociation products, H·, and OH·, and H_2O^+ plus e^-.

Further Radical Chemistry

The *immediate products* of the combination of the preceding with water molecules or hydrogen ion are as follows: The ionized water molecule H_2O^+ dissociates to H^+ and OH·. The free electron interacts with water to produce H_2O^-, which, in turn, dissociates to H· and OH^-; or it reacts with water to produce e_{aq}^- or with hydrogen ion to produce H·. The existence of e_{aq}^- was widely disputed for many years, and the predominant reducing radical was variously thought to be H·, H_2O^-, and several other more exotic forms. In 1953, Platzman described the solvated, or hydrated, electron, e_{aq}^-, and that description has now come to be widely accepted. Its identity was principally confirmed by kinetic measurements. The hydrated electron is stable with respect to its reaction with water shown in Figure 6.2, the reaction rate constant for this reaction being about 200 mol s^{-1}.

To recapitulate, the products formed in the first 10^{-11} s are H·, OH·, and e_{aq}^-. Also formed are H^+ and OH^-, but since they are ubiquitous in their distribution as the normal products of the dissociation of water, their formation is of little importance.

These early reactions in the chain of events leading to the final products of water radiolysis occur without any dependence on diffusion. The products formed by the initial radiation event are able to react with water molecules and/or hydrogen ions, which are, to a first approximation, immediately accessible.

Some further remarks are necessary with regard to the aqueous electron, e_{aq}^-. This entity is also sometimes called the *hydrated* electron. Electrons emitted as the result of water ionization lose their kinetic energy by multiple interactions with their surrounding environment, finally approaching the Fermi energy. After electrons are sufficiently slowed down, when their kinetic energy has fallen to 100 eV or less, and if they have not been captured by either of the processes outlined in Figure 6.2, that is, reaction with water or hydrogen ion, they attract the permanent dipoles of several water molecules to form the hydrated electron, e_{aq}^-. This new entity has a lifetime in neutral water of about 2×10^{-4} s. The hydrated electron is more stable than the free electron.

In the first of the ionization–excitation events of the fast electron, sometimes called a spur by radiation chemists, possibly only 2 to 5 water molecules are ionized or excited. These radicals or excited molecules are formed within a few ångstroms of each other. From 10^{-12} to 10^{-10} s, the active species undergo the reactions given previously, lose their kinetic energy, and diffuse from their point of formation.

Recombination

One immediate reaction that can occur while the reactive species are still very close to each other is recombination of the early products and conversion of their energy to thermal energy in the vicinity. Radicals, as well as ion pairs, can recombine. However, since with the passage of time diffusion separates them and effectively dilutes them in the water environment, recombination is over in 10^{-11} s, after which diffusion-controlled processes become important.

Chemical Stage

After the thermalization of the various species formed in the early reactions, the third, or chemical, stage begins. It lasts until all the chemical reactions have ended. These reactions occur in the vicinity of the track and are nonhomogeneous; that is, the concentrations controlling the reaction rates are not uniform. These reactions are enormously complex. A complete listing of these reactions can be found elsewhere, and many of these reaction products are of little importance. For neutral water the list of products can be summarized in the following way (not balanced equations of course):

$$H_2O \text{ (irradiated)} \Rightarrow H\cdot, e_{aq}^-, OH\cdot, H_3O^+, H_2, OH^-, HO_2\cdot, O_2,$$

$$O_2^-, HO_2^-, H_2O_2, O_2^-$$

In acidic water, that is, in a high concentration of H^+, hydrated electrons react in less than 10^{-10} s with protons to produce a hydrogen radical (see the reaction diagrams in Figure 6.2).

G VALUE: EXPRESSION OF YIELD IN RADIATION CHEMISTRY

The G value for reactions in radiation chemistry was defined as a convenient means to compare relative yields of various chemical species as the result of deposition of energy from ionizing particles. The explicit definition of G value is given as follows, noting that there are two forms of G, the integral yield G, and the differential yield G'. G is defined as the number of molecules formed or destroyed as the result of the deposition of 100 eV of energy from the ionizing secondary electron. The differential G value, G', is related to G as follows:

$$G'(E) = \frac{dE\, G(E)}{dE} \tag{6-1}$$

where $G(E)$ is the integral yield for energy E.

The initial G values for the primary radiolytic product of water at neutral pH are given in Table 6.1.

REACTIONS IN THE TRACK; THE ROLE OF SCAVENGERS

The reaction of water radicals with water is an interesting process in itself, but to have a biological outcome from an irradiation event, there must be a chemical interaction of the reactive species with important biomolecules to lead ultimately to an interference in normal cellular func-

TABLE 6.1	Yields of Primary Radiolytic Products of Water at Neutral pH[a]
Products	G Value
e_{aq}	2.6
H	0.6
OH·	2.6
H_2	0.45
H_2O_2	0.75

[a] From data by J. K. Thomas (1967).

tion. Radiation chemists have come to use the term *scavenger* for the chemical species that interact with the radicals and other active forms present in irradiated water. The scavenger, which can be any molecular species capable of interaction, reacts with the radicals to bring the water chemistry to an end. Interestingly, it has been noted that for low concentrations of any scavenger molecule the fraction of the radicals that interacts with the scavenger is nearly constant over a wide range of scavenger concentration. Why is this so? If you picture a diffusing reactive species, its reaction with the scavenger is a "default" reaction to the other water reactions. At lower concentration of the scavenger, the active molecule (radical) simply diffuses a little further before the scavenging takes place.

Fricke Dosimeter

The Fricke dosimeter system to be described, the oxidation of ferrous ion (Fe^{2+}) to ferric ion (Fe^{3+}) as the result of the irradiation of an aqueous solution of the ferrous ion, was originally developed as a dose-measuring device. It has only limited use for that purpose now, but it is an enormously valuable tool to examine the kinetics and chemistry of the interaction of active water species with any scavenger.

For the Fricke dosimeter system, ferrous ion, Fe^{2+}, acts as the scavenger for the active products of the radiation chemistry of water. The chemical equations for the events of the Fricke are as follows:

$$H\cdot + O_2 \Rightarrow HO_2\cdot \tag{6-2}$$
$$HO_2\cdot + Fe^{2+} \Rightarrow HO_2^- + Fe^{3+} \tag{6-3}$$
$$HO_2^- + H^+ \Rightarrow H_2O_2 \tag{6-4}$$
$$OH\cdot + Fe^{2+} \Rightarrow OH^- + Fe^{3+} \tag{6-5}$$
$$H_2O_2 + Fe^{2+} \Rightarrow OH^- + Fe^{3+} + OH\cdot \tag{6-6}$$
$$H\cdot + H_2O \Rightarrow OH\cdot + H_2 \tag{6-7}$$

Reaction 6-7 occurs in absence of oxygen. Careful examination of these equations shows that, in the presence of oxygen, each molecule of hydrogen radical H· causes the oxidation of three molecules of ferrous ion. Reaction 6-2 produces the HO_2· radical, which, in turn, oxidizes *one* ferrous ion and produces HO_2^-. The HO_2^- reacts with hydrogen ion to produce H_2O_2, which oxidizes the *second* ferrous ion. The OH· produced in reaction 6-6 oxidizes the *third* ferrous ion. Finally, it is clear that each initial hydrogen radical has led to the oxidation of three ferrous ions.

The H_2O_2 radiolytically produced (not through reaction 6-2) will oxidize *two* ferrous ions, one directly and one as a result of production of the OH· as described in reaction 6-6. Finally, the radiolytically produced OH· radical itself is responsible for the oxidation of *one* ferrous ion according to reaction 6-5.

From these reactions, then, we can predict the overall G value for the oxidation of ferrous ion in an oxygenated medium at acidic pH. The overall G value for the production of ferric ion, then, is

$$G(Fe^{3+}) = 2G(H_2O_2) + 3G(H) + G(OH) \qquad (6\text{-}8)$$

Inspection of equations 6-2 to 6-6 will demonstrate that both oxygen (6-2), and hydrogen ion (6-4) enter into the reactions when these are present in the solution. In the absence of oxygen the G value is

$$G(Fe^{3+}) = G(H) + G(OH) + 2G(H_2O_2) \qquad (6\text{-}9)$$

It is left to the reader to show that equation 6-9 is indeed correct, following the same reasoning as before for the fully oxygenated circumstance. Note that reaction 6-7 accounts for the contribution $G(H)$. Using the generally accepted values of G for the various reactants, we may reconstruct the G value for oxidation of ferrous ion (the same as for production of ferric ion). Note that the G values used are those for the production of the products of radiolysis in acidic solution, so they are different from the values in Table 6.1.

For low LET radiation (for example, ^{60}Co gamma rays), $G(H)$ is 3.65, $G(H_2O_2)$ is 0.75, and $G(OH)$ is 3.15. With equation 6-8, the G value for ferric ion production is 15.6. In the absence of oxygen, the $G(Fe^{3+})$ is 8.30 (equation 6-9).

Interpretation of the Fricke

What does the Fricke model tell us that is relevant to biological systems?

1. Scavenger molecules such as Fe^{2+} will react with the radical species produced by the irradiation of water. By inference, appropriately

reductive molecular species in the cell will act in the same way, being in turn oxidized in the process.

2. Other molecular species that are present in the cell's environment will modify the chemical outcome by their presence, notably oxygen.

3. It can be demonstrated that not all the energy deposited in ionization and excitation ultimately creates chemical change in the scavenger.

To give an example of the latter point, if it is given that, on average, approximately 17 eV is required to produce a radical pair (H· and OH·) and if 100 eV is deposited in the Fricke dosimeter, then we would expect 5.88 pairs of radicals (the G value would be 5.88 for H· and OH·), and the Fricke G value would be

$$G(\text{Fe}^{3+}) = 0 + 3 \times 5.88 + 5.88 \quad \text{(in the presence of } O_2)$$
$$= 23.53$$

The observed G value for the Fricke dosimeter is 15.6. The conclusion is that a significant proportion of the chemistry represents interactions between the initial radicals to cause their disappearance so that they are not effective in oxidizing ferrous ion. To reiterate this very important point, the initial products of radiolysis will react with each other to remove themselves as effective agents of further chemical change.

4. We might also infer that competing scavengers would reduce the yield of Fe^{3+}; that is, there would be protection of Fe^{2+} if other chemical species were present to competitively react with the radiolytic products.

DIRECT AND INDIRECT ACTION

Emphasis so far has been entirely on the radiation chemistry of water as it relates to possible effects on biological systems. In addition to the reactivity of the products of radiolysis of water that causes them to interact with important biomolecules in the cell, there is also the possibility that energy deposited by the high-kinetic-energy electron might be deposited directly in the biological molecule of interest. In this case the initial physicochemical reactions take place in the molecules of the important cell constituents such as DNA, rather than in water. The result would be ionization and/or excitation in atoms of these molecules and radical formation from the important biological molecule. These radicals can undergo reactions similar to those seen for the radiolysis of water and its subsequent chemistry, including recombination. Two general classes of interactions of radiation can be identified.

Direct Action

For this process the energy is directly deposited in the target molecule of biological interest, without the intervention of radical species derived from water radiolysis. The dose–response relationship will generally be log-linear as expressed in the relationship

$$E = E_0 e^{-kD} \qquad (6\text{-}10)$$

where E_0 is the starting number of undamaged molecules, E is the number remaining after dose D, and k is a constant (the inactivation constant, the units for which are rad^{-1} or $gray^{-1}$). The reciprocal of k is the dose required to reduce E_0 by $1/e$, that is, to 37% of its original value. This dose is called the D_{37} or, more generally, the D_0.

The relationship in equation 6-10 is based on an assumption that a single direct ionization or excitation event in the biological molecule will cause a change that leads to loss of biological activity in the molecule. The complete derivation and significance of this expression is given in the target theory discussions in Chapter 7.

Molecular Weight by Direct Action

Hutchinson and Pollard (1961) showed that a relationship existed between the molecular weight of the molecule and the target size; that is, if single-event kinetics were correct, the target size would be the molecular weight. If an assumption can be made about the average energy deposited per ionization event, the target size may be calculated from equation 6-10 by the evaluation of the inactivation constant. Assume the average interaction is 75 eV (known empirically from experiment). The target size in daltons is related to k, the inactivation constant, as follows. The dose unit, either the rad or gray, is converted to eV/g. The constant for this conversion is 6.2×10^{13} eV g^{-1} rad^{-1}. If the dose is in gray, the constant is 6.2×10^{15} eV g^{-1} Gy^{-1}. The target size in grams is

$$\text{target size (g)} = k \times \left(\frac{75 \text{ eV per event}}{6.2 \times 10^{13}} \right) \qquad (6\text{-}11)$$

and in daltons,

$$\text{target mol. wt.} = N_a \times (\text{target size in grams}) \qquad (6\text{-}12)$$

where N_a is Avogadro's number, 6.02×10^{23} molecules/mole. Equations 6-11 and 6-12 can be rewritten in the following form:

$$\text{target mol. wt.} = \frac{7.28 \times 10^{11}}{D_{37}} \qquad (6\text{-}13)$$

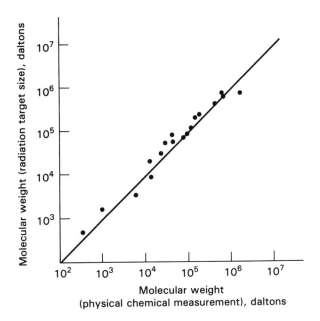

Figure 6.3 Molecular weights for a wide range of biologically active molecules were determined independently by physicochemical methods and by radiation target size determination. The data are plotted from the results of Hutchinson and Pollard (1961).

Remember that the units in which the dose is expressed must be appropriate for the constant used in equation 6-11.

The data shown in Figure 6.3 permit us to conclude that a wide range of bioactive molecules loses its biological activity as a result of a single ionizing interaction. Indirectly, the data also support the value of 75 eV per interaction for the average ionization event. Implicit in the straight-line relationship of Figure 6.3 is a constant value of G for the inactivation of any of these different molecules. G is 1.35, since one molecule is inactivated by the deposition of 75 eV.

The experiments of Figure 6.3 and the discussions of direct action depend on the irradiation of the target molecules in the absence of water, that is, in the dry state. If the molecules are irradiated in aqueous solution, a very different state of affairs exists. In this case the indirect action of the products of water radiolysis will predominate for low LET radiations.

Indirect Action

Indirect action is defined as the outcome that results from the action of the radiolytic products of water on the target of biological importance. The important characteristic of indirect action is that damage mediated through this mechanism is damage that can only occur when the active radical species developed through water radiolysis arrive at and interact with the biologically important molecules. The process will therefore usually be diffusion limited, and the dose–response relationship will often be complex. Indirect action is explicitly defined as the interaction of solute molecules and the reactive species of solvent molecules formed by the direct action of radiation on the solvent. Since water is the main solvent in living systems, most indirect action will be the result of reactive species formed from water molecules.

Before leaving the subject of indirect action, it is important to indicate which of the radical species produced in water radiolysis are predominant in producing biological damage. This subject may be examined in model systems by irradiating in the presence of scavengers, which are specific in their reactivity with a single water radiolysis radical. These experiments have shown, beyond doubt, that the hydroxyl radical, OH·, is the predominant species for reaction with biological molecules. It has recently been suggested that another radical, the superoxide radical, $O_2 \cdot ^-$, might play an important role, but the data are not convincing and are based on experiments with cells in which the enzyme that converts the superoxide radical, superoxide dismutase, is absent or in low activity (see, for example, Samuni and Czapski, 1981).

If we examine the G values for radiolytic species in water in Table 6.1, we will observe that the hydrated electron, e_{aq}, has a G value as great as that for the hydroxyl radical, OH·. In spite of this fact, it appears from experimental evidence that the hydrated electron is of lesser importance for damage in biological systems.

RECOMBINATION, RESTITUTION, AND REPAIR

Restoration of molecules to the preirradiation condition is possible through three mechanisms. One of these, recombination, has been mentioned previously. *Recombination* can occur in the very early stages after the irradiation event, while the water radiolysis species produced are still very close together. The time scale for recombination is less than 10^{-11} s. Recombination is simply the coming together of ion pairs or radical pairs to form the molecule from which they originated. As diffusion increases the initial separation distance of the radical species produced in the spur, recombination becomes less and less likely.

Restitution is a chemical restoration of the altered molecule to its original state without the intervention of enzymatic or other biocatalytic steps. The time scale for restitution is milliseconds and less. Chemical restitution can proceed by a number of pathways, some of which are not very well understood. An example will be useful to understand this process.

Suppose, as the result of either indirect action or direct action, that DNA has undergone a chemical event leaving a radical at some location in the molecule. This radical can, in turn, undergo a reaction that will more or less permanently fix the damage (for example, reaction with oxygen), or it may react with another molecule in a radical exchange reaction that leaves the DNA molecule restored to its preirradiation condition and another radical is created. A typical example of this latter process could be the interaction of a DNA radical with a sulfhydryl-containing molecule, leading to restoration of the DNA molecule and the conversion of a sulfhydryl molecule to a radical compound.

On a longer time scale, enzymatic *repair* of damage can occur. These processes will be discussed in some detail later. The time scale for such processes may be minutes to hours.

MACROMOLECULAR TARGET IN THE CELL

The principal target for ionizing radiation-induced chemical transformation, where these changes can be biologically important for cell survival, is well known by now to be DNA. The molecule carries the genomic information required for self-replication, biochemical renewal in the cell, and cell division. Part of the radiation-induced damage will be the result of indirect action of the products of water radiolysis. Part of it will be the result of direct action on the DNA molecule. While DNA is the principal target, it is clear that other bioactive molecules in the cell are also undergoing deactivation as the result of both direct and indirect radiation damage. The cell can sustain very large losses in the biological activity of most of the molecular species in the cell other than DNA without serious functional deficit. Enzymes, for example, are being continuously synthesized, and damaged molecules will be replaced.

Does the DNA molecule have a special or unique sensitivity to radiation damage that leads to it being the central molecular change associated with radiation damage in the cell? No, as just asserted in the previous paragraph; every molecule in the cell will be susceptible to radiation damage by both the direct and indirect effects. The most important characteristic of DNA that makes its destruction or alteration crucial for cell survival and self-replication is that the total genome is unique in each cell. There is very limited redundancy of information in the molecule, and irreversible damage, when it occurs, may lead to loss of genetic coding vital to cellular function and survival. To consider this issue in a different way, examine the consequence of molecular damage to a vital intracellular enzyme. If such an enzyme is a large protein molecule, it will certainly suffer significant radiolytic damage as the result of direct or indirect action. The loss of vital function of even a significant proportion of the molecules available does not cause the loss of the function itself. Furthermore, the cell continually replaces or "turns over" important biological molecules. For this latter function, the cell depends ultimately on coding information stored in the DNA molecule.

Evidence for DNA as the Target Molecule

What is the experimental evidence that DNA is the target molecule, the alteration of which leads to irreversible changes in cell function?

1. For simpler organisms, such as bacteriophage and viruses, we can establish a quantitative relationship between the damage to DNA and biological function.
2. For higher organisms the relationship between DNA damage and loss of biological function is not so easy to establish quantitatively, but certainly loss of function has been shown to correlate with single-strand and double-strand breaks in DNA.
3. The repair of DNA damage has been shown in many organisms to relate closely to cell survival as measured by the ability of the cell to divide.
4. Cells that lack the ability to repair DNA as a result of a genetic alteration are exquisitely more sensitive to radiation exposure than are their normal phenotypes.
5. Chemical agents that are known to block the repair of DNA damage increase the sensitivity of cells to irradiation.
6. Finally, on physical chemical grounds, DNA is the largest molecule in the cell, and at doses that will cause a cell line to lose its ability to further divide, it is highly unlikely that significant damage can have been done to the smaller molecules in the cell.

The direct actions of ionizing radiations on DNA and other bioactive molecules in the cell are easily visualized as bond disruption or radical formation at the site of energy deposition, leading to inactivation of the target molecule. The indirect actions through the products of water radiolysis are more complicated, and several possible reaction pathways have been established.

REACTIONS OF THE PRODUCTS OF WATER RADIOLYSIS

There are a number of possible reactions of the radicals produced in water radiolysis with biologically important molecules. Not all of these are ultimately important in cellular alterations following irradiation, but the relative importance of each is not yet completely understood.

1. Extraction of hydrogen atoms:

$$R—H + H· \Rightarrow R· + H_2$$
$$R—H + OH· \Rightarrow R· + H_2O$$

2. Dissociative reactions:

$$R—NH_3^+ + e_{aq}^- \Rightarrow R· + NH_3$$
$$R—NH_2 + H· \Rightarrow R· + NH_3$$

3. Addition reactions:

$$R—CH{=}CH—R + OH· \Rightarrow RCHOH—CH·—R$$

Two other reactions are important in understanding radiolytic alterations. One of these has been mentioned earlier, *restitution*, the non-enzymatic restoration of the original altered molecule at the expense of a second molecule. An example is

$$R· + R—SH \Rightarrow R—H + R—S·   \text{(restitution)}$$

The other important reaction class represents the role that oxygen can play in damage fixation as opposed to restitution.

$$R· + O_2 \Rightarrow R·—peroxide   \text{(damage fixation)}$$

The peroxidated radical cannot undergo restitution and is more stable than the original radical. For this reason the reaction with oxygen will generally lead to irreversible changes in the target molecule, but certainly the peroxidated form is not the final chemical product. The reactions of restitution, sometimes called chemical repair, and damage fixation are examined in greater detail in Chapter 9, dealing with modification of the radiation response.

Reactions with DNA

All the preceding reactions are, in principle, capable of occurring with DNA as the result of interaction with radiolytic products of water. Since the reaction constants for OH· and e_{aq}^- are very nearly identical, we would presume that each would react equally efficiently with DNA. H· has a significantly smaller reaction constant, so its efficiency would be expected to be less. For reasons that are not at all clear, OH· appears to be very much the predominant reactive species. As mentioned earlier, it is possible to measure the relative contributions from each radical species by introducing other scavengers that react at a much faster rate with the species in question than does DNA. For example, the use of efficient OH·

scavengers significantly reduces the effectiveness of irradiation for several end points in which it is believed that DNA damage plays an important role. All these experiments show that the OH· radical is the key reactant in DNA damage.

The measured reaction rate constants are

$$OH· + DNA, \quad 3 \times 10^8 \, mol \, s^{-1}$$
$$H· + DNA, \quad 8 \times 10^7$$
$$e_{aq}^- + DNA, \quad 1.4 \times 10^8$$

The damage sustained by DNA can be of several types.

1. Functional groups on the purines or pyrimidines may be irreversibly altered, leading to the presence of an incorrect nucleotide.
2. Damage to purines or pyrimidines may be so extensive as to cause them to be lost to the DNA molecule. The result is apurinic or apyrimidinic sites in the DNA molecule.
3. A radical transfer mechanism is known, in which a radical formed in a base site is transferred to the sugar-phosphate backbone, leading to loss of the base and chain scission.
4. Damage to the deoxyribose-phosphate backbone may cause scission of the backbone and eventuate in breakage of a single strand. This very important damage class is referred to as a *single-strand break* (SSB).
5. Damage to the deoxyribose-phosphate backbone in two or more nearby locations can lead to scission of the molecule, and this damage is referred to as a *double-strand break* (DSB).

Of the mechanisms listed, attacks on the sugar-phosphate backbone of the DNA molecule appear to be of the greatest importance. Attacks on the bases are generally believed not to cause strand scission, and the enzymatic repair mechanisms for replacement of damaged bases are efficient and fast. From the experimental work on viruses and bacteriophage, it has been shown that the OH· attacks the 1', 2', 5', and 4' hydrogens on the deoxyribose moiety of the DNA molecule and abstracts one of them to form water, a mechanism described previously for a generic action of the OH·. The preferential site for this action is the 4' position. There is a significant supply of —SH containing compounds in the cell, for example, glutathione, which can, in turn, engage in the restitution reaction described earlier. If this restitution reaction takes place, the DNA molecule will be returned to its intact state. If, on the other hand, oxygen is present and reacts with the DNA radical, fixation of the damage is the result and leads to chain scission. The lifetime of the OH· is such that it can move only on the order of 120 to 150 Å before it reacts (the diffusion distance for the molecule when its rate constant is 3×10^8 mol s^{-1}). DNA single strands are separated by about 20 Å in the helix, and interbase distance is, on average, about 3.4 Å. Clearly, the hydroxyl radical must come into existence very close indeed to the target site in the DNA molecule in order to diffuse and interact in its short lifetime.

CHAIN SCISSION IN DNA

Chain scission of one strand of a DNA molecule is not a particularly lethal event for the cell. Repair processes are fast, efficient, and relatively free of errors. The close association of the two strands allows precise reading of the appropriate missing bases from the intact complementary strand and the correct reinsertion of these bases. It is generally believed that single-strand breaks are a common event in a cell, with or without radiation, and that their repair is uneventful and efficient. Until relatively recent time the single-strand break was believed to play an important role in radiation biology and in the loss of cellular function for replication and cell division; but it is now generally accepted that the SSB only infrequently contributes to loss of function in the cell.

Double-strand breaks (DSB) are far more serious in their consequences for the cell. A suitable template for restoration of the damage is not readily at hand, and repair of DSBs is an error-prone process that will frequently lead to mutation in the genome and/or loss of reproductive capacity.

Double-strand breaks can result from either the scission of both strands simultaneously, and close together, as the result of direct action that deposited a significant amount of energy in the region in which the damage occurred; or they can result from cooperative interaction of two neighboring SSBs that are close enough together to allow separation of the molecule. It is presently a point of some contention as to how far the two breaks must be from one another for interaction. Since the two strands between the breaks are held together by hydrogen bonding, the total energy in these hydrogen bonds will determine whether the molecule will separate. Estimates range from six to twelve base pairs for the maximum separation of two SSBs that can lead to a DSB. The mechanisms for repair of DSBs is complicated and, as mentioned, error prone.

For ease of understanding of the mechanisms by which DSBs can occur, Figure 6.4 outlines the two mechanisms. The processes modeled are shown to be the result of a greatly simplified direct action, that is, the passage of a high-kinetic-energy electron that results in the local deposition of energy and local radiolysis products, as shown. Some liberties are taken with the diagrams, in that appropriate alteration of the course of the incoming electron is not indicated. All these processes can also occur as the result of indirect action, with appropriate distribution of molecular

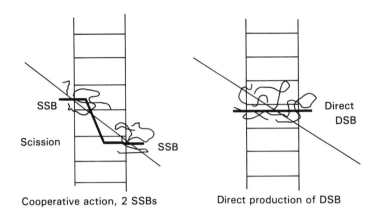

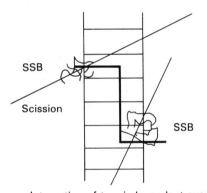

Figure 6.4 Mechanisms by which chain scission (double-strand breaks) can be caused in the DNA molecule. The electron tracks are shown as straight lines. The irregular lines represent the diffusion paths of the products of water radiolysis that are produced at that location.

species from water radiolysis, but with the energy transfer reaction occurring outside the immediate vicinity of the DNA backbone.

REFERENCES

BURTON, M., and MAGEE, J. L. (1956) Einige chemische Aspecte der Strahlenbiologie. *Naturwissenschaften* **43,** 433–442.

HUTCHINSON, F., and POLLARD, E. (1961) In *Mechanisms in Radiobiology*, Volume 1, M. Errera and A. Forrsberg, eds. Academic Press, Inc., New York, p. 85.

MOZUMDER, A., and MAGEE, J. L. (1966) Model of tracks of ionizing radiations for radical reaction mechanisms. *Radiation Res.* **28,** 203–214.

PLATZMAN, R. (1953) In *Conference on Basic Mechanisms in Radiation Biology*. National Research Council Publication No. 305. National Academy of Sciences, Washington, D.C.

SAMUNI, A., and CZAPSKI, G. (1981) Radiation induced damage in *Escherichia Coli* B: the effect of superoxide radicals and molecular oxygen. *Radiation Res*. **76**, 624–632.

THOMAS, J. K. (1967) In *Radiation Research*, G. Silini, ed. North Holland Publishing, Amsterdam.

SUGGESTED ADDITIONAL READING

ALTMAN, K. I., GERBER, G., and OKADA, S. (1970) *Radiation Biochemistry*, Volume I, *Cells*. Academic Press, Inc., New York.

SCHWARZ, H. A. (1974) Recent research on the radiation chemistry of aqueous solutions. In *Advances in Radiation Biology*, Volume 1, L. G. Augenstein, R. Mason, and H. Quastler, eds. Academic Press, Inc., New York.

VON SONNTAG, C. (1987) *The Chemical Basis of Radiation Biology*. Taylor and Francis, London.

PROBLEMS

1. A 10-cm^3 air-saturated Fricke dosimetry solution in a 1-cm-diameter tube is irradiated for 10 min in a ^{60}Co source of gamma rays. The optical density measured at 3040 Å in a 1-cm light path at 30°C was 0.260 after the completion of irradiation and 0.003 before irradiation.

   (a) What is the total dose in rad absorbed by the solution?

   (b) If 10 cm^3 of methanol is irradiated in the same tube for the same 10 min, what is the absorbed dose in rad? The Z/A for methanol is 0.562 and that for the Fricke dosimeter is 0.553. Explain any assumption you make to solve the problem.

2. For water radiolysis by ^{60}Co gamma irradiation, the steady-state G values for the active intermediates, assuming the solution is 0.8 N in H_2SO_4 are

$$G_H = 3.65, \quad G_{OH} = 2.95, \quad G_{H_2O_2} = 0.85, \quad G_{H_2} = 0.45$$

   (a) Calculate the air-saturated ferric ion yield in the Fricke dosimeter.

   (b) If a scavenger is present that reacts with OH radical at such a fast rate that no other reaction with OH is possible, what will be the ferric ion yield in the same situation?

   (c) The diffusion constant for H radical is 8×10^{-5} cm^2/s in water. Calculate the time it takes to diffuse a distance of 10 Å from the point of its origin.

3. There is little or no variation in response of a Fricke dosimeter for dose rates up to 10^7 rad/s. Between 10^7 and 10^{10} rad/s the yield (G value) falls from 15.6 to 8. Explain this observation.

4. Consider a spur that contains 100 eV of energy in water reactants. If 17 eV of energy is needed to create a radical pair (H and OH), and it is assumed that no molecular products are formed, what should the ferric ion yield be for irradiation in either air or nitrogen? Compare your answer with the value of 15.6 in air for the observed value.

7

Theories and Models for Cell Survival

INTRODUCTION

Most modern radiobiological theory is based, to some extent, on the cell survival curve. The cell survival curve describes the relationship between the fractional survival of a population of radiated cells and the dose of radiation to which the cells have been exposed. The survival curve experiment is done in vitro; that is, cells are irradiated outside the animal and then their survival is measured in petri dishes. Until Puck made his remarkable discoveries in the 1950s (Puck and Marcus, 1955) about how to grow mammalian cells in vitro, the survival curve experiments were restricted to those that could be done with standard microbiological technique, that is, with bacteria, yeast, and other simple organisms. The Puck methods opened up new vistas for exploration of the response of mammalian cells. Such diverse responses of cellular systems as radiotherapeutic killing of cancer cells and the rate of development of cancer after irradiation depend in part on the cell survival models for their interpretation.

CLONOGENIC SURVIVAL

The end point for survival in such experiments as those just described must be clearly defined. The term *clonogenic survival* is defined as the ability of a single cell to give rise to a colony of cells on a petri plate. In other words, clonogenic survival does not describe the continued existence of a single cell, but rather describes the ability of a cell to reproduce. This end point is sometimes referred to as *reproductive death* as distinguished from true survival, which is the continued functional and metabolic existence of one cell. It will be seen that cellular respiration and metabolism are very difficult indeed to affect with ionizing radiation. Doses one hundred or more fold higher than that necessary for reproductive death fail to completely suppress the metabolic and respiratory activities of a cell. In all discussions of survival curves and survival dose–response relationships in radiation biology it must be emphasized that it is clonogenic survival, the ability of a cell to reproduce itself, that is being measured and analyzed.

Clonogenic or reproductive death is defined for purposes of radiation biology as the loss of the ability of a single cell to act as a progenitor (clone) for a significant line of offspring. The word significant in the definition is essential, since a cell is often able to divide two or three times before finally failing to divide further. For this reason, an operational definition of clonogenic survival has become common to radiation biology. If a cell can produce at least 50 offspring, it is considered to be a clonogenic survivor.

For the study of proliferative death, the dependence of survival on dose remains the principal tool available to the investigator in radiation biology. Unfortunately, the radiation sensitivity of a wide range of mammalian cells is not amenable to this tool, since they divide infrequently, if at all, and, if they can be induced to divide, have a limited division potential. These cell lines include many of the well-differentiated cells of the tissues of the mature organism and include the differentiated cells of brain, muscle, nerve and the various other nonstem-cell, differentiated lines in the bone marrow. This is an important shortcoming of the clonogenic survival end point, which limits our ability to interpret and understand entirely the action of radiation at the organismic level.

In Chapter 10 we will describe a number of methods that have been developed for the quantitative assessment of the radiosensitivity of normal tissues and tumors growing in their in vivo environment. All these methods have been invaluable in verifying that studies with cellular systems in vitro can be validated by in vivo studies.

LEA'S TARGET THEORY MODEL

One of the earliest interpretive models for cell killing was that proposed by Lea (1955). It has come to be widely known as the *target theory* of cell killing. The target theory was developed by Lea using data on cell killing of microorganisms and from radiation inactivation data on bioactive molecules. In spite of this, the model has been found, within limits, to have wide applicability to mammalian clonogenic survival.

Basic Assumptions

Several important assumptions are the basis for all mathematical models that attempt to describe the loss of clonogenic potential resulting from irradiation. These are outlined next.

1. The killing of a cell (loss of its reproductive capacity or clonogenic killing) is the result of a multistep process.
2. The absorption of energy in some critical volume (or volumes) in the cell is the necessary first step.
3. The deposition of energy as ionization or excitation in the critical volume will lead to the production of molecular lesions in the cell.
4. The expression of these molecular lesions at the biological level causes the loss of the ability of the cell to carry out normal DNA replication and cell division.

Of course, Lea, the author of the first definitive model for cell killing, was not aware of the crucial role of DNA in this process, but he did predict the need for specific inactivating lesions in important molecules of the cell.

The absorption of energy is relatively well understood, the mechanisms of the chemical or molecular lesion less so; but in light of our modern knowledge about the important role of DNA in cell replication, the process of molecular lesion production in this critical molecule is reasonably modeled by the hypotheses for the direct and indirect action of radiation at the chemical level. Finally, the expression of the molecular lesion as a biological end point is not understood in great detail, but is now known to be clearly tied to the knowledge of the irreversible defects that can be produced in DNA by the direct and indirect actions of ionizing radiation and that still exist after all repair processes have been exercised.

The essential assumptions made by Lea (1955) for his target theory model of radiation action are the following:

1. There exists within the cell a discrete, physically describable but not yet identified target for radiation action. The volume of sensitive substance (target or targets) has physical meaning; that is, the target is a true space-occupying entity.
2. There may exist a multiplicity of these targets in a single cell, and the inactivation of n of them leads to loss of reproductive capacity. The exact number of these targets is not known in advance.

3. Deposition of energy is a discrete and random process (stochastic) both in time and in space. In Chapter 6 the discreteness and randomness of the energy deposition process was described in great detail.

4. There are no conditional probabilities for interaction of the radiation with the individual targets. That is, the order of inactivation is not important. In the simple case, if there are two targets, after the first target is inactivated, the inactivation of the second target is neither more nor less difficult or likely.

BIOLOGICAL SURVIVAL CURVES

Before proceeding with the development of the model, it is first necessary to describe the general nature of the dose–response relationships for clonogenic survival that are generally seen. Biological cell survival data, whether for microorganisms or mammalian cells, are generally describable by two types of relationships, either the exponential survival curve or the *shouldered* (sigmoid) survival curve.

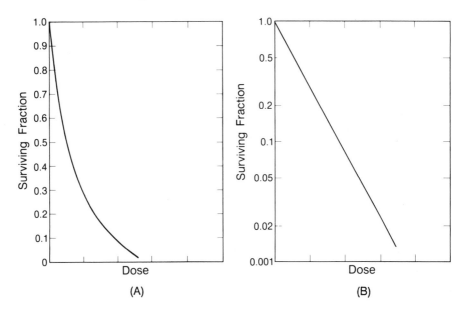

Figure 7.1 The exponential survival curve. (A) is a plot of the surviving fraction as a function of dose. (B) is a plot of the surviving fraction on a logarithmic scale as a function of dose. The notation S/S_0, for surviving fraction, is general in radiation biology and is taken to be the number of clonogens surviving any dose (S), divided by the number of clonogens surviving (S_0) after zero dose. In laboratory shorthand, S has come to be known as the surviving fraction, standing in for S/S_0.

Exponential Survival Curve

The exponential type of curve (Figure 7.1) is found with haploid yeast, some bacteria, and, in mammals, for spermatogonia. The linear exponential curve of Figure 7.1B is also characteristic of postirradiation survival of biological function for a wide range of bioactive molecules. The reader will recall that this property of the linearity of log survival against dose permitted the determination of molecular weight for a wide range of bioactive molecules for which a single inactivation event was all that was required for the molecule to lose its bioactivity. The implication of the shape of this curve is that the loss of clonogenic potential is related to dose in an exponential fashion, and, presumably, this relationship indicates that simple, single events are responsible for the biological outcome. Note that the arithmetic (nonlogarithmic) plot of S/S_0 (Figure 7.1A) is nearly linear for higher surviving fractions. This is still not a mathematically useful property. The logarithmic plot is linear at all values of surviving fraction and it is amenable to mathematical analysis.

Threshold–type Survival Curve

The plots in Figure 7.2 are characteristic of diploid yeasts and nearly all mammalian cells. They are characterized in the logarithmic plot by a

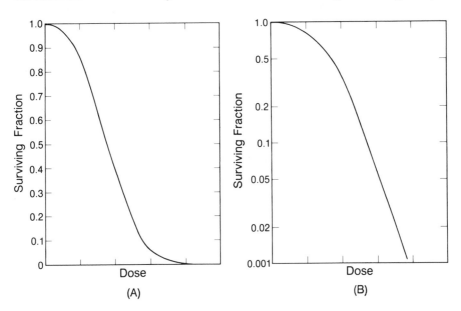

Figure 7.2 The sigmoid or threshold type survival curve. (A) is a plot of the surviving fraction as a function of dose. (B) is a plot of the surviving fraction on a logarithmic scale as a function of dose.

shoulder and a more or less straight portion reminiscent of the low-dosage portion of the first-order plot of Figure 7.1A. It could be inferred from the shoulder portion of the curve that there must be an accumulation of damage before there is loss of the cell's ability to reproduce itself. Lea's inference was that there existed a number of targets, of which n had to be inactivated before loss of clonogenic potential. With this inference it is clear that some minimum dose would have to be administered before there were any loss of clonogenic potential.

DEVELOPMENT OF THE TARGET THEORY MODEL

The target theory was developed by Lea following the generalized assumptions made earlier in this chapter. For the formal development of the model, the following assumptions, notation, and limits were formulated.

Assumptions

1. Assume that a population is exposed to low LET radiation, so interaction of ionizing events is rare. That is, the likelihood of multiple events in a single target is small if the target volume is reasonably small itself.

2. Assume that the cell contains a sensitive volume or volumes or target(s) of size v.

3. Designate those energy absorption events that can (but need not) produce the necessary biological damage as *active* events. This excludes events such as ion recombination and radical recombination, as well as chemical restitution events.

4. Those active events registered within the volume v will be called *hits*.

5. Let V stand for the total cell volume of the population exposed. It will be the product of the average cell volume and the number of cells at risk in the radiation field.

6. Let D represent the density of active events; that is, D is proportional to dose, and it will be called *dose* in the development of the model. This term can be made equal to the real dose in any unit the observer wishes to use by the choice of the appropriate proportionality constant. *Note*: D is the events occurring per unit volume.

Derivation

Now, among the various energy absorption events that occur in condensed media from ionizing radiation, the nature of active events is unknown, but certainly, for most purposes, it must correspond with the very small amount of energy associated with the interaction of ionization and excitation events with either water (the indirect effect) or directly with the molecule that makes up the target. For even a small dose, the number of active events is bound to be very large. For example, if an active event corresponds to about 60 eV, the average energy transfer from a secondary electron, then for a dose of 1 rad (1 cGy) (100 erg/g), there will be about 4×10^3 events in a 20-μm-diameter cell.

Returning to the development of the model, it is clear that the probability that an active event will be scored anywhere in the population exposed is given by the ratio v/V. This follows from the fact that, of the total number of events VD (events per unit volume $\times$ total volume), vD of these (event in the sensitive volume $\times$ events per unit volume) are registered in the sensitive volume. The ratio of these terms is v/V. This ratio is a dimensionless parameter to which we assign the name *hit probability*, ρ. It is a small number, since, in the limit, it is the reciprocal of the number of cells when v approximates the cell volume. The meaning of hit probability is a simple geometrical concept. The probability of a hit is directly measured by the fraction of the total volume that is the sensitive volume.

Now also set $\mathfrak{D}$ equal to the total number of events scored in the total population of cells. That is, $\mathfrak{D} = VD$. The total number of events is the product of total cell volume exposed and the dose (events per unit volume). Now, the *probability* that a cell will be hit h times can be expressed, using the simple binomial theorem, by the product of three terms as follows:

$$(\rho^h)\,(1 - \rho)^{\mathfrak{D}-h}\,(_{\mathfrak{D}}C_h) = P(\rho, h, \mathfrak{D}) \tag{7-1}$$

where the first term is the probability that the cell will be hit h times, the second term is the probability that the remaining events will not be hits, and the third term is the binomial coefficient that expresses all the ways that h hits and $\mathfrak{D} - h$ misses can be assigned for $\mathfrak{D}$ events.

A reminder: The binomial coefficient for $\mathfrak{D}$ things taken h at a time is

$$(_{\mathfrak{D}}C_h) = \frac{\mathfrak{D}!}{h!(\mathfrak{D} - h)!} \tag{7-2}$$

So far the arguments in developing the model have been purely physical. We must now introduce the concept that a function can be described that represents the probability that a cell will survive h hits. This function, $H(h)$, is called the *hit survival function*. The nature of this function and its shape are generally unknown.

The net probability function corresponding to h hits can be adjusted for the biological hit survival function to give P a new meaning of survival probability for a given ρ, h, and $\mathfrak{D}$. P is now the likelihood of surviving if it is hit h times during $\mathfrak{D}$ events with a hit probability of ρ.

$$P(\rho, h, \mathfrak{D}) = (_{\mathfrak{D}}C_h)(\rho^h)(1 - \rho)^{\mathfrak{D}-h} (H(h)) \tag{7-3}$$

Since, depending on the nature of the survival function $H(h)$, a cell may have a nonzero survival probability for $h = 0, 1, 2, 3, \ldots , \mathfrak{D}$, the total survival probability for the cell is the sum of the survival probabilities for all values of h to the maximum, $\mathfrak{D}$.

General Survival Equation

The total survival probability per cell is given as

$$S(\rho, \mathfrak{D}) = \sum_{h=0}^{h=\mathfrak{D}} P(\rho, h, \mathfrak{D}) \tag{7-4}$$

where, since we have summed the P's for all the h's, S depends only on the hit probability and the dose. The $H(h)$ is accounted for for each value of h; that is, $P(\rho, h, \mathfrak{D})$ is the function of equation 7-3.

Equation 7-4 is the general survival equation for the target theory model. As with many mathematical formulations, it has limited utility in this general form. Fortunately, several special limiting values for the parameters and functions provide useful special solutions of the general equation. One of these is the *single-hit inactivation model*, which provides the classic first-order relationship of log survival on dose that was described earlier in this chapter and diagrammed in Figure 7.1.

Single-hit Model

Assume that a cell will survive *only* if it has received no hits at all and that it will always die (that is, lose clonogenic function) if it has received one or more hits. Remember that ultimately in the application of these models dying and surviving are taken to be loss of or maintenance of the cell's ability to divide. Stating this assumption in terms of the *hit survival function*, for the single-hit model,

$$\text{for } h \geq 1, \quad H(h) = 0$$

$$\text{for } h = 0, \quad H(h) = 1$$

The interpretations of these limiting values of the hit survival function are just as given in the preceding paragraph. That is, when the hit number is 1 or more the cell will die (that is, lose function), and when the hit number is zero the cell will survive.

Since $H(h)$ is zero for all h's equal to or greater than 1, the general survival equation reduces to the following form for this special case:

$$S(\rho, \mathfrak{D}) = P(\rho_0, h = 0, \mathfrak{D}) = (1 - \rho_0)^{\mathfrak{D}} \tag{7-5}$$

And from the identity $(1 - \rho_0)^{\mathfrak{D}} = e^{\mathfrak{D} \ln (1 - \rho_0)}$

$$S(\rho, \mathfrak{D}) = P(\rho_0, h = 0, \mathfrak{D}) = e^{\mathfrak{D} \ln (1 - \rho_0)} \tag{7-6}$$

And since, for small values of ρ_0 (which should always be the case for the reasons discussed earlier in this chapter), $-\ln (1 - \rho_0) \cong \rho_0$, then set $\rho_0 = p_0$ and the survival equation for this special case can be rewritten as in equation 7-7.

$$S = e^{-p_0 \mathfrak{D}} \tag{7-7}$$

This is the familiar expression for the single-hit survival curve, and it is identical with the formulation in Chapter 6, in which the molecular weight was determined from survival relationships. It is still not in terms that are directly useful, since the dose equivalency of $\mathfrak{D}$ has not been stated.

In equation 7-7 replace $\mathfrak{D}$ by VD, where D is the active events per unit volume. Let a new constant, p, be equal to $p_0 V$. The exponent is then replaceable by pD. D, as mentioned earlier, is dose, in whatever units are appropriate. If D is expressed in rad, then p has the dimensions of rad^{-1}, and if D is expressed in gray, the p has the dimensions $gray^{-1}$. The parameter p is given the name *inactivation coefficient*.

The *mean lethal dose*, D_0, is the reciprocal of the inactivation coefficient, and it is equal to the dose required to reduce the surviving fraction to $1/e$ (that is, 37%) of its initial value.

MULTITARGET SINGLE-HIT SURVIVAL

The single-hit inactivation model is of great classic interest, but it is of only limited applicability in radiation biology. More frequently, curves of the sigmoid type of survival versus dose, such as those described in Figure 7.2, are seen. The general extension of the target theory model that finds the widest applicability is the multitarget single-hit model (MTSH). In this general extension of the single-hit model, it is assumed that a number of targets are present in each cell. One or more of these must be inactivated. Each target has an equal probability of being hit. More explicitly, it is assumed that all v's (target volumes) are the same. This latter assumption may be one of the greatest weaknesses of the multitarget single-hit formulation. It is unlikely indeed that all targets in the cell will have the same volume. Fortunately, the target theory model is relatively insensitive to all but large differences in target volume when the number of targets is reasonably large. For example, if the number of targets is 10 or more, the sensitivity to variation in target volume is not very great.

Assumptions

1. There are n targets in the cell.
2. Each target has the same probability, q, of being hit.
3. One hit is sufficient to inactivate each target (but not the cell).
4. In the general case, there will be a hit survival function as in the single-hit formulation. This function will be $B(b)$.

With these assumptions and the application again of the binomial theorem, the probability that a cell will survive with b targets hit is given as

$$P(q, b, n, D) = (1 - e^{-qD})^b (e^{-qD})^{n-b} (_nC_b) (B(b)) \qquad (7\text{-}8)$$

where $(e^{-qD})^{n-b}$ is the probability of a miss, and $(1 - e^{-qD})^b$ is the probability of a hit, each for the given values of q, b, n, and D. The term $(_nC_b)$ is, again, the binomial coefficient for n things taken b at a time.

The *hit survival function* $B(b)$ for the multitarget single-hit model can be assigned the following limiting conditions in a fashion similar to that used for the single-hit inactivation model.

1. For $b < n$, $B(b)$ assumes a value such that $P(q, b, n, D) = 1$. That is, for any number of hits b that is less than n, the probability of survival is 1.
2. For $b \geq n$, $B(b) = 0$, and $P(q, b, n, D) = 0$. That is, for any number of targets hit that is less than n, the cell will survive, and for any number of targets hit that is equal to or more than n, the cell will die.

Evaluating the general survival equation, equation 7-4, is straightforward, given these limiting conditions. However, a more insightful and intuitive way to evaluate the expression is as follows: the survival probability for all values of b less than n is 1. The survival probability for the nth hit is zero. The probability of the nth hit is given by equation 7-8 as $(1 - e^{-qD})^n$. Since this nth hit assures nonsurvival, it can be subtracted from the survival probability for all hits up to $n - 1$, which is 1.0, to give the overall survival probability:

$$P(q, b, n, D) = 1 - (1 - e^{-qD})^n \qquad (7\text{-}9)$$

Note in this formulation that D was used initially, following the same reasoning that was used for the single-target formulation. Other approaches to the development of appropriate limits of the hit survival function are also possible, but the hit survival function just described allows a useful formulation of the general survival equation. The final formulation for the *multitarget single-hit model* (MTSH) is as follows:

$$S(q, n, D) = 1 - (1 - e^{-qD})^n \qquad (7\text{-}10)$$

For an elegant and extensive derivation of these equations, the reader is directed to the work of Dertinger and Jung (1970) in the Suggested Additional Reading.

Properties of the MTSH Function

What are the properties of the function described in equation 7-10 when the logarithm of the surviving fraction is plotted as a function of dose?

1. Each log curve becomes a straight line for surviving fractions below about 0.1, at which time the shoulder region has no more effect on the curvature.
2. The extrapolation of this straight-line portion of the plot back to the zero dose ordinate yields a value for n, the target multiplicity.
3. Except for the special case of $n = 1$, in which circumstance the relationship degenerates to the first-order (single-hit) form, every curve has a shoulder that increases in breadth with the multiplicity n.
4. The curves for any n greater than 1 will have zero slope at zero dose.
5. q has the dimensions and nature of the inactivation coefficient as in the first-order (single-hit) case.
6. $1/q$ is the dose for $1/e$ (37%) survival in the *linear* portion of the survival plot, and it is called the D_0. The D_0 has the dimensions of dose.

Quasi-threshold Dose

Another special parameter of the multitarget single-hit relationship has been established through usage and because of its special utility in estimating the parameters of the relationship from laboratory data. If the linear portion of the curve of the logarithm of the surviving fraction versus dose is extrapolated back to the point where it crosses the 1.0 surviving fraction ordinate, that intercept is the dose that has come to be called the *quasi-threshold dose*. It is labeled the D_q.

A special and useful relationship among D_0, D_q, and n can be easily proved from the preceding relationships:

$$D_q = D_0 \ln n \qquad (7\text{-}11)$$

Single-target Multihit Model

In addition to the two models discussed, there is a third class of function called the *single-target multihit model*. In this inactivation scheme there is a single target that must receive n hits for inactivation. The mathematics is more complicated, but solutions are available (see Dertinger and Jung, Suggested Additional Reading). The model appears to have little application in biology and it will not be considered here except to note why it is different from the *multitarget single-hit* formulation. A conditional probability enters into the single-target multihit model, since whether the single target has already been hit must be considered. To compare the two in simple terms, with the multitarget single-hit model, a hit in the target removes it from further consideration, since it is inactivated. Further hits do no further damage. For the single-target multihit model, the probability of an encounter remains unchanged until the cell is inactivated and removed after multiple hits.

MOLECULAR MODEL FOR CELL DEATH

Need for an Alternative Model

Lea's target theory has been the principal tool for the interpretation of cell survival curves ever since its origins in 1946 and its detailed publication in 1955, in spite of some known shortcomings in the use of the model for the interpretation of real laboratory data on cells. Lea developed his model to explain the actions of ionizing radiations on molecules of biochemical importance and on microbiological organisms. For these purposes the target theory models seemed to be entirely adequate. As experience grew with mammalian cell lines grown in tissue culture, it was clear that not all cell lines had survival data that were adequately described by the target model. Starting in the early 1960s, several investigators reported that their data were better explained by using a function in which dose appeared both to the first power and the second power. One of the most impressive and earliest of these data sets was that of Neary, Preston, and Savage (1967), regarding chromosome aberrations, in which a linear–quadratic formulation was clearly preferable to the multitarget single-hit model.

What are the deficiencies of the multitarget single-hit model in describing loss of reproductive capacity in mammalian cells? There are two important discrepancies that have theoretical implications.

1. Is the straight portion of the curve really straight? The MTSH model predicts that the logarithm of cell survival fraction will relate linearly to dose for survival fractions below about 0.1 or 0.2. If this is so, it is possible to estimate the D_0 (the reciprocal of q, the inactivation coefficient) from the slope of the straight-line portion of the curve. Remember D_0 is defined as the dose required to reduce survival by $1/e$ on the exponential portion of the survival curve. As frequently as not, with mammalian cell lines the determination of the value of D_0 shows it not to be constant as it is measured at lower and lower survival levels. Generally, the D_0 tends to decrease (q increases) as the survival level decreases. The most convincing evidence for the changing D_0 is that of Hall and others (1972), where they found a significant and continuous decrease in D_0 as survival was followed through many decades of survival for a mammalian cell line.

   2. Is the initial slope zero? The initial slope of the survival curve for the MTSH model predicts that a zero slope will be found at zero dose. This is a very difficult value to establish, but evidence continues to accumulate that, for most systems, the initial slope is significantly less than zero.

Role of Enzymatic Repair

   The target theory makes no assumptions about either time, in general, as an important variable or about time-dependent enzymatic repair of DNA. The repair of DNA damage is now known to be central to the effectiveness of radiation in causing loss of clonogenic potential.

   Within the last 5 to 10 years, a number of alternatives to target theory have been developed that respond to the objections raised. At least two of these models are of a linear–quadratic form. That is, the surviving fraction is described by a relationship that embodies a linear term in dose and a quadratic term in dose. The theory of dual radiation action proposed by Kellerer and Rossi (1971) has temporal precedence, but it tends to be an explanation of empiric observations rather than development of new approaches *ab initio*.

MOLECULAR THEORY OF RADIATION ACTION

A model proposed by Chadwick and Leenhouts (1981) attempts to correct the deficiencies of the target theory models. The new model, called the *molecular model* by its authors, has come to be called the *linear–quadratic model* by workers in the field.

Assumptions

1. It is taken that there are certain critical molecules in the cell, the integrity of which is essential for the survival of the reproductive function of the cell.

2. These critical molecules are assumed to be double-stranded DNA, and the critical damage is assumed to be a double-strand break in the DNA molecule.

3. The action of radiation, either direct or indirect, is considered to be the rupturing of molecular bonds in the DNA strands (lesions).

4. These lesions in DNA are capable, under certain conditions, of being repaired, and modifications of radiobiological effects can be expressions of varying degrees of repair.

5. Repair processes include physicochemical recombinations, charge transfer processes, chemical restitution, and enzymatic repair.

This model clearly attempts to bridge the gap between the physical processes of radiation energy deposition and damage and the biological outcomes expressed through DNA repair or lack of repair. There is no explicit role for time in the molecular model, but since enzymatic repair of DNA is so central to it, time certainly has an implicit role.

Development of the Molecular Model

Let N_0 be the total number of critical bonds per unit mass, the rupture of which can lead to a single-strand break under appropriate conditions (that is, N_0 is the total number of bonds available for rupture in the target cell). N is the number of these bonds remaining intact after any dose. Let K be the probability constant for rupture of a single bond per unit dose, and let D be the dose. Then

$$\frac{-dN}{dD} = KN \quad \text{and} \quad N = N_0 e^{-KD} \tag{7-12}$$

The number of bonds broken per unit mass per unit dose is the difference of N and N_0.

$$N_0 - N = N_0 - N_0 e^{-KD} \tag{7-13}$$

or

$$N_0 - N = N_0(1 - e^{-KD})$$

Now a correction must be introduced for the *repair* of these bond breaks.

Let r be the proportion of damaged bonds that is repaired, and $f = 1 - r$ is the proportion that is *not* repaired. Introducing the repair constant (or nonrepair constant if you choose) f in equation 7-13 gives the number of *effective* DNA strand breaks, that is, those that can lead to strand scission.

$$N_0 - N = f N_0(1 - e^{-KD}) \tag{7-14}$$

Two Mechanisms for DNA Damage

Clearly, the discussion to this point is about single-strand breaks, not molecular scission of the DNA molecule. The latter is what is taken in the model to be the cause of loss of reproductive capacity in the cell. According to Chadwick and Leenhouts, DNA can be envisioned as undergoing a double-strand break as the result of two different mechanisms.

i. Both strands are broken by one event.

ii. Each strand is broken independently, and the breaks are close enough in time and space for the molecule to rupture.

These mechanisms are diagrammed in Figure 6.4.

It is necessary to formulate an expression that allows for double-strand scission by either of these processes. For the first class of rupture, only a single event leads to strand scission. For the second mechanism, it is necessary to account for strand rupture in either strand and the likelihood that the ruptures will be close enough together for strand scission.

Derivation of the Molecular Model

Let

n_1 = number of critical bonds on strand 1

n_2 = number of critical bonds on strand 2; n_1 is taken equal to n_2

k = probability constant for bond rupture per bond per unit dose, analogous to K in equation 7-12; the constant k will be the same for each strand

f_1, f_2 = *unrestored* fractions of bonds in strands 1 and 2, respectively

A new parameter now needs to be introduced that will identify the fraction of double-strand breaks (DSBs) that are produced by mechanisms i and ii. Let Δ be the fraction of dose D that acts through mechanism i, both strands broken in a single event, and $1 - \Delta$ be the fraction of the dose acting through mechanism ii. Then, respectively, the number of single strands broken on each of strands 1 and 2 by mechanism ii can be developed as follows.

Let q be equal to the number of broken bonds per cell. Broken bonds on strand 1 per cell will be

$$q_1 = f_1 n_1 \{1 - e^{[-k(1-\Delta)D]}\} \qquad (7\text{-}15)$$

and broken bonds per cell on strand 2 will be

$$q_2 = f_2 n_2 \{1 - e^{[-k(1-\Delta)D]}\} \qquad (7\text{-}16)$$

It seems obvious that for the two SSBs to lead to a DSB they must be effectively associated in both space and time. A new parameter must be introduced that is somewhat analogous to hit survival function in target theory. The term *effectiveness factor* is introduced for this parameter.

The effectiveness parameter expresses the likelihood of a rupture occurring from two SSBs appropriately associated in time and space. If this parameter is named E, an expression can be written for the mean number of double-strand breaks per cell occurring as the result of mechanism ii, as the product of the number of SSBs in strand 1, the number of SSBs in strand 2, and the necessary effectiveness factor. Let Q be equal to the number of DSBs per cell. There must also be another correction for the number of DSBs that are repaired, analogous to the SSB case. Let f_0 be the fraction of unrepaired DSBs.

The number of unrepaired DSBs formed by mechanism ii is

$$Q_{ii} = En_1n_2f_1f_2f_0\{1 - e^{[-k(1-\Delta)D]}\}^2 \tag{7-17}$$

Using a similar notation, where n_0 is the number of sites that can sustain a DSB (obviously equal to or less than n_1 or n_2), k_0 is the hit probability constant, and f_0 is the unrestored fraction of DSBs, from the initial equation we can write, for mechanism i, the rupture of both strands of DNA at the same time.

$$Q_i = n_0f_0 [1 - e^{-k_0\Delta D}] \tag{7-18}$$

Now, the mean number of unrepaired double-strand breaks per cell as the result of both mechanisms i and ii acting is

$$Q = n_0f_0[1 - e^{-k_0\Delta D}] + En_1n_2f_1f_2f_0\{1 - e^{[-k(1-\Delta)D]}\}^2 \tag{7-19}$$

Finally, still one more parameter must be introduced, which is the proportionality constant between cell death and double-strand breaks. Call this parameter p. Lumping constants but retaining p for reasons that will become clear later, equation 7-19 is rewritten as 7-20.

$$Qp = p\{[\chi[1 - e^{(-k_0\Delta D)}]] + [\phi[1 - e^{(-k(1-\Delta)D)}]^2]\} \tag{7-20}$$

The constant χ is the lumping of n_0 and f_0. The constant ϕ is the lumping of E, n_1, n_2, f_1, f_2, and f_0.

Equation 7-20 now states the number of *lethal* double-strand breaks that occur as the result of dose D of radiation of quality characterized by Δ. It does not state the fraction of cells killed or surviving. Clearly, a cell can only be killed once, and further action on the remaining cells is constrained to that smaller number of cells. This is the usual Poisson-type cell killing that is common in biology. To convert to fraction of cells inactivated, equation 7-20 is put in the Poisson form, where all of equation 7-20 is the exponent. The result is the formulation for the fraction of cells killed:

$$F_d = 1 - e^{-Q_p} \qquad (7\text{-}21)$$

Linear–Quadratic Formulation

If we reasonably assume, as in the target theory model, that k and k_0 (the hit probability parameters) are quite small, then we can simplify equation 7-21 and restate the equation in the form of a surviving fraction:

$$S = \exp[-p(\alpha D + \beta D^2)] \qquad (7\text{-}22)$$

or, in logarithmic form,

$$\ln S = -p(\alpha D + \beta D^2) \qquad (7\text{-}23)$$

where

$$\alpha = (f_0, n_0, k_0, \Delta) \quad \text{and} \quad \beta = [f_0, E, n_1, n_2, f_1, f_2, k^2, (1 - \Delta)^2]$$

Note again that p has not been lumped in the constants. The reason is that p is a biological effectiveness factor for double-strand breaks, and this factor can be examined experimentally and independently of the other constants. The molecular model has found widespread usage, and p has not always been maintained separate from the other lumped constants. Frequently, in the literature, formulations of the model lump p into the constants α and β. It is clear from equation 7-23 why the molecular model has been widely called the linear–quadratic model. Dose appears in a linear (first-order) term and in a quadratic (second-order) term.

Another useful form of the linear–quadratic relationship is one that is linearized. If we divide through equation 7-23 by the dose D, the following form results, in which $\ln S/D$ is plotted against dose, D. The result is a linear relationship in which the y intercept is $-p\alpha$ and the slope is $-p\beta$.

$$\ln \frac{S}{D} = -p(\alpha + \beta D) \qquad (7\text{-}24)$$

Has the linear–quadratic model been found to be more useful biologically and a better fit mathematically? On the whole, the fit of the data for survival of mammalian cells has been significantly improved by using a formulation with dose to the first and second power. The greatest improvement is that there is no longer the conceptual difficulty of zero slope at zero dose. The L–Q formulation has a limiting slope at low doses that is equal to $-p\alpha$, since $\lim d(-\ln S)/dD$ as D approaches zero is $-p\alpha$. The coefficient for the dose squared term, $-p\beta$, is always very small compared to $-p\alpha$, and the dose squared term makes little contribution at low dose.

The L–Q model is widely used by experimental radiobiologists since it fits the data well and has descriptive utility; but, for reasons that are not at all clear, the fundamental assumptions of the model are not widely accepted. The dual radiation action model of Kellerer and Rossi (1971) ultimately finds expression in the same linear–quadratic formulation as the Chadwick and Leenhouts molecular model.

THEORY OF DUAL RADIATION ACTION

The theory of dual radiation action was proposed by Kellerer and Rossi (1971), partly as an explanation for an empirical observation that, when the relative biological effectiveness (RBE) of neutron radiations was examined as a function of the dose delivered, the RBE increased as the dose was reduced in such a way that Kellerer and Rossi were convinced that the neutrons were acting in such a fashion that the effect was determined by dose to the first order, while for the comparison x-rays there appeared to be a second-order term. Neutrons are a high linear energy transfer radiation and would be described in the L–Q model as having most of their dose from the Δ component, while the comparison x-rays would act principally through a $1 - \Delta$ component. Kellerer and Rossi made no such comparison since their work predated that of the authors of the L–Q model by nearly 10 years.

Assumptions

The *dual radiation action model* proposes that cell inactivation is through the formation of lesions in critical sites, and the authors of the model developed an approach from their observations that, at low doses, high LET radiation had an effectiveness relative to low LET radiation such that the logarithm of the ratio of the equieffective doses (RBEs) as a function of the high LET radiation dose followed a relationship the slope of which was $-\frac{1}{2}$. The assumption was that, at low doses, only a single neutron (high LET radiation) track would traverse a cell, leading to inactivation of the cell. Under this assumption then, the yield of lesions was proportional to absorbed dose.

$$\epsilon = k_n D_n \qquad (7\text{-}25)$$

ϵ is the yield of lesions, k_n the proportionality constant for the neutron radiation used, and D_n the dose. Since the RBE (ratio of equieffective doses) is described by a slope of $-\frac{1}{2}$, then the following expression describes the relationship:

$$\frac{D_x}{D_n} = \text{RBE} = \sqrt{\frac{\lambda}{D_n}} \qquad (7\text{-}26)$$

And the equieffective neutron dose can be written in terms of the equivalent x-ray dose D_x.

$$D_n = \frac{D_x^2}{\lambda} \tag{7-27}$$

Derivation

If equation 7-27 is substituted in equation 7-25, the expression obtained is

$$\epsilon = kD_x^2, \qquad \text{where } k = \frac{k_n}{\lambda} \tag{7-28}$$

The authors' conclusion was that cell inactivation was by a single particle event for high LET radiation and that low LET radiation effects were described by the square of the dose term, suggesting an interaction of two events. The relationship to the Δ term of the molecular model is obvious. They further suggest that the two equations

$$\epsilon = kD_x^2 \quad \text{and} \quad \epsilon = k_nD_n \tag{7-29}$$

are both approximations to a more general expression:

$$\epsilon = k(\lambda D + D^2) \tag{7-30}$$

where λ depends on radiation quality and has such a small value for low LET radiation that the linear term in dose can be neglected as long as the dose is not too small.

To explain the experimental observations, as well as to incorporate the new ideas of microdosimetry at the cellular level, Kellerer and Rossi proposed the dual radiation action model using the following assumptions:

1. The exposure of a biological cell to ionizing radiation leads to the production of *sublesions* in the cell, and the number of such sublesions is directly proportional to the energy imparted (dose).
2. A biological *lesion* is formed through the interaction of two such sublesions. The interaction of sublesions to produce a lesion is possible at significant distances on a subcellular scale.
3. Once the lesion is formed, there is a fixed probability of the lesion leading to a deleterious biological effect (reminiscent of the "hit-survival function" of target theory).
4. All pairs of sublesions within a specific distance of one another have equal probability of interacting. If the sublesions are farther apart than the *sensitive site* dimension, then the probability of interaction is zero. In other words, the interaction probability is a step function of site dimension or separation, with the probability of interaction

equal to 1 for distances less than the sensitive site dimension and a probability of 0 for distances greater than this dimension.

Detailed derivation of the Kellerer and Rossi model can be found in their original paper, but a briefer and possibly more intuitive presentation can be found in Goodhead (1982).

The mean number of lesions (not sublesions) as a function of dose is (after Goodhead)

$$\mathcal{E}(D) = \int_0^\infty \mathcal{E}(z) \cdot f(z, D) \, dz \tag{7-31}$$

where $f(z, D) \, dz$ is the probability that for dose D the specific energy lies between z and $z + dz$. The specific energy is the energy imparted per ionization event per unit mass. It will be a property of the radiation type and bears some similarity in purpose to the Δ function of the L–Q model.

The mean number of lesions within a sensitive site may be written as

$$\mathcal{E}(z) = kz^2 \tag{7-32}$$

where k is a biological property of the system. The specific energy, z, is a direct measure of the number of sublesions, since one of the assumptions was that sublesions were directly related to the specific energy. The term z is squared, since sublesions are required to interact in pairs. Then

$$\mathcal{E}(D) = \int_0^\infty kz^2 \cdot f(z, D) \, dz = k\overline{z^2}(D) \tag{7-33}$$

where $\overline{z^2}(D)$ is the mean value of z, squared, for dose D. Kellerer and Rossi (1972) have rewritten this equation in terms of the means of the spectra of specific energies from single events from their microdosimetry experiments.

$$\overline{z^2}(D) = \frac{\overline{z_1^2}}{\overline{z_1}} D + D^2 = \frac{\displaystyle\int_0^\infty z^2 \cdot f_1(z) \, dz}{\displaystyle\int_0^\infty z \cdot f_1(z) \, dz} D + D^2 \tag{7-34}$$

where $f_1(z)$ is the distribution of specific energies z_1 of each single ionization event. Substituting ζ for the ratio of the two integrals in equation 7-30, we arrive at the final expressions of the dual action theory.

$$\mathcal{E}(D) = k(\zeta D + D^2) \tag{7-35}$$

Survival Equation

The survival equation describes effect, much as the expression of the L–Q model describes double-strand scissions of DNA. The usual Poisson conversion provides the equivalent statement for cell survival.

$$\frac{S}{S_0} = e^{-k(\zeta D + D^2)} \qquad (7\text{-}36)$$

Significance of the Dual Radiation Action Model

Both the dual radiation action model and the L–Q model have been criticized widely for various reasons. Those related to the L–Q model have been discussed. The criticisms of the dual radiation action model have, until recently, been centered around the idea that the empirical evidence cited earlier on the relationship of neutron RBE to dose was inadequate to justify the development of the model. In 1978, Kellerer and Rossi introduced a generalized formulation of the dual radiation action theory that introduced the concept of a "site size" and a probability of two sublesions interacting (Kellerer and Rossi, 1978). More recently, Goodhead (1982) has criticized the revised model on the basis of experimental evidence that the "sensitive site" concept was untenable. It will be some time before the various controversies are resolved.

REPAIR–MISREPAIR MODEL OF CELL SURVIVAL

A modeling approach distinctly different from all those previously discussed in this chapter is that of Tobias and others (1980). Rather than considering the geometrical identity and location of lesions in DNA, this model more explicitly deals with the DNA repair processes themselves. In approaching the model, the authors have started with some assumptions about the biochemical processes, which can be summarized as follows.

Assumptions

1. There is an initial process of physical energy transfer, followed by migration of the deposited energy and ultimately the production of long-lived molecular species as the result of the radiation chemistry of the system.
2. The radiation chemical stage is followed by one characterized by biochemical processes including repair or damage enhancement or fixation, accompanied by cell progression through various physiological states.
3. The cells, if they survive, may express permanent alteration in their phenotype.

The repair–misrepair (RMR) model (Figure 7.3) describes the yield of relevant macromolecular lesions per cell as a function of dose (D); there is a time (t) dependent transformation of these lesions, and accompanying

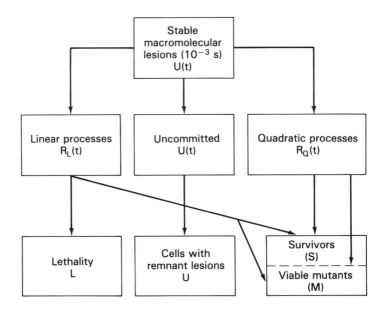

Figure 7.3 Steps in the repair–misrepair (RMR) model. (After Tobias and others, 1980)

time- and dose-dependent probabilities of survival (S), lethality (L), and mutation (M). The model also postulates a class of lesions, (U) lesions, which Tobias suggests might stand for "uncommitted" lesions; and there are various repair states (R), which are the result of the transformation of (U) lesions.

Formulation

In the simplest form of the model, it is assumed that there are two repair (R) states; R_L is the yield per cell resulting from a monomolecular repair reaction that is linear on the concentration of U lesions, and R_Q is the yield per cell of a repair process that is quadratic, where the rate is proportional to the square of the density of U lesions. If the U lesions are homogeneously distributed in the reaction volume, then the rate, R_Q, is proportional to $U(U - 1) \cong U^2$, if $U \gg 1$.

Assuming the delivery of a dose of low LET radiation in a time that is brief compared to repair rates, then

$$\frac{dU}{dt} = -\lambda U(t) - kU^2(t) \qquad (7\text{-}37)$$

where λ and k are the rate constants for the linear and quadratic repair processes, respectively. Integration of equation 7-37 for the time period 0 to t yields

$$U(0) - U(t) = \int_0^t \lambda U(t)\, dt + \int_0^t kU^2(t)\, dt \qquad (7\text{-}38a)$$

Defining R_L and R_Q as

$$R_L = \int_0^t \lambda U(t)\, dt \quad \text{and} \quad R_Q = \int_0^t kU^2(t)\, dt \qquad (7\text{-}38b,c)$$

then

$$U(t) + R_L(t) + R_Q(t) = U(0) \qquad (7\text{-}39)$$

The authors propose solutions for the "decay" of uncommitted U lesions and the growth of the R states with the assumptions that, for a given cell and a specific state, λ and k are constants independent of time and dose. If $U(0)$ is U_0, $R_L(0) = R_Q(0) = 0$, $U(\infty) = 0$, and $\epsilon = \lambda/k$ is the repair ratio, then the following equations hold:

$$U = \frac{U_0 e^{-\lambda t}}{1 + (U_0/\epsilon)(1 - e^{-\lambda t})} \qquad (7\text{-}40)$$

$$R_L(t) = \epsilon \ln\left[1 + \frac{U_0}{\epsilon}(1 - e^{-\lambda t}) \right] \qquad (7\text{-}41)$$

$$R_Q(t) = \frac{U_0\,[1 + (U_0/\epsilon)](1 - e^{-\lambda t})}{1 + (U_0/\epsilon)(1 - e^{-\lambda t})} - \epsilon \ln\left[1 + \frac{U_0}{\epsilon}(1 - e^{-\lambda t}) \right] \qquad (7\text{-}42)$$

Equations 7-40, 7-41, and 7-42 are directly derivable from equation 7-39 by insertion of the appropriate solved integrals of equations 7-38b and c. The algebra is of no great interest and is not developed here. Do not lose sight of the fact that U is determined by the dose; therefore, these equations simply represent rates of change of uncommitted lesions and the rate of appearance of repair products of first- and second-order repair processes. To this point, dose is not a variable, although it is implicit in U_0.

The formulation of the RMR model is uniquely different from either the molecular model or the dual radiation action model in that the rate of production of initial lesions is linear with dose, and, especially, time is explicitly part of the formulation. The RMR model shares with the molecular model the assumption that linear and quadratic terms for rates of repair are possible. However, Tobias and others (1980) introduced an additional concept that was indeed unique.

Eurepair and misrepair.   Both the linear and the quadratic repair processes are assumed by the authors to be capable of correct repair of the macromolecular damage, *eurepair*, or incorrect repair of the lesions, *misrepair*. Two new parameters are introduced. Let ϕ represent the probability that the linear repair is correct (eurepair), and let δ represent the probability that quadratic repair is correct (eurepair). The probabilities of misrepair, then, are $1 - \phi$ and $1 - \delta$ for the two processes, respectively. We can then formulate the number of lesions per cell repaired by the linear process as R_{LE} (eurepaired) and R_{LM} (misrepaired) and the same for the quadratic process as R_{QE} and R_{QM}. The original presentation of Tobias and others presented derivations for a number of limiting cases for the RMR model. Only two of these cases, the simplest, will be examined here.

Case I: Linear Eurepair, Quadratic Misrepair

Assumption. All linear repair is eurepair, $\phi = 1$, and all quadratic repair is misrepair and lethal, $\delta = 0$. Remnant U lesions existing at time t are also lethal. Using the Poisson statistic in the same fashion as for all the foregoing models, the survival at time t, $S(t)$, is expressed in the following equation by applying equations 7-40, 7-41, and 7-42 to state the fractions of linear and quadratic repair and the fraction of remnant lesions, $U(t)$.

$$S(t) = e^{-U_0}\left[1 + \frac{U_0}{\epsilon}(1 - e^{-\lambda t})\right]^{\epsilon} \tag{7-43a}$$

There is, again, no explicit statement of dose in 7-43a, but U_0 is a function of dose. Various sources, as well as the earlier models discussed here, suggest that the relationship between U_0 and dose could be described by either of the following expressions:

$$U_0 = aD \quad \text{or} \quad U_0 = a_1 D + a_2 D^2 \tag{7-43b, c}$$

where a, a_1, and a_2 are constants.

For a practical case survival equation to serve as a description for case I, an additional constraint must be introduced. This is a factor, T, that is related to the maximum time for repair, t_{max}, which is presumably the time from irradiation to mitosis or to time for replication of a portion of the DNA that cannot be replicated in the unrepaired state. T is defined by the following expression:

$$T = 1 - e^{-\lambda t_{max}} \tag{7-44}$$

Then the surviving fractions, S, can be written

$$S = e^{-aD}\left(1 + \frac{aDT}{\epsilon}\right)^{\epsilon} \tag{7-45a}$$

And when $\lambda t_{max} \gg 1$, then $T \cong 1$:

$$S = e^{-aD} \left(1 + \frac{aD}{\epsilon} \right)^{\epsilon}$$ (7-45b)

This survival expression has been found to be a generally good fit to the experimental data and, on the whole, to be a better fit than either MTSH or L–Q. The survival in the shoulder region is significantly better than that for MTSH, and the fit to the exponential portion of the survival curve at high doses is a better fit than the L–Q model.

The power of the RMR model is that very few assumptions are general and rigid. The nature of the dose–response relationship for the production of U lesions is not fixed and can be adjusted to fit experimental data. For the development of the expressions leading to equations 7-45a and b, equation 7-43b was assumed to describe the yield of lesions. There is also a great deal of flexibility in the model to deal with fractionated, divided dose, and continuous exposure conditions.

Comparison with Conventional Multitarget Single-hit Model

Conventional target theory, as discussed in earlier sections, describes survival for m number of hits and an inactivation constant of a with the following equation:

$$S_m = e^{-aD} \sum_{i=0}^{m-1} \frac{(aD)^i}{i!}$$ (7-46)

If equation 7-45a is expanded as a power series, setting ϵ equal to $m - 1$, we obtain

$$S (\epsilon = m - 1) = e^{-aD} \sum_{i=0}^{m-1} \frac{(m-1)!}{i! (m-1-i)!} \left(\frac{aD}{m-1} \right)^i$$ (7-47)

Tobias and others (1980) point out that these expressions only differ by the constant in i and m preceding the dose term in equation 7-47 and in the constant divider of the dose term. For the single-hit survival curve, $m = 1$, both equations are identical in the familiar form, $S = e^{-aD}$. For the two-hit MTSH survival curve, $m = 2$, the same form of survival curve is produced by the RMR model for the state where $\epsilon = 1$. The significance of this latter value is that $\lambda/k = 1$, and thus the coefficients for the linear repair and quadratic repair are equal.

Case II: Linear Repair Is Not Always Eurepair

When linear repair is assumed to not always be eurepair, then the constant ϕ has a value different from 1.0. This case will not be derived here, and the reader is referred to the initial reference for details. However, the form of the final expression is important. It is identical with the expression for survival under case I, but an additional exponent, ϕ, is added to the dose term. The expression for case II survival is

$$S = e^{-aD}\left(1 + \frac{aD}{\epsilon}\right)^{\epsilon\phi} \tag{7-48}$$

An important consequence of the form of equation 7-48 is that it will have a finite negative slope at low doses. As the surviving fraction, S, approaches zero, the first derivative of equation 7-48 is $-a(1 - \phi)$.

LETHAL–POTENTIALLY LETHAL (LPL) MODEL

The LPL model of Curtis (1986) is based, in part, on the structure proposed by Pohlit and Heyder (1981) as outlined in Figure 7.4. The number of B and C lesions are linearly related to dose by the proportionality constants η_{AB} and η_{AC}. The B lesions are repaired by a first-order process to return them to the undamaged A state. The alternative is that the B lesions are permanently converted to the irreparable or lethal state C by a second-order process.

Assumptions

The assumptions of Curtis's LPL model are as follows:

1. The B lesions of Figure 7.4 are spatially distributed and long-lived (minutes), potentially repairable lesions created in the nucleus of the cell by low LET radiation. The lesions are reparable to return

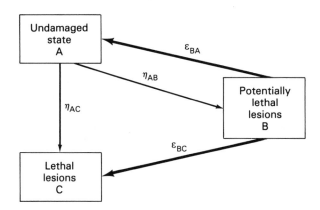

Figure 7.4 Hypothetical model for the induction of potentially lethal and lethal lesions by radiation and the possible pathways for repair or damage fixation. (After Pohlit and Heyder, 1981)

the B lesions back to the A state by an enzymatic process that is first order in the number of B lesions. If the B lesions are not repaired, they will interact with each other to form C lesions, which are irreparable and lethal. The interaction process is second order in the number of B lesions. It is explicit in the model that the interaction of B lesions does not depend on the spatial separation of these lesions, but only on the total number of B lesions.

2. The lesions can be formed at the time of irradiation on a very short time scale if they are created simultaneously or nearly so, and they are very close together. The constant η_{AC} describes the yield per unit dose for these directly formed lethal lesions.

3. The constants η_{AB} and η_{AC} are the rate constants for the production of both types of lesions, and the rate of formation of each lesion will be equal to the product of the dose rate, D', and the constant. For example, $\eta_{AC}D'$ is the rate of formation of lethal lesions for dose rate D'. The constants ϵ_{BA} and ϵ_{BC} are the rate constants for the formation of restored lesions per unit time and for the formation of irreparable lesions per unit time, respectively. Since ϵ_{BC} is the result of a bimolecular interaction of B lesions, it will be a second-order reaction in the concentration of B lesions.

4. It is further assumed that the rate of repair of the lesions does not depend on the number of lesions; that is, there is no saturation of the repair process.

During Irradiation

The two differential equations for the rate of change of the number of lesions of both types can be written as

$$\frac{dB(t)}{dt} = \eta_{AB}D' - \epsilon_{BA}B(t) - \epsilon_{BC}B^2(t) \tag{7-49}$$

$$\frac{dC(t)}{dt} = \eta_{AC}D' + \epsilon_{BC}B^2(t) \tag{7-50}$$

When the initial conditions are that $B = C = 0$, in other words, the number of either type of lesions is zero at time zero, the integration of equations 7-49 and 7-50 are

$$B(t) = \frac{2\eta_{AB}D'\,(1 - e^{-\epsilon_0 t})}{\epsilon_0 + \epsilon_{BA} + (\epsilon_0 - \epsilon_{BA})e^{-\epsilon_0 t}} \tag{7-51a}$$

where

$$\epsilon_0 = \sqrt{\epsilon_{BA}^2 + 4\epsilon_{BC}\eta_{AB}D'} \tag{7-51b}$$

and

$$C(t) = \eta_{AC}D + \epsilon \ln \left\{ \frac{2\epsilon_0}{\epsilon_0 + \epsilon_{BA} + (\epsilon_0 - \epsilon_{BA})e^{-\epsilon_0 t}} \right\} + \frac{(\epsilon_0 - \epsilon_{BA})^2 t}{4\epsilon_{BC}}$$

(7-52a)

where

$$\epsilon = \frac{\epsilon_{BA}}{\epsilon_{BC}}$$

(7-52b)

After Irradiation

After the irradiation is complete, there is no longer a source term for the production of new lesions, but repair continues with the same rate constants for the repair of the remaining lesions. The reader is referred to Curtis for the appropriate differential equations, which are easily set up from the model conditions. The integrations yield equations 7-53 and 7-54 for the number of B lesions and the number of C lesions as a function of time. The initial conditions for these integrations are the following, assuming the irradiation stops at time, T: $B(T)$ is the solution of equation 7-51a for time T; $C(T)$ is the solution of equation 7-52a for time T.

$$B(T + t_r) = \frac{N_{PL}e^{-\epsilon_{BA}t_r}}{1 + \dfrac{N_{PL}(1 - e^{\epsilon_{BA}t_r})}{\epsilon}}$$

(7-53)

and

$$C(T + t_r) = \frac{N_L + \left[1 + \left(\dfrac{N_{PL}}{\epsilon}\right) \right](1 - e^{\epsilon_{BA}t_r})}{1 + \dfrac{N_{PL}(1 - e^{\epsilon_{BA}t_r})}{\epsilon}}$$
$$- \epsilon \ln \left[1 + \frac{N_{PL}(1 - e^{\epsilon_{BA}t_r})}{\epsilon} \right]$$

(7-54)

where $N_{PL} = B(T)$; $N_L = C(T)$ and t_r is the available repair time. As before, ϵ is the ratio of the repair constants.

Survival Equation

To calculate the survival at time $t = T + t_r$, it is assumed that the sum of all lesions, both lethal (C) and potentially lethal (B), are now indeed lethal, since there is now no longer time to repair B lesions. With the assumption that the lesions are distributed according to the Poisson statistic, the probability that a cell has *no* lethal lesions, where n_{tot} is the total of both C lesions and B lesions at time $T + t_r$, is

$$S = e^{-n_{tot}(T+t_r)} \qquad (7\text{-}55)$$

Substituting from equations 7-53 and 7-54 for $B(T + t_r)$ and $C(T + t_r)$ and simplifying,

$$S = e^{-(N_L + N_{PL}) + \epsilon \ln\left[1 + \dfrac{N_{PL}(1 - e^{-\epsilon_{BA} t_r})}{\epsilon}\right]} \qquad (7\text{-}56)$$

or

$$S = e^{-N_{tot}}\left[1 + \dfrac{N_{PL}}{\epsilon(1 - e^{-\epsilon_{BA} t_r})}\right]^{\epsilon} \qquad (7\text{-}57)$$

where N_L = total number of lethal lesions at the end of the exposure time

N_{PL} = number of potentially lethal lesions at the end of the exposure time

$N_{tot} = N_L + N_{PL}$

ϵ = ratio of the two rate constants, $\epsilon_{BA}/\epsilon_{BC}$

t_r = total repair time available after the end of the exposure

As in the case of several of the previous models, although equation 7-57 appears excessively complicated, there are some very useful limiting conditions and simplifications.

Low-dose-rate Approximation

If low dose rate is defined such that further decrease in the dose rate has no effect on the survival curve, then potentially lethal (B) lesions will be repaired before they have a chance to interact. Lethality will be determined by the rate of direct formation of lethal (C) lesions. The formalism is the following: $D' <<< \epsilon_{BA}^2/(2\eta_{AB}\epsilon_{BC})$. Then expanding the equation for ϵ_0 (see equation 7-51b), and with extensive derivations, the equation for survival at low dose rate becomes

$$S = e^{-\eta_{AC} D} \qquad (7\text{-}58)$$

This equation is a form with which the reader is very familiar from all the previous models, including the single-hit form of the Lea hit theory model.

High-dose-rate Approximation

If high dose rate is defined such that the shape of the survival curves is no longer a function of dose rate, another useful approximation is obtained. It can be shown that, if $T \lll 2/\epsilon_0$, then the survival equation will reduce to

$$-\ln S = (\eta_{AC} + \eta_{AB}) D - \epsilon \ln [1 + \eta_{AB}D/\epsilon(1 - e^{-\epsilon_{BA}t_r})] \quad (7\text{-}59)$$

For the conditions of this approximation the dose rate constraints are not particularly onerous. Exposure times of less than 5 min will usually satisfy the criterion with most mammalian cell lines. It will be noted that equation 7-59 is the same as the general survival equation, but with the assumption that all the lesions created by the radiation are still present at the end of the exposure period.

Linear–Quadratic Approximation

A special form of equation 7-59 can be developed for the region of the survival–dose response curve for low doses. For details of the derivation the reader is directed to Curtis (1986), but the general procedure is to rewrite the survival function as a logarithmic relationship and to expand this relationship as a power series. If only the first two terms, those in dose to the first and second power, are retained and higher-order terms are neglected, the following survival expression is developed:

$$-\ln S = (\eta_{AC} + \eta_{AB}e^{-\epsilon_{BA}t_r}) D + \frac{\eta_{AB}^2}{2\epsilon} (1 - e^{-\epsilon_{BA}t_r})^2 D^2 \quad (7\text{-}60)$$

This formulation is identical with the survival expression for the molecular model, with rather complicated constants, including the exponentials in the constants. To visualize this expression in the form of the linear–quadratic expression of the molecular model, the constants α and β can be extracted from equation 7-60.

$$\alpha = \eta_{AC} + \eta_{AB} e^{-\epsilon_{BA}t_r} \quad (7\text{-}61)$$

$$\beta = \frac{\eta_{AB}^2}{2\epsilon} (1 - e^{-\epsilon_{BA}t_r})^2 \quad (7\text{-}62)$$

The linear-quadratic form under the low-dose approximation has been analyzed by Curtis against the survival data for C3H-10T $\frac{1}{2}$ cells as published by others. The values of α and β for these data, developed by others, are $\alpha = 0.1366 \ \text{Gy}^{-1}$ and $\beta = 0.02 \ \text{Gy}^{-2}$. The linear–quadratic approximation holds for about 22 Gy for this set of data, a value at which the surviving fraction has been reduced by four log cycles.

SUMMATION

As we progress through the historical development of biophysical models for cell survival, we are impressed that there is significant convergence as we develop through more and more sophisticated models. A generalization would be that the linear–quadratic model is an adequate approximation for survival fractions greater than 10^{-3}, but that further sophistication is necessary for detailed description of survival at surviving fractions less than this value.

Another important observation is that as we progress through the historical model development, there is increasing attention to the significance of irradiation time, LET, repair rates, fractionation, protraction, and other subtle aspects of the exposure paradigm.

REFERENCES

CHADWICK, K. H., and LEENHOUTS, H. P. (1973) A molecular theory of cell survival. *Phys. Med. Biol.* **18,** 78–87.

—— (1981) *The Molecular Theory of Radiation Biology.* Springer Verlag, Berlin–Heidelberg.

CURTIS, S. B. (1986) Lethal and potentially lethal lesions induced by radiation—a unified repair model. *Radiation Res.* **106,** 252–271.

GOODHEAD, D. T. (1982) An assessment of the role of microdosimetry in radiobiology. *Radiation Res.* **91,** 45–76.

HALL, E. J., and others (1972) Survival curves and age response functions for Chinese hamster cells exposed to x-rays or high LET alpha-particles. *Radiation Res.* **52,** 88–98.

KELLERER, A. M., and ROSSI, H. H. (1971) RBE and the primary mechanism of radiation action. *Radiation Res.* **47,** 15–34.

—— (1972) The theory of dual radiation action. *Curr. Topics Radiation Res. Quart.* **8,** 85–158.

—— (1978) A generalized formulation of dual radiation action. *Radiation Res.* **75,** 471–488.

LEA, D. E. (1955) *Actions of Radiation on Living Cells,* 2nd ed. Cambridge University Press, New York.

NEARY, G. J., PRESTON, R. J., and SAVAGE, J. R. K. (1967) Chromosome aberrations and the theory of RBE. III. Evidence from experiments with soft x-rays, and a consideration of the effects of hard x-rays. *Intl. J. Radiation Biol.* **12,** 317–345.

POHLIT, W., and HEYDER, I. R. (1981) The shape of dose–survival curves for mammalian cells and repair of potentially lethal damage analyzed by hypertonic treatment. *Radiation Res.* **87,** 613–634.

PUCK, T. T., and MARCUS, P. I. (1955) A rapid method for viable cell titration and clone production with HeLa cells in tissue culture: the use of x-irradiated cells to supply conditioning factors. *Proc. Natl. Acad. Sci. U.S.* **41,** 432–437.

TOBIAS, C. A., and others (1980) The repair–misrepair model of cell survival. In *Radiation Biology and Cancer Research,* R. E. Meyn and H. R. Withers, eds. Raven Press, New York, pp. 195–230.

SUGGESTED ADDITIONAL READING

DERTINGER, H., and JUNG, H. (1970) *Molecular Radiation Biology*. Springer Verlag, Heidelberg, Berlin (English Edition, in particular Chapters 2 and 3).

ELKIND, M. M., and WHITMORE, G. F. (1967) *The Radiobiology of Cultured Mammalian Cells*. Gordon and Breach, New York and London.

MEYN, R. E., and WITHERS, H. R., eds. (1980) *Radiation Biology in Cancer Research*. Raven Press, New York.

PROBLEMS

1. Show by simple algebraic or graphical means, explaining your steps, that the following relationship holds for the multitarget single-hit model. $D_q = D_0 \ln n$.

2. The following data are provided for the survival of a hamster cell line grown in vitro immediately after irradiation.

Dose (cGy)	Surviving Fraction
0	1.00
50	0.85
100	0.52
150	0.33
200	0.16
250	0.064
300	0.024
400	0.0035

   (a) Plot the data graphically on three-cycle semilog paper and determine the MTSH parameters D_0, D_Q, and n; write the MTSH expression for these parameters.

   (b) Repeat part (a) using the linear–quadratic model to fit the data. What are the values of $p\alpha$ and $p\beta$?

3. In developing the expression for single-hit survival, it is assumed that the target volume v is constant. This is biologically unlikely. Let us see how sensitive to variation in target volume the shape of the survival curve may be. Take a value for the inactivation constant at, for example, 0.01, and:

   (a) Plot the survival function for the simple exponential on semilog paper as a function of dose.

   (b) Repeat this procedure assuming a mixture in which half the cells has v's that are 80% of that for the remainder.

   (c) Repeat for a mixture in which half the population has v's that are 10% of that for the remainder. (*Hint*: What are the dimensions of the inactivation constant?)

8

Survival Curve
and Its Significance

INTRODUCTION

In Chapter 7, several models were discussed and developed that are intended to give insight into the mechanism of radiation action on living cells. A reminder is in order. Survival or nonsurvival is defined for the purpose of radiation biology as the ability of a single cell to reproductively proliferate to form a colony of cells. This ability is defined as clonogenic potential. The clonogenic survival definition has ancient roots in microbiology, and many of the methods for quantitation arise from that source. It is still the practice to plate individual microbial cells, which are then observed for their ability to form a visible colony. The same methods of colony counting are now widely used in scoring the survivability of mammalian and other cells after exposure to ionizing radiation.

Until the late 1950s it was not possible to use in vitro cell culture methods for mammalian cell lines. The quantitative radiation biology of the era before this time is characterized by studies on the classic microbiological organisms that could be grown on petri plates and could have survival measured. Typical organisms on which clonogenic survival studies were done at that time are *Escherichia coli*, *Bacteroides subtilis*, *Saccharomyces* sp., and some other lower organisms, such as *Tetrahymena* sp. or other protozoa. The classic studies of Puck and Marcus (1955) that led eventually to the development of methods for clonogenic plating and growth of mammalian cells had a dramatic impact on radiation biology.

After this time it was possible to examine radiation clonogenic survival for a number of mammalian cell lines grown in vitro.

The nature of the mammalian cell lines that can be grown in culture is important. Most of the lines are "immortal" cells, derived generally from cancer tissue that has been removed from a human or an animal donor. These lines are not necessarily representative in their radiation response of the radiation response of noncancer cells, and it is a continuing goal of radiation biologists to develop what are known as untransformed cell lines for clonogenic survival studies. Some of the more common cell lines that are used in radiation biology are HeLa cells, derived from a human cervical cancer; V79 and CHO cells, derived from hamster lung or ovary, respectively; 9L cells derived from a rat gliosarcoma; T1 cells derived from a human kidney; and a few others that will be referred to throughout the remainder of the text. All these cell lines have the potential for immortal growth and are considered to be transformed cells.

TECHNIQUE OF THE CLONOGENIC SURVIVAL CURVE

What is eventually graphed for the survival curve for cells irradiated and grown in vitro is the logarithm of the fraction of the plated cells that survive and produce a colony of offspring, as ordinate, against the dose as abscissa. The criterion of clonogenic survivability is usually taken as the production of 50 or more cells at the minimum. Some cells divide only a few times before failing to divide further, and these are not counted as clonogenic survivors. One of the more crucial points of a successful clonogenic survival experiment is to establish the clonogenic potential of the unirradiated cells. Generally, mammalian cell lines are not completely effective in establishing growing colonies. The denominator of the surviving fraction is the fraction of unirradiated cells that will produce colonies. This fraction generally will range from 0.5, or less, to as high as 0.9. Rarely, if ever, is the unirradiated cell line capable of 100% generation of new colonies. The variability in the control surviving fraction, which is called the *plating efficiency*, is one limit on the precision of the survival curve technique.

Determining the Surviving Fraction

Samples of the cells are irradiated under carefully controlled laboratory conditions, such that the plating efficiency would remain unchanged and as high as achievable if no irradiation took place; and appropriate numbers of the cells are transferred to plastic containers for growth. A defined medium in which the cells are known to grow well is added to the container, and they are put in the incubator at 37°C with an appropriate atmospheric environment of oxygen and carbon dioxide.

One of the more crucial aspects of this process is to choose the correct number of cells to be added to the culture plate. One would prefer to have from 20 to 50 colonies on the plate when they are examined after 10 to 20 days from seeding. When the surviving fraction is very low, this will mean seeding large numbers of cells to achieve the minimum number of colonies. When the survival curve is not established, it is necessary to make educated guesses as to the appropriate surviving fraction. The reliability of the estimate of surviving fraction follows the usual Poisson statistic, just as in the counting of radioactive disintegrations. The standard deviation of the number of colonies counted will be the square root of that number. For instance, if 50 colonies are counted, the standard deviation is 7.07, or 14%. If 5 colonies are counted, the standard deviation is 2.24, or 45%.

Feeder Cells

The pioneer experiments of Puck and Marcus were successful because they developed the ingenious idea that an underlayer of metabolizing but nondividing cells under the cells of experimental interest would provide essential growth factors for support of division in the experimental cell line. This *feeder cell* method has since then been widely used. A large number of cells are preirradiated to a dose that will ensure no division. The cells are plated into the dishes and allowed to attach. The experimental cells are then plated in their small numbers over the top of the feeder layer. In recent years, media have been developed that allow the omission of a feeder layer for the growth of some cell lines.

More recently, developments in cell culture methods allow us to examine the radiation sensitivity of some cells other than the "immortal" lines (also called *established cell lines*). It is now possible to undertake survival analysis in vitro from fresh explants of normal and tumor tissues, as well as with nonimmortal lines that divide for only a few tens of generations.

CHARACTERISTICS OF THE MAMMALIAN CELL SURVIVAL CURVE

The mathematical analysis of survival curves in Chapter 7 demonstrated that it is often a plot of the natural logarithm of the surviving fraction against dose that is the most useful format for examination of the data. That is the format in which the data are usually plotted. What parameters are most useful to examine? It was suggested in Chapter 7 that the linear–

quadratic model for clonogenic survival was one that effectively represented the data for a reasonably restricted range of dose. The parameters
for that relationship, $p\alpha$ and $p\beta$, are certainly mathematically useful, but
they are indeed hard to use to visualize the shape of the survival curve.
For this reason, the multitarget single-hit model (MTSH) still finds wide
utilization in the radiobiology literature. The reason for this is that knowledge of the parameters n, D_0 and D_q allows us to carry a very effective
mental picture of the shape of the survival curve. We shall follow that
practice in this chapter, using the linear–quadratic parameters only as
supplemental information.

The general characteristics of mammalian survival curves are outlined in Figure 8.1. In this representation, the formalism of the multitarget
single-hit model for the surviving fraction is used. The logarithm of the
surviving fraction is plotted as the ordinate, and the dose is plotted as
the abscissa. The parameters that describe this curve are the inactivation
constant k and the extrapolation number, or target multiplicity, n. The
inactivation constant is measured as the slope of the linear portion of the
survival curve. More usually, it is determined by estimating the dose
required to reduce the surviving fraction by $1/e$ in the linear portion of
the curve. The target multiplicity n is determined by back-extrapolation

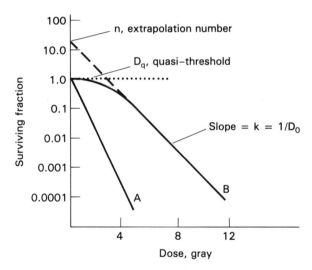

Figure 8.1 General characteristics of the mammalian survival curve as plotted
using multitarget single-hit mathematics. The curves A and B are, respectively,
representative of cell killing for high LET radiation, showing typical single-
hit kinetics, and for low LET radiation, showing typical MTSH kinetics. The
back-extrapolation to determine the target multiplicity n is shown for curve
B. D_q, the quasi-threshold dose, is measured as the dose intercept of the back-
extrapolate at a surviving fraction of 1.0 of the line used for estimation of n.

of the linear portion of the survival curve to the zero-dose abscissa. Curve A in the figure is for the special case where the target multiplicity is 1. Such a curve is generally associated with survival following high LET radiation. It is, in principal, a degenerate form of the general curve, B, for the special case of target multiplicity of 1.

An additional derived parameter is also shown in Figure 8.1. That is the D_q, generally called the *quasi-threshold dose*. The D_q is determined by estimation of the dose at the point where the back-extrapolate of the exponential portion of the curve crosses the surviving fraction ordinate of 1.0. The reader is referred to Chapter 7, where the special relationship between D_q, D_0, and n is given. As a reminder, that relationship is given again:

$$D_q = D_0 \ln n \qquad (8\text{-}1)$$

The general characteristics of the in vitro cell survival curve as outlined describe quite satisfactorily the survival for nearly all mammalian cell lines that have been studied. The parameters of the multitarget single-hit equation do indeed vary among cell lines, but not to great extremes. The D_0 shows the least variation among all the many lines studied since the first mammalian cell survival curve was published in 1956. It varies in the range of 1 to 2.5 Gy (100 to 250 rad), with most lines having D_0 values around 1.2 to 1.5 Gy (120 to 150 rad). The extrapolation number or target multiplicity varies much more than that, but, still, most lines studied have extrapolation numbers between 1.5 and 20.

Figure 8.2 shows typical survival curves for three cell lines that have had wide utilization in cell biology. These three lines are HeLa, CHO, and T1. The sources of these cell lines are described earlier in this chapter. As a point of historical interest, the original work by Puck on human cell culture, as well as the studies of radiation survival, were done on the HeLa line, a cell line derived from a human cervical carcinoma. The survival data for HeLa and CHO fit well to the multitarget single-hit model, and the parameters are easily and reliably estimated. On the other hand, the survival curve for the T1 line, a cell line of human kidney origin, does not conveniently fit the multitarget single-hit model, and it is not possible to get single-valued estimates of n and D_0. For this latter cell line, it is necessary to fit the data to the linear–quadratic function described in Chapter 7. Represented in the data in Figure 8.2 is the usual range of variability seen among mammalian cell lines. The HeLa line is significantly more radiosensitive than either of the other two lines, with an extrapolation number of about 2. The CHO line has an extrapolation number of about 5 to 6, and these parameters have little meaning for the T1 line, which is intermediate in radiosensitivity.

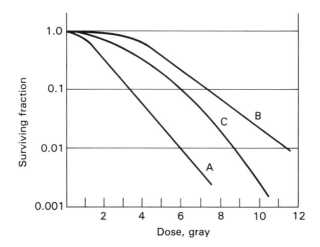

Figure 8.2 Typical postirradiation survival curves for three mammalian cell lines grown in tissue culture. Curve *A* is for HeLa cells, curve *B* for CHO cells, and curve *C* for T1 cells.

SIGNIFICANCE OF THE SHOULDER ON THE SURVIVAL CURVE

Prior to 1956 the interpretation of the significance of shoulders on survival curves was pretty well confined to considerations of target multiplicity. When the target theory of radiation action was discussed in Chapter 7, the point was made that the model contained only one significant biologically based function. This function, the hit survivability function, described the likelihood of cell survival if a designated number of hits were sustained by the cell. This approach was strongly reinforced by the totality of experience on microbiological organisms and on direct radiation of large bioactive molecules, where target theory adequately explained the data. Of particular importance was the observation that shoulders on survival curves (or in this case remaining biological activity) for bioactive molecules irradiated in solution were certainly also either of the single-hit-type curve or, occasionally, MTSH-type curve. No molecular repair is possible for these large biomolecules in solution, and the MTSH-type curves were perfectly adequately explained by target theory.

In 1959 a landmark study was published by Elkind and Sutton that set out to show that the shoulder of a survival curve in mammalian cells was indicative of repair of damage to whatever was the radiation sensitive system in the cell. The experimental data of Elkind and Sutton will be examined in some detail, since they represent such a significant step forward in our understanding of radiation biology.

The observations made by Elkind and Sutton were as follows. (No attempt has been made here to present their data quantitatively. It is more important to recognize the significance of the experiment rather than to pay attention to the details of the doses used or the surviving fractions.) The upper portion of Figure 8.3 is the normal survival curve for this cell line for a surviving fraction down to about 0.1. The important point of this portion of the figure is that D_q can be established as shown. For this mythical cell line, D_q is taken to be about 2 Gy (200 rad). Having established D_q, it is then possible to irradiate a stock of cells to a level that would give a surviving fraction of about 0.1, or 4.4 Gy (440 rad), for the hypothetical case shown. The surviving fraction to which the cells are irradiated is irrelevant as long as the surviving fraction chosen is in the

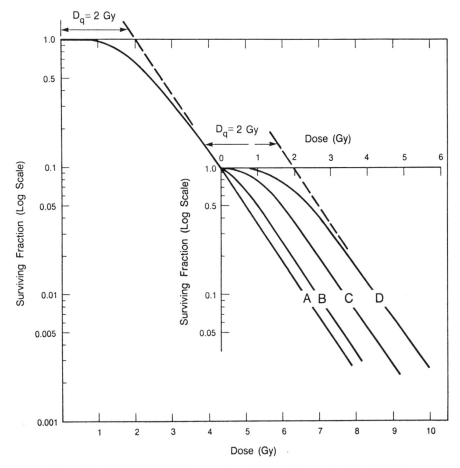

Figure 8.3 The Elkind–Sutton experiment. The cells are first irradiated to a surviving fraction of about 0.1. These preirradiated cells are then used as stock for further irradiations after holding the cells for various times. Curve A is for cells in which the irradiation is continuous with no interruption. Curves B, C, and D are for cells irradiated after 15 min, 2 h, and 6 h, respectively.

exponential portion of the survival curve. Having preirradiated a stock
of cells, a new survival curve is constructed as shown by the lower coordinates. The surviving fraction for the preirradiated cells that receive
no further irradiation becomes a surviving fraction of 1.0 for the new
survival curve. The remaining preirradiated cells are then reirradiated
with a range of doses to establish a survival curve after various time
intervals of holding the cells without treatment.

Curve A in Figure 8.3 is the survival curve obtained if there is no
interval of time between the first dose, sometimes called the *conditioning
dose*, and the second dose. This is the expected result, since the conditions
are identical with those under which a complete survival curve is normally
developed. It is as if the dose were given in a single session.

Curves B, C, and D are the survival curves obtained for the preirradiated cells when various time periods are allowed to lapse before the
conditioning dose is given, according to the schedule described in the
figure. It is apparent that as time passes the cells become less radiosensitive; and after an adequate time period has passed, the radiosensitivity
of the cells is the same, or nearly the same, as that for previously unirradiated cells. The direct proof of the last statement is that D_q for the
preirradiated cells is the same as D_q for previously unirradiated cells.

The interpretation given these data by the authors was that repair
processes were at work to restore the cell to its previous condition of
radiosensitivity and that these processes were relatively fast.

It is important to note that the slope of the linear portion of the log
survival curve remains the same, no matter what the period of delay
between the conditioning dose and the final dose.

REPAIR OF SUBLETHAL INJURY

The concept of sublethal injury and its repair was developed by Elkind
and Sutton from the just described experiments. The conditioning radiation exposure certainly kills a large number of cells as evidenced by the
low surviving fraction if no other radiation exposure is made after the
conditioning dose. In the surviving cells, however, there remains a significant injury that will not be revealed unless additional radiation exposure occurs. In this sense the injury in the surviving cells is *sublethal*.
In other words, the injury is not lethal to the cells unless something else
happens. It is also clear that this sublethal injury is repaired at a rapid
rate. The measured half-time for the repair of sublethal injury reported
by others is in the range of 12 to 15 min. Elkind and Sutton concluded
that, at least by 18 h after a conditioning dose of 505 rad, surviving cells
fully repaired their sublethal damage, since survival following the second
exposure was the same as for cells that had received no prior irradiation.
They rightly point out that inferences can only be drawn about the fraction

of the population that survives the first dose. Some repair of sublethal damage is certainly occurring in the population of cells destined to die, but it is inadequate to assure survival.

Interpretation of SLD

An attractive interpretation of these findings can be made in terms of the molecular model of radiation damage described in Chapter 7. DNA is the critical target, and it is well established that DNA can undergo changes that will cause the cell to lose its reproductive capacity. The molecular model proposes that this damage to DNA can occur by either of two mechanisms: (1) a direct hit on both strands of the DNA molecule leading to a double-strand break, or (2) individual strand breaks occurring in the separate strands that can interact, if conditions are appropriate, to lead to a double-strand break. If we visualize sublethal damage as a single-strand break in DNA that can be rapidly and efficiently repaired, then the addition of a second single-strand break at a nearby region of the opposite DNA strand could lead to interaction of these lesions to produce a double-strand break that could be lethal to a cell's reproductive capacity. Rapid repair of these single-strand lesions would lead to restoration of the cell's integrity and resistance to additional radiation.

The evidence for the hypothesis just formulated is certainly not at all conclusive, but there is much supporting data. One of the more significant findings is that the kinetics of the repair of single-strand breaks is very well matched to the kinetics of repair of sublethal injury. It is also well known that ionizing radiation is an efficient producer of single-strand breaks in DNA. It must be said, however, at the time of this writing there is an increasing body of opinion and experiment that denies the relevance of DNA single-strand breaks to cell survival. Those who raise such concerns do agree that the evidence is strongly supportive of the presence of some kind of "sublesion" in the DNA that can lead to major and irreversible damage to the DNA if it interacts with another nearby sublesion.

One of the more convincing arguments against the single-strand break hypothesis is the experimental evidence collected with ultrasoft x-rays (Goodhead, Thacker, and Cox, 1978; Thacker, Goodhead, and Wilkinson, 1980). These authors irradiated cellular systems with ultrasoft x-rays generated as the 280 eV $K\alpha$ x-rays from carbon. The electron tracks produced from these x-rays can be no more than a few nanometers in length, and, therefore, lesions produced in DNA must be at least that close together. In spite of this, the extremely soft carbon x-rays are more effective at cell killing and chromosome abberation production than are normal hard x-rays. With x-rays of such low energy, the separation of single-strand breaks caused by independent events could not be more than a few nanometers. This separation of strand breaks by only one or two base pairs in the DNA molecule could not, under normal circumstances of higher-energy x-irradiation, be expected to happen with any significant frequency. Only time and further experiment will clarify this issue. For interpretation of sublethal injury, the DNA single-strand break model is useful and illuminating. If in the future it is revealed that the sublesion is physically not a strand break, the interpretation remains valid.

High LET/Low LET and SLD

Remembering the concepts of the molecular model, it should be possible to shed additional light on sublethal injury and its repair by comparing the effects of high and low LET radiations on the phenomenon. The high LET radiation, represented by the Δ fraction of the dose in the molecular model, it is suggested, produces its damaging effect by production of double-strand breaks as single events, while low LET radiation is thought to produce the preponderance of its damage through interaction of two sublethal events. The latter is the $1 - \Delta$ fraction of the dose in the molecular model. Data have been published on the two possible combinations (Ngo, Blakely, and Tobias, 1981). In one case a low LET conditioning dose is given to the cells, followed by a high LET test dose, and the second case is one in which the conditioning dose is given with high LET radiation and the test dose is given with low LET radiation. Figures 8.4 and 8.5 show a schematic representation of these data for hypothetical cells and hypothetical low and high LET radiations. The data are not taken from Ngo, Blakely, and Tobias, but they are prototypical of their findings.

Figure 8.4 shows that high LET radiations are, as expected, more effective at cell killing than low LET radiations. There is no shoulder, nor would there have been one if the cells had never been irradiated. Whatever repair time is allowed to lapse, even up to the 6 h used by Elkind and Sutton in their first experiment, the survival curve for cells pretreated with low LET radiation and then irradiated with high LET radiation indicates that no repair can be revealed by treatment with high LET radiation. Certainly, the same sublethal lesions are present from the pretreatment with low LET radiation, but they are irrelevant to the survival when the cells are exposed to high LET radiation. The high LET killing is by events that break both strands of DNA simultaneously, and this mode of cell killing is not facilitated by the presence of preexisting single-strand breaks. Whether the sublethal damage repair can be revealed by a postconditioning dose exposure to high LET or not, the repair of this damage must be continuing, since the cells are not privy to what test irradiation will be used for the second challenge dose.

The intepretation of the data in Figure 8.5 is as follows: It is reasonably certain that the high LET irradiation kills the cells by single events in which lethal double-strand breaks occur, and that these double-

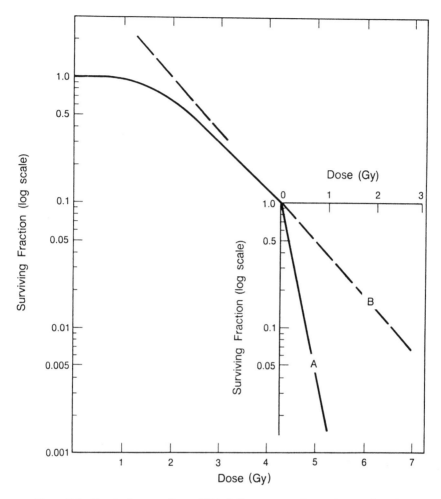

Figure 8.4 Survival curves for an Elkind–Sutton type of experiment when the conditioning dose is undertaken with low LET radiations, and the test exposure is done with high LET exposure. Curve *B* is the survival that would have occurred if the second irradiation were done with low LET radiation and no repair time had been allowed before the administration of the test irradiation. Curve *A* is the expected result for the second irradiation being administered with high LET radiation. The survival curve, *A*, is the same whether a time interval has been allowed between irradiations or not, and the survival curve after the high LET test dose is the same as if it were given to previously unirradiated cells.

strand breaks are repaired only slowly and with difficulty by the cell. However, while these double-strand breaks are the direct cause of loss of reproductive capacity, many single-strand breaks are also caused, which are generally irrelevant to the death of the cell because of the predomi-

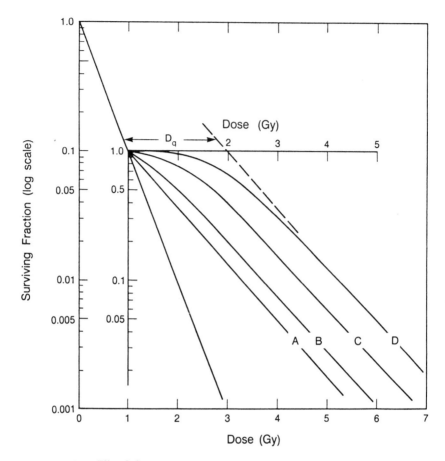

Figure 8.5 Elkind–Sutton experiment for a conditioning dose of high LET radiation followed by irradiation at low LET. Curves *A*, *B*, *C*, and *D* are for test exposures after the same time for repair as in Figure 8.3. The unlabeled curve is for continued high LET radiation.

nance of the double-strand event. These sublethal lesions are, however, revealed when the cells are irradiated with low LET radiation, which may produce single-strand events that can interact with the single-strand events produced by the earlier high LET radiation. As curves *A*, *B*, *C*, and *D* show, as time for repair is allowed, the single-strand sublethal lesions produced by the high LET radiation can be removed, and the D_q associated with low LET radiation is reestablished.

Before moving on to another type of repair of radiation injury, the operational definition of sublethal injury is important to know and understand. *Sublethal injury* produced in cells by ionizing radiation is a class of injury that would not normally lead to cell death unless additional radiation exposure occurs. The *challenge*, or second irradiation, produces additional injury that, through interaction with the existing sublethal lesions, will lead to cell death.

REPAIR OF POTENTIALLY LETHAL INJURY

Potentially lethal injury was first proposed by Phillips and Tolmach (1966). Lyman and Haynes (1967) shortly thereafter also demonstrated potentially lethal injury and its repair in yeast. This type of cellular damage is operationally very different from that associated with sublethal injury. The latter requires an additional exposure to ionizing radiation to produce additional injury that, through interaction with the existing lesions, will lead to cell death. Expressly stated, sublethal injury or damage may only be revealed by additional radiation exposure, and indeed that exposure must be from low LET radiation. *Potentially lethal damage* (or *injury*), usually abbreviated PLD, is defined as injury that will inevitably kill the cell unless intervention occurs that alters the outcome. Recently it has come to be realized that this intervention (repair process) is probably starting in the cell immediately when radiation damage is incurred, whether there is external intervention or not.

The earlier studies were all performed on cells grown in vitro, but recently it has been possible to show that potentially lethal damage is present and is also repaired in organized tissue systems.

Expression of PLD Repair

How is potentially lethal damage revealed and how is its repair demonstrated? These can be demonstrated by appropriate modulation of the environmental conditions under which the cell is growing in vitro. Cell survival can be increased under certain conditions after exposure to ionizing radiation, and this is generally identified as *repair of potentially lethal injury*; or cell survival can be decreased under other conditions, and this is called *expression of potentially lethal injury*, or possibly this latter phenomenon could better be called *prevention* of repair of PLD.

Fixation of PLD with Hypertonic Saline

PLD can be shown to be repaired by placing cells in environments that are not conducive to the continued progress of the cell through the cell cycle of DNA replication and mitosis, but that are, at the same time, hospitable to the vegetative functions of the cell. A medium that serves this cause admirably is *depleted medium*, a medium in which cell growth has taken place to the point of exhaustion of the energy source. At that point the medium is separated from the cells growing in it, and it is used as a replacement medium for the irradiated cells. An alternative is a balanced salt solution that contains no energy source; but this environment is more hostile to the cells, and the plating efficiency may be quite low. Other procedures are also available. For example, if cells are grown on the petri dishes to confluency, cells tend to go out of their normal cell cycle. If they are irradiated in that condition and then subcultured immediately or some hours later, a differential survival is seen, with protection provided by continued maintenance of the cells in the confluent state. This effect is also attributed to repair of PLD. In this latter case, repair of PLD is probably allowed to continue for a longer period owing to the longer time that the cells have before entering DNA replication when they are harvested from plateau phase cultures.

Expression of PLD is demonstrable in a somewhat reverse way. If cells are placed in hypertonic saline ($0.5\ M$ NaCl) (Raaphorst and Kruuv, 1976) immediately after they have been irradiated and survival is measured, the saline can be shown to decrease survival. This has been taken to be a demonstration that some PLD repair is proceeding in cells after irradiation whether or not the environment is altered. The hypertonic saline, it has been suggested, irreversibly fixes PLD so that it cannot be repaired.

Interpretation of PLD

Figure 8.6 embodies the elements of both repair and expression of PLD. As mentioned previously, holding in depleted medium allows time for PLD repair, and exposure to hypertonic saline fixes the PLD irreversibly. The times for PLD repair are significantly longer than for repair of sublethal injury. Instead of a halftime in minutes, it appears that PLD repair has a half-time more on the order of 1 to 2 h. It is tempting to interpret PLD repair in terms of repair of double-strand breaks in DNA, and that those environments that promote PLD repair are holding the cell from progression through the cell cycle and DNA replication while the double-strand breaks are being repaired. At this time, such an idea can only be called a hypothesis, but it is indeed attractive. In any case, it is now known that most cell lines can repair double-strand breaks in DNA, even though the process is slow and error prone (Resnick and Martin, 1976).

One of the more telling arguments against the hypothesis that repair of DNA double-strand breaks is responsible for PLD is that such repair should not depend on the nature of the radiation that produces the double-strand breaks. Unfortunately, it has not been possible to demonstrate repair of PLD in cell systems that have been irradiated with high LET radiation sources. It is just a possibility that the nature of the double-strand lesion after high LET exposure is significantly different from the

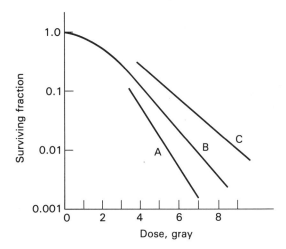

Figure 8.6 Repair of potentially lethal injury (PLD). Curve B is the normal survival curve for the cell line represented. If the cells are held some hours in depleted medium after irradiation (usually 6 to 24 h) and then plated, curve C will be obtained. If the cells are placed in 0.5 M saline immediately after irradiation and held there for some hours before plating, curve A will be obtained.

lesion produced by the cooperative interaction of single-strand breaks or sublesions in DNA.

CELL SURVIVAL AND CELL AGE

Life Cycle of the Cell

Since the classic studies of Howard and Pelc (1953) demonstrated that replicating cells have a "life history" that repeats itself in a cyclic way, interest has been very great in the relative sensitivity of the cell at various stages in its life cycle. The Howard and Pelc model is now part of the everyday experience of the cell biologist, but a brief recapitulation is useful. The marker that has been known since the earliest days of cell biology is mitosis. This particular phase of the cell cycle is morphologically evident and easily recorded. Mitosis is the creation of two daughter cells from the parent. The contribution of Howard and Pelc was to define several additional separate stages in the life cycle of the cell from mitosis to mitosis. These stages are only identifiable by biochemical rather than morphological markers. The most important of these stages for the dividing cell is the replication of its DNA prior to the distribution of two complete copies of the genome to two daughter nuclei at mitosis. This stage, which for all cell lines studied occurs at approximately the midpoint

TABLE 8.1   Cell Cycle Times in Two Representative in Vitro Cell Lines

Cell Cycle Phase	CHO (Hamster), h	HeLa (Human), h
T_c (generation time)	11	24
T_m (mitotic time)	1	1
T_s (synthesis time)	6	8
T_{G-2} (postsynthetic pause)	3	4
T_{G-1} (presynthetic pause)	1	11

of the time period between mitoses, occupies, in general, a goodly portion of the intermitotic time, and it has been named the *S phase*, or *synthetic phase*. The S period is preceded by a stage in which biochemical preparation for the start of DNA replication is the predominant activity. This stage is labeled the G-1 period and is often called the *presynthetic pause*. After replication of DNA is complete, the cell enters another period of preparation for mitosis. This period has been labeled the G-2 period, or *postsynthetic pause*.

In summary, the life cycle of the cell can be pictured as a continuous cyclical activity, usually timed from mitosis, and made up of a G-1 period, followed by the S period of DNA replication, followed by the G-2 period of postreplication activity, followed, again, by mitosis. The relative portion of the cell cycle taken up by each of these phases is highly variable among cell lines, both in culture and in vivo. The G-2 period is rarely more than a small proportion of the cell cycle time, usually only 1 to 2 h in a typical cell line with a 12- to 14-h time between mitoses. The S period is generally on the order of 4 to 8 h or so in the most extreme cases. The largest variability is seen in the G-1 period, which can range from almost nonexistent in, for example, bone marrow cells in culture, to a very large fraction of the intermitotic time. Table 8.1 gives the periods of the cell cycle for two cell lines widely used in laboratory experimentation, the HeLa human line and the CHO Chinese hamster cell line.

Cell Age and Radiosensitivity

Probably the earliest demonstration of a cell cycle or cell age dependence of radiation sensitivity were the indirect observations of Elkind and Sutton (1959) during the course of their landmark experiments on the repair of sublethal injury. They observed in one set of experiments in which the dose was given in two divided installments separated by a variable time that, as the time between doses was increased, the surviving fraction increased monotonically for upward of 2 h. These results for the first 2 h of dose separation were consistent with their hypothesis for repair of sublethal injury. However, after the first 2 h they reported "fluctuations in the radiosensitivity" marked by a decreased survival, followed by an increased survival. They furthermore noted that if the incubation of the cells between radiation doses was at 20°C rather than at 37°C, as in the just described experiment, the surviving fraction increased monotonically to the expected value of full repair of sublethal injury, without the oscillations seen at 37°C. Their interpretation was that repair could proceed at 20°C, but that changes in the physiological state at 37°C led to the fluctuations.

The modern interpretation of this pioneer work of Elkind and Sutton would be that at 37°C there is progress through the cell cycle, with partial synchrony, leading to a change of the distribution of cells in the cycle, with increases or decreases in the fraction of the cells in more sensitive portions of the cycle. At 20°C the sublethal damage is repaired, but progress through the cycle was prevented.

These early studies led to extensive reports on the radiosensitivity of cells at various stages in the cell cycle. Of these, two of the most important were the reports of Terasima and Tolmach (1963) on the HeLa cell line and that of Sinclair and Morton (1966) on the hamster V79 cell line.

The data of Sinclair and Morton are shown in Figure 8.7. In their experiments, cells were synchronized by mitotic shake-off so that, at time $t = 0$, the cells were in the mitotic phase of the cell cycle. At various times as shown after the mitotic cell collection, the cells were either exposed to a dose of 660 rad (6.6 Gy) of x-rays and their clonogenic survival

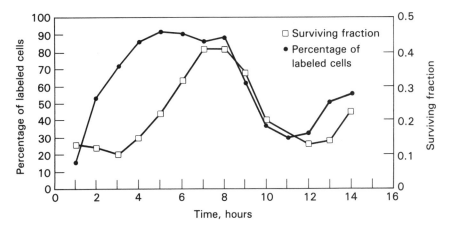

Figure 8.7 Data (modified from Sinclair and Morton, 1966) show, in the solid circles, the percentage of the cell populations that take up a DNA synthetic label as a function of time and that are presumably in the S phase of the cell cycle. The open squares show the surviving fractions from a clonogenic assay after 660 rad of 250 kVp x-rays.

was determined, or the cells were exposed to tritium-labeled thymidine in the culture medium. The latter procedure will label all cells that are in the stage of replication of their DNA. From these data it is evident that there is a wide fluctuation in the radiosensitivity of cells, depending on their position in the intermitotic cell cycle. Cells that are in mitosis at the time of irradiation (at time zero) and cells that are in G-2 and reentering mitosis (at a time 10 to 12 h after collection) are by far the most radiosensitive, the surviving fraction at these times being less than a quarter of that seen for the most resistant cells. The most resistant cell population is the population of cells that is nearly complete in its cycle of DNA replication, the late S phase. The data presented in Figure 8.7 on labeling with a DNA synthesis marker show the progress of the cell through the cell cycle. The period during which the labeling percentage is at its maximum marks the period during which DNA synthesis is occurring.

Figure 8.8 shows complete survival curves for synchronized cells irradiated at various times after mitotic harvest. Two striking features about the curves in Figure 8.8 are the lack of any shoulder for the mitotic and G-2 cells and the extreme resistance and broad shoulder for cells that are very late in the DNA synthetic cycle (late S). The extent of the sensitivity differences are also striking. For example, the survival of late S cells at 1000 rad (10 Gy) is two log cycles greater than the survival of mitotic and G-2 cells.

The understanding of cell age effects on radiosensitivity are markedly complicated by the great variability in the relative lengths of the

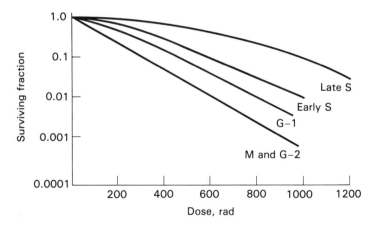

Figure 8.8 Survival curves are plotted as a function of dose for Chinese hamster cells (V79 line) for various times after synchronization by mitotic selection. The M and G-2 curve represents data for cells exposed and plated immediately after collection and for cells irradiated 12 h after collection when they are in G-2 and reentering mitosis. (Data from the experiments of Sinclair, 1968)

various portions of the cell cycle as outlined in Table 8.1. Terasima and Tolmach's (1963) work with HeLa cells demonstrated the cell cycle dependence of that cell line with a similarly mitotically harvested synchronized cell population. Their results, while looking superficially different, are, after all, much like those for hamster cells. The HeLa cell line has a very long G-1 period (see Table 8.1), during which there is a beginning radiosensitive phase associated with cells in mitosis or just leaving mitosis, followed by a high degree of radioresistance in mid G-1. Again, the HeLa cells acquire marked radioresistance near the end of the S phase.

A wide variety of cell lines of the mouse, hamster, rat, and man has been tested for the changes in radioresistance with cell cycle age. Some generalizations are possible. It is clear that, for all lines and species tested, cells are the most sensitive just before, during, and just after mitosis. Most, but not all, cell lines show an increased radioresistance near the end of the S phase of the cycle. There are some significant exceptions. A rat cell line developed from a gliosarcoma, which has been designated the 9L cell line, shows almost no age sensitivity at all during the various phases of the cell cycle. Even for this line, however, there is a small but significant increase in radioresistance near the end of the S phase. Those few studies that have attempted to evaluate the cycle-dependent radiosensitivity of cells growing in in vivo tissue systems have demonstrated a similar age response function. Withers and others (1974) showed that gastrointestinal crypt cells had the same high radioresistance near the end of the S phase of the cell and increased radiosensitivity during at least part of the G-1 portion of the cycle. In this study the cells were synchronized by the administration of hydroxyurea, a drug that blocks cells in the S phase and thus produces at least partial synchrony in the S phase.

RADIATION-INDUCED CELL PROGRESSION DELAY

Delay of cell division induced by radiation is a very well known and widely demonstrated phenomenon. Cells appear to suffer a transient block in their progress through the G-2 phase of the cell cycle, and the duration of this block is somewhat dependent on the radiation dose and the age of the cells after last mitosis at which the radiation is delivered (Leeper, Schneiderman, and Dewey, 1973). This study of mitotically synchronized CHO hamster cells identified two different types of blocks to the progression of the cell through the cell cycle. The delay caused for the cells in their progression through G-2 and mitosis was more or less linearly related to the time in the cell cycle at which the cells were irradiated. For example, a dose of 150 rad (1.5 Gy) in the middle of G-1 led to a division delay of approximately $\frac{1}{2}$ h, while the same dose delivered in late S or early G-2 led to a delay of 2 to 3 h. For the same conditions of irradiation in late S, 300 rad (3 Gy) led to a 4-h division delay, and 600 rad (6 Gy) caused a 6- to 7-h division delay.

In addition to the division delay reported by many, Leeper, Schneiderman, and Dewey also identified a small delay of entry of cells into the S phase if they are irradiated in early to mid G-1. This type of delay does not seem to be very dose sensitive and probably contributes little to the overall delay in progression through the cell cycle.

A mechanism has been proposed for the delay of the progress of cells through mitosis. This mechanism, which has been put forward by a number of investigators, has been suggested to be a defect in translation of structural and control protein or proteins essential for mitosis.

In organized cell systems, it has been suggested (Sweigert and Alpen, 1987) that there is a significant difference in the extent of division delay caused by radiation. These authors found a reliable linear regression of division delay on the administered radiation dose. In the case of monolayer cells, the delay was seen to be 13 min/Gy. For cells grown in multicellular spheroid geometry, the division delay was again linear, with the delay being 32 min/Gy, two and a half times as long as for monolayer cells. The authors offer no explanation for this finding, but they do observe that the cell cycle time for cells in the spheroids was almost twice as long as for monolayer cells, with nearly all the additional time in an extended G-1 period.

REFERENCES

ELKIND, M. M., and SUTTON, H. (1959) X-ray damage and recovery in mammalian cells in culture. *Nature* **184,** 1293–1295.

GOODHEAD, D. T., THACKER, J., and COX, R. (1978) The conflict between the biological effects of ultrasoft x-rays and microdosimetric measurements and application. In *Sixth Symposium on Microdosimetry, Brussels, Belgium,* J. Booz and H. G. Ebert, eds., pp. 829–843. Commission of the European Communities. Harwood, London.

HOWARD, A., and PELC, S. R. (1953) Synthesis of deoxyribonucleic acid in normal and irradiated cells and its relation to chromosome breakage. *Heredity* (Suppl.) **6,** 261–273.

LEEPER, D. B., SCHNEIDERMAN, M. H., and DEWEY, W. C. (1973) Radiation induced cycle delay in synchronized Chinese hamster cells: comparison between DNA synthesis and division. *Radiation Res.* **53,** 326–337.

LYMAN, J. T., and HAYNES, R. H. (1967) Recovery of yeast after exposure to densely ionizing radiation. *Radiation Res.* (Suppl.) **7,** 222–230.

NGO, F. Q. H., BLAKELY, E. A., and TOBIAS, C. A. (1981) Sequential exposures of mammalian cells to low and high LET radiations. *Radiation Res.* **87,** 59–78.

PHILLIPS, R. A., and TOLMACH, L. J. (1966) Repair of potentially lethal damage in x-irradiated HeLa cells. *Radiation Res.* **29,** 413–432.

PUCK, T. T., and MARCUS, P. I. (1955) A rapid method for viable cell titration and clone production with HeLa cells in tissue culture: the use of x-irradiated cells to supply conditioning factors. *Proc. Natl. Acad. Sci. U.S.* **41,** 432–437.

RAAPHORST, G. P., and KRUUV, J. (1976) Effect of tonicity on radiosensitivity of mammalian cells. *Intl. J. Radiation Biol.* **29,** 493–500.

RESNICK, M. A., and MARTIN, P. (1976) The repair of double-strand breaks in the nuclear DNA of *Saccharomyces Cerevisiae* and its genetic control. *Mol. Genetics* **143,** 119–129.

SINCLAIR, W. K. (1968) Cyclic x-ray responses in mammalian cells in vitro. *Radiation Res.* **33,** 620–643.

——, and MORTON, R. A. (1966) X-ray sensitivity during the cell generation cycle of cultured Chinese hamster cells. *Radiation Res.* **29,** 450–474.

SWEIGERT, S. E., and ALPEN, E. L. (1987) Radiation induced division delay in 9L spheroid vs. monolayer cells. *Radiation Res.* **110,** 473–478.

TERASIMA, T., and TOLMACH, L. J. (1963) Variations in several responses of HeLa cells during the division cycle. *Biophys. J.* **3,** 11–33.

THACKER, J., GOODHEAD, D. T., and WILKINSON, R. E. (1980) The role of localized single track events in the formation of chromosome aberrations in cultured mammalian cells. In *Eighth Symposium on Microdosimetry*, J. Booz and H. G. Ebert, eds., pp. 587–595. Commission of the European Communities. Harwood, London.

WITHERS, H. R., and others (1974) Response of the mouse intestine to neutrons and gamma rays in relation to dose fractionation and division cycle. *Cancer* **34,** 39–47.

SUGGESTED ADDITIONAL READING

ELKIND, M. M., and WHITMORE, G. F. (1967) *The Radiobiology of Cultured Mammalian Cells.* Gordon and Breach, New York.

PROBLEMS

1. The following survival data are provided on a hamster cell line irradiated and then immediately plated for clonogenic potential.

Dose (cGy)	Surviving Fraction
0	1.00
50	0.85
100	0.52
150	0.33
200	0.16
250	0.064
300	0.024
400	0.0035

(a) Plot the data graphically on three-cycle semilog paper.

(a) Plot the data graphically on three-cycle semilog paper.

2. With the MTSH parameters from Problem 1, generate a split dose analog to the Elkind–Sutton experiment with the hypothetical cell line given. Use a conditioning dose of 200 cGy and a recovery time of 10 h (maximum sublethal repair).

3. Chinese hamster cells irradiated with ^{60}Co gamma rays in the fully oxygenated condition can be assumed to have a D_0 of 150 rad and an extrapolation number of 5.

   (a) Using MTSH kinetics, compute the surviving fraction at 10, 100, 200, 300, 400, 600, and 1000 cGy, and plot these data on semilog paper. Show your MTSH calculations.

   (b) Using the linearized form of the molecular model for cell killing (linear–quadratic model), plot the data obtained in part (a) in such a way that the two parameters $p\alpha$ and $p\beta$ may be determined. Plot this line on the results of part (a).

   (c) What does this tell you about the sensitivity of the survival data to the form of survival equation that is used?

   (d) Now compute the MTSH surviving fractions for 1200, 1500, 1800, and 2000 cGy. Again, determine $p\alpha$ and $p\beta$ from this plot and overlay the L–Q line on the MTSH line. What conclusions can you draw about the effect of high doses on the goodness of fit of either model, MTSH or L–Q?

9

Modification
of the Radiation Response

INTRODUCTION

Many substances are known to affect the outcome of the radiation exposure of biological materials. Some of these enhance the detrimental
effect of the radiation, while some will reduce the effectiveness of the
radiation. In nearly all, if not all, cases the action of agents that modify
radiation effects is mediated through alterations of the chemical effects
of the radiation. None are known to interfere in the earlier physical and
physicochemical processes that were described in some detail in Figure
4.1. The earlier processes are those included in the initial interaction
event of the incident radiation and the excitation and ionization of molecules of the medium in which the energy is deposited. Other reactions
that might interfere with the final outcome of the radiation interaction
with the target molecules will commence in the diffusion phase of the
process, when either the radical products of irradiation are diffusing in
the medium, or possible interaction reagents are diffusing to the site of
deposition of the radiation energy.

ROLE OF WATER

Water is not often thought of as a modifier of radiation action, but, indeed,
its absence or presence has the most significant action of all the agents
that are at our disposal to modify the radiation response. The importance

of the radiation chemistry of water has been emphasized many times in this text and elsewhere. The whole of the indirect action mechanism in aqueous media depends on the production of the hydroxyl and hydrogen radicals and the hydrated electron e_{aq}^-, and on the many subsequent chemical reactions that take place with these species, leading to further possible mediators of macromolecular damage. Among the many examples of these other reactive chemical species are H_2O_2, $O_2 \cdot^-$, and the superoxide radical $HO_2 \cdot$. The relative importance of the various species is still not well known, but there is general agreement that the initial $OH \cdot$ radical is the predominantly harmful species.

For biological materials, biomolecules or cells irradiated in the anhydrous or nearly dry state require that most of the damage incurred as the result of ionizing radiation exposure must be related to the direct effect, the direct action of high-kinetic-energy electrons themselves on the target system. There is a large body of published data on the comparative effects of irradiation of biomolecules and living systems in various states of hydration. An extensive review is that of Augenstein, Brustad, and Mason (1964). A particularly interesting example of the effects of the aqueous medium on radiation sensitivity is that of Augustinsson, Jonsson, and Sparrman (1961) in which they studied the effect of the state of hydration on the radiosensitivity of two enzymes, cholinesterase and arylesterase. These rather old but very informative data are chosen as an example of the modifying actions of water because several different modulating actions can be seen to be at work. If we refer to Figure 6.2, the simpler interactions of water are outlined. As the principal chemical component of hydrated cell systems, water enters into the dynamics of radiation action in two ways. The initial ionization and/or excitation events will be principally in water in the usual hydrated cellular systems. Following these initial events, water interacts with the molecular products of ionization or excitation to produce further highly reactive species, which proceed to interact with vital macromolecules in the cell. In the absence of water the initial ionization–excitation events will take place directly in the macromolecules of the target system. This direct effect will usually be less effective than the indirect effect mediated through the products of radiolysis of water, but occasionally there might be enhancement of radiation effectiveness because of special molecular characteristics of the target molecule.

The data in Figure 9.1 for arylesterase, an enzyme that hydrolyzes aromatic esters of simple aliphatic acids, represent the "expected" response modulation of water. There is a linear increase in the effectiveness of the radiation for the inactivation of this enzyme as the concentration of water is increased. This increased effectiveness is shown by a decrease in the D_{37}. There is no explicit experimental proof that the increased sensitivity of the enzyme is the result of indirect action through water radicals, but the evidence is very strong indeed. As will be seen later, the presence of highly active scavengers or sinks for water radicals, such as NO or R—SH compounds will reverse this sensitization. The enzyme, arylesterase, depends on the presence of a functional mercapto group (—SH) for its enzymatic action, and protection of this functional group by removal of water radicals through competing processes that scavenge the water radicals from the medium will be effective in protecting the biological activity of the enzyme. It was just noted that extraneous R—SH compounds are indeed active reactors with water radicals, particularly with OH·, and NO is an effective reactant to remove OH· as well.

The complex response seen for acetyl cholinesterase in Figure 9.1 has no simple explanation; however, the authors of the report suggest that for this enzyme, which contains nonfunctional mercapto groups upon which the enzymatic activity does not depend, the presence of water in low concentration promotes the transfer of the deposited energy to the nonfunctional —SH groups as a preferred route for the removal of the initial ionization and excitation energy. As the —SH groups are exhausted and as the water concentration increases, the efficient indirect action of water as a mediator of radiation action is expressed as increased radiosensitivity of the target molecule, cholinesterase.

It is not generally possible to alter the water concentration for eukaryotic cells to demonstrate the modifying action of water on radiation damage, nor is it of any practical significance; but for all organisms that have been examined, from bacteria to seeds, the general response is the same. Water increases the effectiveness of ionizing radiation in a general way, even though for some special circumstances, such as for cholinesterase in the experiments just described, there may be a restricted range of water concentrations where protection is provided by the presence of

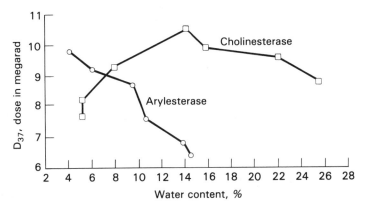

Figure 9.1 Relative radiosensitivity of two enzymes when they are irradiated under different conditions of hydration. (Augustinsson, Jonsson, and Sparrman, 1961)

water. These special circumstances are always thought to be the result to unusual relationships of the reactants in the same fashion as for the nonfunctional —SH groups of cholinesterase.

TEMPERATURE AND RADIATION DAMAGE

The Lea target theory, as well as any of the more modern theories of the action of ionizing radiation, do not explicitly identify temperature as an independent variable; but the activation energies for the destruction of a wide range of bioactive molecules, phages, and viruses have been measured and found to be nonzero. Since activation energies can be shown to not be zero, there should be effects of temperature on those steps in the reaction process for which these activation energies are significant. Augenstein, Brustad, and Mason (1964) is a major contribution to the understanding of the role of temperature in the direct action of ionizing radiation on biomolecules, and Brustad (1964) has formulated a model for the effect of temperature.

It would be predicted that any portion of the damage to a molecule resulting from the direct ionization of a sensitive site, leading to irreversible destruction of the biological properties of the molecule, would be relatively insensitive to temperature. Enough energy is deposited to completely separate an electron pair involved in a covalent bond, and this amount of energy is large enough that no activation energy would be required. To the extent that bond rupture in the same site is the result of excitation rather than ionization, it could be predicted that activation energy might be required for complete bond destruction. The formula proposed by Brustad (1964) is of the following form, where C is the inactivation coefficient for the enzyme from irradiation as a function of the absolute temperature, T.

$$C = \frac{1}{D_{37}} = C_0 e^{-\epsilon_0/RT} + C_1 e^{-\epsilon_1/RT} + C_2 e^{-\epsilon_2/RT} \tag{9-1}$$

In this formulation, $1/D_{37}$, which is the inactivation constant in the single-hit inactivation model of hit theory, is the reciprocal of the dose for inactivation of 63% of the molecules and is equal to $1/D_0$ when inactivation is first order in dose. The approach to the determination of the appropriate activation energies follows the usual method of the Arrhenius plot. The log of C of equation 9-1 is plotted as a function of the reciprocal of the absolute temperature in degrees Kelvin. In the usual Arrhenius plot, the logarithm of yield of the product or loss of reactant is plotted against absolute temperature. In the case at hand, the D_{37} is reciprocally related to yield, and the larger the D_{37} is the smaller the yield per unit dose. For simple reactions the Arrhenius plot is a straight line with a slope equal to ϵ/RT. The formulation of equation 9-1 is generalized, where several reactions are occurring simultaneously. Typical data for the inactivation of enzymes, of which one of the more comprehensive is that of Fluke (1966), fit a form of equation 9-1 in which the first coefficient, C_0, is zero or close to it. The second coefficient, C_1, was generally in the range of a few percent. The third coefficient, C_2, accounted for the largest part of C. The typical data, which can be written as follows, using as an example Fluke's (1966) data for lysozyme, are as follows:

$$C = \frac{1}{D_{37}} = 0.005 + 0.02e^{-620/RT} + 0.7e^{-2540/RT} \qquad (9\text{-}2)$$

The significance of the formulation of equation 9-2 is that indeed there is a term in the expression, C_0, for which there is no activation energy and for which, for this enzyme, lysozyme, irradiated in the dry state, there are second and third terms that contribute to the inactivation coefficient and are temperature dependent.

Augenstein, Brustad, and Mason (1964) have proffered a model explanation based on this generally observed multicomponent modulating action of temperature. Their explanation includes as one aspect the thermal facilitation of charge transfer from a site at which ionization has occurred to another critical site, the destruction of which leads to inactivation. In addition, they suggest that the localization of *excitation* energy at critical sites can lead to potential inactivation of the biomolecule if sufficient additional thermal energy is provided or is already present to cause disruption of the critical bond. They divide the temperature domain into three regions for which the effects might be due to either mechanism, thermal activation of charge transfer or thermal enhancement of bond breakage at the site of radiation energy deposition.

$T < 100\,K$

Inactivation at this low temperature is temperature insensitive, and the conclusion might be that the initial ionizing inactivations must occur directly in the critical site of the molecule without charge or energy migration. Charge migration at these temperatures is not likely, and therefore the initial event must occur in the critical target volume. Since that volume is generally a small fraction of that of the whole molecule, we would predict a very high inactivation dose, and indeed that is what is seen, as, for example, in the sample data of Fluke quoted previously.

$100 < T < 170\,\text{K}$

In this temperature range, Augenstein, Brustad, and Mason suggest that exciton migration, coupled with energy contained in the vibrational and/or rotational modes of the target molecule, is part of a process that causes the excitation energy to become localized at a critical site. The small activation energies for this process, on the order of RT, are in accord with such a hypothesis.

$170 < T < 420\,\text{K}$

For this temperature range, more complex processes are involved, with associated increased activation energies. One possibility is a cooperative interaction between migrating charge and involvement of hydrogen atoms mobilized from hydrogen bonds and their participation in torsional modes, leading to inactivation of critical sites. Several experiments have suggested that frequently inactivation of bioactive molecules is associated with disruption of —SH·HS— configurations, which are usually stabilized by the associated hydrogen bonds.

A simpler explanation of the effect of temperature on inactivation of enzymes would be the assignment of the nontemperature-dependent portion of equation 9-2 to inactivation by *ionization* events, which should be independent of temperature except at very high temperatures, and the assignment of *excitation* events to those that have temperature dependence. The result of such an interpretation would be that at moderate temperatures a very small fraction of inactivations would be by ionization. Excitations that become ionizations when adequate activation energy is provided would be hard to differentiate from ionizations that occurred directly, other than through the temperature effect considered here. This hypothesis is attractive but it has not yet been testable.

In aqueous media, where a large part of the damage might be by the indirect effect related to water radiochemistry, activation energies for the conversion of excited water molecules would be small indeed, and little temperature dependence would be expected. There are very few experimental data to either prove or disprove this point.

OXYGEN EFFECT

Some of the very earliest experimenters in radiation biology discovered that the presence or absence of oxygen had a strong influence on the effectiveness of radiation in producing damage, either to bioactive molecules or to living materials. As far as can be discovered, the earliest published report of the oxygen effect is that of Schwarz (1909). It was not until the earlier 1950s, however, that the central importance of oxygen as a radiation response modifier came to receive the attention that it deserved. The landmark papers of Gray and others (1953) and of Thomlinson and Gray (1955) not only alerted the radiation biology community to the basic significance of the presence or absence of oxygen, but, furthermore, it was pointed out in these papers that there was practical importance related to oxygen sensitization of the radiation response of malignant tissues undergoing radiotherapy. The latter paper emphasized the poorly oxygenated condition of most tumor tissues and pointed out that it was likely that some tumor tissues were protected from radiation killing by the anoxic levels of oxygen in them.

It is artificial and contrived to attempt to discuss the modifying effects of oxygen without looking at the larger picture of the interactions of other chemical substances, either those endogenously present in living systems or those that can be added to the system. The close interrelationship becomes evident when we examine the *competition model* for the interaction of sensitizing and protecting chemicals, which was proposed by Alexander and Charlesby (1955) and by Ormerod and Alexander (1963). The model suggests the following steps as part of the usual chemical changes that occur after production of radicals by either the direct or the indirect action of radiation.

1. The primary events are the production of radicals of organic molecules as the result of either direct or indirect action.

$$RH \rightarrow RH\cdot^+ + e^- \rightarrow R\cdot + H^+ + e^- \quad \text{(direct action)}$$

$$RH + OH\cdot \rightarrow R\cdot + H_2O \quad \text{(indirect action)}$$

2. The radicals in biomolecules may be restored by interaction with sulfydryl compounds either endogenous to the system or added. The most prominent of these compounds is reduced glutathione, which is abundantly present in most cells. The result of the reaction has been termed *chemical repair*.

$$R\cdot + R'SH \rightarrow RH + R'S\cdot \quad \text{(chemical repair)}$$

3. A competing reaction may take place with oxygen that fixes the state of the radical in the biomolecule and presumably leads to chemically irreparable damage. This reaction is central to the sensitizing effect of oxygen.

$$R\cdot + O_2 \rightarrow RO_2\cdot \quad \text{(damage fixation by oxygen)}$$

4. Many electron-affinic compounds are capable of reactions that mimic those of oxygen in their outcome. These radiation sensitizers have come to have practical importance, and they will be discussed in detail in later portions of this chapter. The reaction given is typical but not exclusive of the possible transformations caused by these oxidizing and electron-affinic compounds. Note that the product in

which the damage has been fixed by the sensitizer is not the same product that results from damage fixation by oxygen.

$R\cdot + X\,(\text{sensitizer}) + H_2O \rightarrow$

　　　　$ROH + X\cdot^{-} + H^{+}$　　　　(damage fixation by radiation sensitizers)

5. Two competing reactions with the preceding reactions are the scavenging of water radicals by —SH compounds or other added chemicals. This reaction is analogous to the chemical repair reaction, in which the reactant with the water radical is RSH. Molecular oxygen reacts competively with the thiols in the intracellular medium to reduce the concentration of either or both to levels where their concentrations are reduced to the level that their participation in the preceding reactions is limited. The rate constants for the reaction of the water radicals either with each other or with the sulfhydryl compounds are very high, but in this nonhomogeneous reaction medium the concentration of the R—SH scavenger plays a complex role.

The preceding reactions impute a greater understanding of the processes of oxygen sensitization of the radiation response than is known from experimental or theoretical bases. The actual mechanism of oxygen interaction is still an unresolved question. Many proposals have been put forward, but none has completely withstood the test of time and experiment.

Oxygen Enhancement Ratio

The effect of oxygen has been generally found to be one that can be described as a *dose modifying effect*. That is, there is a constant ratio between the surviving fraction found for irradiation in oxic and anoxic conditions, regardless of the level of the dose. There is also a constant ratio for the doses, under oxic and anoxic conditions, which are found to produce the same level of survival (see Figure 9.2). This later ratio has been defined as the *oxygen enhancement ratio*, commonly abbreviated OER. The expression for the OER is as follows:

$$\text{OER} = \frac{\text{dose in } N_2 \text{ for survival fraction, } S/S_0}{\text{dose in } O_2 \text{ for survival fraction, } S/S_0} \qquad (9\text{-}3)$$

The oxygen effect is ubiquitous. It is found with essentially all living organisms and with nearly all bioactive molecules, as well as with many nonbioactive molecules that are degraded by irradiation. Generally, but not always, the shape of the dose–effect relationship is not changed by irradiation in the presence of oxygen. Dose–effect relationships that follow single-hit kinetics (that is, the dose versus log surviving fraction is linear) are still linear whether irradiated in nitrogen or oxygen. There are, how-

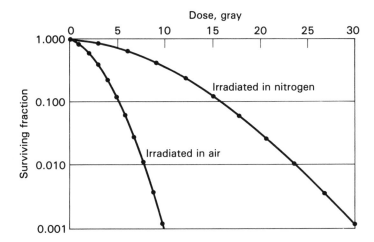

Figure 9.2 Survival data under oxic and anoxic conditions for a hypothetical mammalian cell line. The oxygen enhancement ratio (OER) is taken as 3.0.

ever, some reported instances, especially for mammalian cells grown in vitro, for which the survival curves in nitrogen and oxygen are not in constant proportion. Some investigators have reported, in particular, a relative reduction of the breadth of the shoulder on the survival curve when irradiation takes place under hypoxic conditions. That is, for low doses the constant proportionality of the hypoxic/oxic relationship does not hold. In general these experiences are infrequent.

Effect of Oxygen Concentration

The data in Figure 9.3 are from a landmark study by Howard-Flanders and Alper (1957), which was undertaken by them to define the oxygen concentration relationships for the radiosensitizing effect of oxygen. The data have been plotted here on a logarithmic scale of oxygen concentration and have not been smoothed, as was done by the authors. The data have been so transformed to present more clearly the details of the concentration effect at the lower concentration of oxygen. There is clearly a concentration threshold for oxygen sensitization. To all intents and purposes, the OER is still essentially 1 at a partial pressure of oxygen in the medium of 0.0177% (log pO_2 = −1.75). A relative sensitivity of 2.05 (OER), which is half of the range from 1.0 to 3.08, is reached at an oxygen partial pressure of 0.55% (log pO_2 = −0.26). For this set of data, the oxygen partial pressure associated with the plateau maximum value is not as well demonstrated as for some of the other data sets in their report; but, in general, the maximum oxygen sensitization is seen at partial pressures of 2% to 3%. For the several organisms studied by

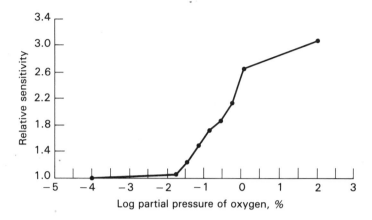

Figure 9.3 Relative sensitivity of the organism, *Shigella flexneri*, Y6R, to kill-ing by 200 kVp x-rays as a function of the partial pressure of oxygen in the medium. Data from Howard-Flanders and Alper (1957). The data have been plotted in logarithmic form from their original table. See the accompanying text for the relevant conversions of log pO_2 to concentration.

Howard-Flanders and Alper, as well as for many other studies done by others subsequently, the general shape of the response is similar. That is, there is a threshold concentration of oxygen for sensitization that is rarely higher than a few hundredths of a percent partial pressure of oxygen, and the maximum sensitization will be seen at partial pressures of 2% to 3% or less.

These authors were able to fit their data for the oxygen sensitization of the radiation killing of two bacteria, *Shigella flexneri*, Y6R, and *Escherichia coli*, B/R, and for haploid yeast to a general expression that has come to be widely used in characterizing the oxygen sensitization of many organisms. Their equation is

$$\frac{S}{S_N} = \frac{m[O_2] + K}{[O_2] + K} \qquad (9\text{-}4)$$

The constants m and K are separately determined for each system ex-amined. The constant m is dimensionless, and the constant K has the dimensions of micromole per liter (μmol/l) when the concentration of oxygen is expressed in the same units. This oxygen concentration is taken as the dissolved oxygen concentration in the medium. S is the surviving fraction for an oxygen concentration of $[O_2]$. S_N is the surviving fraction for irradiation under anoxic conditions, $[O_2] = 0$. Fortunately, the two constants can be evaluated through simple manipulations.

When $[O_2]$ is very much larger than K,

$$\frac{S}{S_N} = \frac{m[O_2]}{[O_2]} = m \qquad (9\text{-}5)$$

The conditions of equation 9-5 are met for the region of the oxygen concentration/sensitization curve where S/S_N is the maximum. In other words, the constant m is equal to the maximum OER; that is, $S_{max}/S_N = m$.

The constant K can be determined by observing that when K is equal to $[O_2]$ the following derivation is possible:

$$\frac{S_K}{S_N} = \frac{m[K] + K}{K + K} = \frac{K(m + 1)}{2K} = \frac{m + 1}{2} \tag{9-6}$$

The evaluation of K, then, is accomplished by determining the maximum OER from a plot similar to that in Figure 9.3, and this derived value of m is inserted in equation 9-6. The result of solving equation 9-6 with the estimated value of m is the value of S_K/S_N (the OER) for $K = [O_2]$. The value of K is then read graphically from the plot of OER versus oxygen concentration as the value of $[O_2]$ for the OER just determined from equation 9-6.

Data obtained by these methods by Howard-Flanders and Alper are shown in Table 9.1.

The significance of equation 9-4 was discussed by Alper (1956) in terms of a competitive model in which there are two types of damage caused by the primary radiation event. Type 1 is a potentially lethal lesion that only becomes lethal through the action of oxygen. The type 2 lesion is taken as always lethal to the target. The type 1 lesion can be chemically restituted by the mechanisms discussed earlier as chemical repair. This chemical repair process is competitive with the damage fixation with oxygen. In terms of this model, the constant m is related to the relative numbers of the lesions of type 1 and type 2 (n_1 and n_2):

$$\frac{n_1}{n_2} = m - 1 \tag{9-7}$$

and K is the ratio of the rate constants for the two processes, chemical repair and oxygen fixation.

TABLE 9.1	Values of the Constants m and K for Several Microorganisms, Determined by Howard-Flanders and Alper (1957)		
	Organism	m	K
	Shigella flexneri, Y6R	2.9	4.0
	Escherichia coli, B/R	3.1	4.7
	Saccharomyces cerevisiae	2.4	5.8

From the data in Table 9.1, taking, for example, *E. coli*, *B/R*, the ratio of the damage of the two types is $n_1/n_2 = 2.11$.

$$K = \frac{k_{\text{rep}}}{k_{\text{fix}}} \tag{9-8}$$

The special case solution for single-hit kinetics has been derived by Dertinger and Jung (1970) (see Suggested Additional Reading, in particular, Chapter 8).

The particular attractiveness of the mathematical model of Alper and Howard-Flanders and Alper is that it provides a sound theoretical basis for the competition theory of the action of oxygen. Furthermore, this information provides an adequate explanation for the oxygen effect associated with high LET radiation. Since it is noted in Chapter 14 that the OER approaches 1.0 as the LET of the radiation increases, an adequate explanation for the lack of oxygen sensitization with high LET radiations is essential. It is certainly known that the direct effect pathway can be equally sensitized by oxygen, and the best example of this is that an oxygen effect may be noted with the low LET irradiation of dry materials. The formulation of equation 9-7 is particularly illuminating when considering high LET radiations. If the lesions produced by high LET radiation are predominantly of type II, nonrepairable, then m will be disappearingly small and no oxygen sensitization will be detectable.

Time Dependence of the Oxygen Effect

The competition model for the effect of oxygen may be tested to some extent by carefully conducted timing experiments for the availability of oxygen either before or after the irradiation takes place. Since the radical species responsible for biological damage would be predicted to have half-lives of only a few milliseconds or less, the time scale is short and the timing must be precise. The irradiation itself must take place in an extremely brief period so that irradiation time is not a variable. The short irradiation times have been accomplished by the use of nanosecond to microsecond pulses from electron accelerators. The experimental procedure has come to be known as *pulse radiolysis*. Several methods have been developed for rapid mixing of the oxygen either before or after irradiation. One of these, the oxygen implosion technique, uses the timed rupture of a diaphragm, allowing access of oxygen to the reaction chamber. A number of experimental data sets are available in the literature. The data illustrated in Figure 9.4 are from Michael and others (1973). The irradiations were accomplished in single 2-μs pulses followed by the admission of oxygen by the implosion method at various times before and after irradiation. The test system used was the bacterium *Serratia*

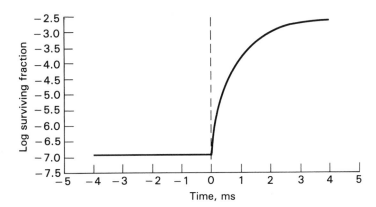

Figure 9.4 Relative sensitivity of the bacterium *Serratia marscesens* to a 28-krad pulse of electron radiation. Oxygen is made available either before irradiation (negative times) or at timed intervals after irradiation. Increasing surviving fraction indicates decreased oxygen sensitization. (From Michael and others, 1973)

marscesens. The time resolution was on the order of 100 μs for the admission of oxygen.

The data show that if oxygen is made available at any time prior to the irradiation, up to that small fraction of a millisecond determined by the time resolution of the oxygen implosion method, the full expression of oxygen sensitization of the organism to radiation is achieved. Bacteria exposed to oxygen at times after the electron pulse showed rapidly decreasing oxygen sensitization. At times approaching 2 ms, the bacteria had the radiosensitivity of fully anoxic cells, indicating that oxygen that was made available at these late times after irradiation was not able to exercise its normal sensitizing action. The authors conclude from their data that, for *S. marscesens* irradiated under anoxic conditions, the oxygen-dependent damage decays with an approximate exponential half-life of 500 μs.

Michael and others (1973) suggest that the extremely short period over which postirradiation addition of oxygen can be effective in influencing the survival of the radiated organism confirms the importance of free radical processes in the oxygen effect. Another deduction of these authors is that the hydroperoxyl radical, $HO_2 \cdot$, and its anion, $O_2 \cdot ^-$, are unlikely candidates for the important radical species produced from water radiochemistry that attack the key biomolecules. $O_2 \cdot ^-$, the superoxide radical, is produced as a result of the interaction of molecular oxygen with the hydrated electron, e_{aq}^-. Since the hydrated electron has a much shorter half-life than the 500 μs observed in the normal intracellular environment, with its large excess of effective hydrated electron scavengers, the role of the hydrated electron and its reactants with molecular oxygen seems to be limited, if not ruled out.

Experiments similar to that shown in Figure 9.4 have been repeated by many investigators, and the results are in substantial agreement. Some studies show structure in the time–response curve for admission of oxygen after irradiation that seem to suggest that oxygen may be working by way of at least two, or possibly three, mechanisms, which are as yet not well defined.

The time dependence of the oxygen effect has been evaluated in a number of systems, but only for bacteria and for thin-film preparations of enzymes and biomolecules have the data been unequivocal. Particularly for mammalian cells and yeast cells, it is difficult to study the effectiveness of the postirradiation availability of oxygen. An important conclusion has been drawn from studies on mammalian cell systems in vitro. Certainly the inability to demonstrate a postexposure effectiveness for these systems with larger dimensions is related to the fact that the diffusion rate for oxygen becomes the limiting factor. We may deduce that, therefore, the site of action of oxygen is not at the membrane surface, but somewhere in the interior of the cell.

Mechanisms of the Oxygen Effect

It has been nearly 80 years since the first reported identification of the sensitizing effect of oxygen on ionizing radiation induced damage, but there is still no unequivocal answer for the mechanism by which this effect is mediated. There is general acceptance for the competition model for the oxygen effect, but little agreement on how the details of the competition occur. In spite of the evidence just cited, which seemed to rule out the importance of the superoxide radical in damage fixation, there continues to be serious interest in this candidate mechanism. The superoxide anion radical is now known to be produced by at least two mechanisms during the radiolysis of water. In addition to the reaction of molecular oxygen with the hydrated electron, there is a competing reaction with the hydrogen radical. The reactions are as follows:

$$e_{aq}^- + O_2 \rightarrow O_2 \cdot^-$$

$$H \cdot + O_2 \rightarrow HO_2 \cdot \rightarrow H^+ + O_2 \cdot^-$$

The radicals $HO_2 \cdot$ and $O_2 \cdot^-$ are in equilibrium, with a pKa of 4.7.

The other important water radical, $OH \cdot$, does not freely react with molecular oxygen, but an alternate pathway to superoxide formation has been proposed via an intermediate reaction with formate residues in the biological media, as follows:

$$OH\cdot + HCO_2^- \rightarrow CO_2\cdot^- + H_2O$$

and

$$CO_2\cdot^- + O_2 \rightarrow CO_2 + O_2\cdot^-$$

Superoxide has received so much attention as a candidate intermediate for the oxygen effect because it is known to play an important role in many biological processes unrelated to radiation damage. The enzyme, superoxide dismutase (SOD), the action of which is mediated by some transition metal ions such as copper and zinc, acts on the superoxide ion radical to catalyze its conversion to oxygen and hydrogen peroxide. This enzyme is ubiquitous in its biological distribution. The data on the role of the superoxide radical are conflicting and controversial. Experiments have been carried out with modulation of the SOD levels, either biochemically or genetically, with the hope that such studies would indicate the importance of the enzyme and its substrate. No clear answer has yet appeared. One interesting study is that of Samuni and Czapski (1978). They showed that the addition of formate to the irradiation medium, which converts practically all the water radicals to superoxide radicals, had only a marginal effect on the OER. Furthermore, the addition of exogenous SOD did not affect the OER, even though the enzyme would be expected to destroy the superoxide radicals.

No other mechanisms are presently proposed that have stood the test of experimental verification, and an elucidation of the mechanism of the competitive reaction of molecular oxygen remains an elusive goal.

THIOLS AND MODIFICATION OF RADIATION RESPONSE

The competition model for the oxygen effect postulates that oxygen is competitively reacting at a site that has sustained radiolytic damage with another class of chemical compounds that can restore the radiolytic damage. This class of compounds is made up principally of the thiols, both those that are endogenous to the cells as well as those that might be added from external sources. The class reaction was identified earlier in this chapter, but to repeat and elaborate, the chemical repair process, as it has come to be called, acts as follows:

$$R\cdot + R'SH \rightarrow RH + R'S\cdot$$

The result is restoration of the radiolytically damaged molecule and the production of a thio radical. The latter can then undergo a bimolecular recombination:

$$R'S\cdot + R'S\cdot \rightarrow R'SSR'$$

The initial thiol compounds are also capable of being effectively removed by molecular oxygen, so competition with oxygen is also going on at a separate level, independent of the presence of radiolytically produced radicals. The thiols can also effectively react with radiolytically produced water radicals; but at the concentrations of both the thiols and the radicals that normally exist in the cell after moderate doses of radiation, this cannot be an effective mechanism for the observed protection against radiation damage provided by the thiols, particularly given the very large rate constants for the consumption of the radiolytically produced water radicals by competing processes.

For intracellular thiols to be effective scavengers of the radiolytic damage to important bioactive molecules according to the equations just given, the thiols must be mobile inside the cell environment. A large proportion of the intracellular thiol content is present as thiol groups in large protein molecules. These can play only a minimal role in the protective action, and we must look to the nonprotein sulfhydryls for the protective action. Of these, the most important is reduced glutathione, a naturally occurring intracellular tripeptide that contains a free thiol group. There is also a small amount of the free amino acid, cysteine, as well as its metabolite, cysteamine, in the intracellular medium, but the concentrations are so low as to be inconsequential relative to that for reduced glutathione (GSH). In addition to the intracellular glutathione, the sulfur-containing polypeptides and proteins are an important source of reduced sulfhydryls. The polypeptide thiols represent better than 80% of the total sulfur content of the cells, but most of this thiol sulfur is in the disulfide form, R—S—S—R, which accounts for much of the structural rigidity of many proteins.

The rate constants for the reactivity of oxygen with radiolytically formed radicals are very fast, and because of this the rates of these reactions are essentially diffusion limited. The ratios of the rate constants for radical fixation by oxygen to the rate constants for chemical repair by the thiols are in the range of 200 to 1000; so, unless the thiol concentrations are very high, the oxygen fixation pathway has a dominant role. Even after all these years of study, the importance of chemical repair by endogenous intracellular thiols remains an unsettled question.

Recently, two compounds have become available that are capable of modulating the level of GSH in the cell, and therefore it has been possible to attempt a direct evaluation of the role of GSH in radiation protection in competition with oxygen. These compounds are DL-buthione-S,R-sulfoxime (BSO), which interferes with the biosynthesis of GSH, and diazenedicarboxylic acid bis(N,N'-dimethylamide) (Diamide), which reduces the intracellular level of GSH by oxidizing it to the disulfide form (GSSG). Several other chemical agents also affect the intracellular level of nonprotein —SH in the cell, but their mechanisms of action are not well understood. In addition to chemical agents, GSH deficient cell lines are also available to test the importance of this compound in radioprotection. The data on the role of GSH and the modification of this role by drugs that affect its concentration were reviewed in detail by Biaglow and others (1983).

Some of the more extensive data dealing with the effectiveness of endogenous intracellular thiols is that of Revesz (1983). This author used a mutant human cell line that was GSH deficient. The cells of the GSH$^-$ line when irradiated in either oxic or anoxic conditions had the same yield of single-strand breaks in DNA, while strand break yield in the GSH$^+$ line showed the typical oxygen enhancement ratio. The author concludes that the data support the competition model for the oxygen effect and demonstrate the effectiveness of intracellular thiols in radioprotection. The results with other endpoints, such as cell survival, are not as definitive. Other studies with BSO treated cells are, again, equivocal. In any case, the data for cell survival indicate that the presence of GSH changes cell survival only marginally. It is possible that, because of the highly favorable reaction rates for oxygen fixation of radicals, as mentioned previously, the thiol protection becomes important only when exogenous thiol compounds are added. In this latter case, protection is unequivocal and widely demonstrated.

Protection by Exogenous Thiols

It has been known for some time, even before the mechanistic explanation of a mode of action was available, that many thiols when added to the cellular milieu just before in vitro irradiation, or even when injected into animals, could provide significant protection against many radiation damage end points. The first recorded observation of this effect was by Patt and others (1949).

Cysteine, a natural amino acid, and its decarboxylated derivative, cysteamine, are both present in normal cells, but in low concentration. Both compounds, when made available in high enough concentrations, reduce the effectiveness of radiation in many systems, ranging from enzymes in solution to the intact animal. Dose modification factors (the factor that the unprotected dose level must be multiplied by to reach the same level of effectiveness in the chemically protected case) on the order of 1.8 to greater than 2.0 are found for these two compounds and a number of related organic thiols. The related thiols with effectiveness as radioprotectors all have features in common. The —SH group is free or is potentially free through hydrolysis in the cell (—SR $\Rightarrow$ —SH) or through reduction of the oxidized form (RSSR $\Rightarrow$ RSH). There also must be a strong basic function removed no more than two carbon residues from the thiol. The mechanisms of action of these compounds, which must be present in relatively high concentration in the cell or near the target molecule is, again, not simple. Certainly these compounds should be effective competitors with molecular oxygen for carbon-based radical sites in important biomolecules. Another pathway for the action of these compounds, partly because they are used at high concentrations, is through their action as radical scavengers in the aqueous medium. By the following reactions, the three important radiolytic products of water are scavenged by high concentrations of thiols.

$Hydroxyl\ radical$: $RSH + OH \cdot \rightarrow RS \cdot + H_2O$
or $RS^- + OH \cdot \rightarrow RS \cdot + OH^-$
    reaction constant, 1 to 2 $\times$ 10^{10} dm^3 mol^{-1} s^{-1}

$Hydrogen\ radical$: $RSH + H \cdot \rightarrow RS \cdot + H_2$
or $\rightarrow R \cdot + H_2S$
    reaction constant, 1 $\times$ 10^9 dm^3 mol^{-1} s^{-1}

$Solvated\ electron$: $RSH + e_{aq}^- \rightarrow R \cdot + SH^-$
or $\rightarrow RS^- + H \cdot$
    reaction constant, ca. 1 $\times$ 10^{10} dm^3 mol^{-1} s^{-1}

All the rate constants shown are near the diffusion-controlled limit of approximately 10^{10} dm^3 mol^{-1} s^{-1}, and therefore these reactions should compete favorably with the oxygen attack on biomolecules, which is also diffusion limited, if the thiols are present in high enough concentration. In addition, we would expect these thiol compounds to effectively react with carbon-centered radicals on the biomolecules to restore them to normal. The relative importance of the two reactions, radical scavenging and chemical repair, are not well known, but theoretical considerations of reaction rates and diffusion control of concentrations would favor the direct water radical scavenging process. The reaction rates for most thiols with carbon-centered radicals, such as those that would be of interest in biological macromolecules, are generally $\frac{1}{100}$ to $\frac{1}{1000}$ of the rates for reaction with water radicals.

The organic thiols have found little practical application in radioprotection, partly because of their toxicity, but also because they must be present before irradiation for effectiveness. Except in the case of planned radiotherapy, there is rarely advance knowledge about accidental exposure to ionizing radiation. An extensive developmental synthesis program produced a number of compounds with improved dose reduction, of which one in particular provided a dose reduction factor in excess of three for some end points, including the killing of bone marrow stem cells. This compound, known by the vernacular name WR-2721, is chemically S-(2-(3-amino propylamino)) ethyl-phosphorothioic acid. This drug, while very effective in certain in vitro circumstances, is known to have little protective effect for some tissues. It has, for example, been shown to be totally ineffective for radioprotection of the central nervous system (Yuhas and Phillips, 1983). The reasons for the wide variability in response with this drug are that it requires chemical transformation in the cell for protective activity, and the ability of the drug to be transported to the site of activity depends strongly on its solubility characteristics. In any case, this drug, as with all the other thiol protectors, is little more than a laboratory curiosity.

NITROAROMATIC RADIATION SENSITIZERS

Thomlinson and Gray (1955) were the first to point out that, because of the ineffective blood supply in fast growing tumors, there would be a radioprotection afforded to tumor tissues by a low oxygen partial pressure in the tissues. Under the conditions proposed by Thomlinson and Gray, there would be an unexpectedly high survival of tumor cells in tumors being irradiated with curative intent, and this protection of cells was due to local anoxia in regions of the tumor poorly supplied with blood.

Since about 1970, there has been a great number of reports of chemical agents that can, through their electronophilic character, sensitize anoxic tissues in ways similar to those associated with molecular oxygen. The earliest of these to receive significant attention was the compound, 2-methyl-5-nitroimidazole-1-ethanol. The generic name for the compound is metronidazole, and it has had wide clinical use as an antiprotozoan, marketed under the trade name Flagyl.

$$NO_2 \overset{CH_2-CH_2-OH}{\underset{N}{\overset{|}{\underset{}{\boxed{}}}}} CH_3$$

2-methyl-5-nitroimidazole-1-ethanol (metronidazole)

The nitroaromatic compounds are the class that has had the most extensive investigation, and some of these are now in clinical trial.

The sensitizing efficiency of the various nitroaromatic sensitizers is related to the redox potential of these drugs, which act through their electron affinity. All this class of compounds have lower reduction potentials than does molecular oxygen, so they must be used at much higher concentrations to elicit their radiosensitizing ability. A characterization of this circumstance is shown in Figure 9.5, in which the data of Whillans and Hunt (1982) are plotted to show the relative effectiveness of molecular

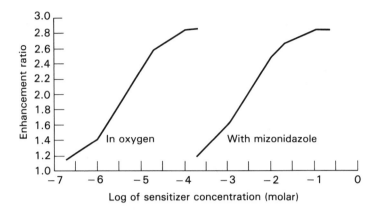

Figure 9.5 Enhancement of radiation sensitivity of CHO cells irradiated with either oxygen (left curve) or mizonidazole (right curve) as the sensitizer. (Recalculated from the data of Whillans and Hunt, 1982)

oxygen and mizonidazole. The concentration of mizonidazole required to effect the same level of sensitization as oxygen differs by approximately three orders of magnitude. The effectiveness of the nitroaromatics also strongly depends on their solubility in nonpolar and polar media, since the compound must be at the site of action for effectiveness.

Mechanism of Action of the Nitroaromatics

The pioneering work on the mechanism of the action of the nitroaromatics was carried out with the compound 5-nitro-2-furaldehyde semicarbazone (nitrofurazone) by Chapman and colleagues (1973). The observation that this class of compounds sensitizes for radiation damage only under conditions of anoxia, as well as the data in Figure 9.5, certainly leads to a prediction that the nitroaromatics are acting in the same role as oxygen in sensitization. That is, the compounds react with radicals on the important biomolecules, presumably DNA, to fix damage caused either by water radiolysis products or by the direct effect. The data by Chapman and others (1973) confirm this hypothesis. They furthermore demonstrated that the nitroaromatics can act just like oxygen in reacting directly with the products of radiolysis of water, presumably principally OH·. Others have shown that the compounds can also react with intracellular thiols such as GSH. In summary, the nitroaromatics can carry out all of the reactions important in the sensitizing and competitive role of oxygen, but with lower efficiency and at much higher intracellular concentrations of the drug.

SENSITIZATION BY 5-HALOGEN SUBSTITUTED PYRIMIDINES

Another class of radiation sensitizers, which acts by mechanisms completely unrelated to the action of oxygen or its mimics, is the pyrimidines substituted in the 5-position with a halogen atom. The principal pyrimidine in this regard is uracil. The significance of the substitution of a halogen atom in uracil is that, with the appropriate choice of halogen substituent, the van der Waals radius of the substituent is so similar to that for a methyl group that the product is treated biologically like methyl-substituted uracil. Uridine substituted in the 5-position with a halogen atom is metabolized competitively with thymidine for incorporation into DNA by the synthetic pathways appropriate for thymidine. This substitution would be so poorly competitive under normal circumstances that very little of the halogenated pyrimidine would appear in DNA. However, inhibition of the thymidylic synthetase of the cells by the halopyrimidine partially blocks the normal pathway for the synthesis of thymidine in DNA and allows the expression of the competitive pathway for incorporation of the halopyrimidine into the cellular DNA. The result is a replicated DNA that contains the halogenated analog in positions normally occupied by thymine. The most frequently used analog is 5-bromouracil, which appears in the DNA product as 5-bromodeoxyuridine in some small fraction of the normal thymine positions.

The sensitizing action of, for example, 5-bromouracil is very different from the action expressed by all the other modifiers of radiation response described previously. The sensitizer in this case must have been present in the cell during a time of DNA replication for incorporation of the sensitizer into the DNA molecule, and only when the analog of thymine is already present in the DNA at the time of irradiation does it have sensitizing action. When 5-bromodeoxyuridine is used as the sensitizing agent, the drug must be available for incorporation for at least one cell number doubling time to achieve maximum sensitization (Raaphorst and others, 1985). The drug itself has no sensitizing action in the absence of incorporation into DNA.

The mechanism of sensitization by incorporation of 5-bromouridine into DNA is thought to be by a special radiosensitivity of the bromine atom in this structure to radiolytic cleavage. The following shows the production of the uracilyl radical:

This direct action pathway for the radiolytic destruction of DNA by the mechanism shown is certainly important, but other indirect actions through water radicals are probably equally important. The only one of such reactions that is well documented is the interaction of the bromo-deoxyuridine residues with the solvated electron according to the following reaction:

REFERENCES

ALEXANDER, P., and CHARLESBY, A. (1955) Physico-chemical methods of protection against ionizing radiations. In *Radiobiology Symposium 1954*, Z. M. Bacq and P. Alexander, eds. Butterworth, London, pp. 241–263.

ALPER, T. (1956) The modification of damage caused by primary ionization of biological targets. *Radiation Res.* 5, 573–586.

AUGENSTEIN, L. G., BRUSTAD, T., and MASON, R. (1964) The relative roles of ionization and excitation processes in the radiation inactivation of enzymes. *Adv. Radiation Biol.* 1, 228–266.

AUGUSTINSSON, K., JONSSON, G., and SPARRMAN, B. (1961) Effects of ionizing radiation on arylesterase and choline-esterase. *Acta Chem. Scand.* 15, 11–15.

BIAGLOW, J. E., and others (1983) The role of thiols in cellular response to radiation and drugs. *Radiation Res.* 95, 437–455.

BRUSTAD, T. (1964) Heat as a modifying factor in enzyme inactivation by ionizing radiations. In *The Biological Effects of Neutron and Proton Irradiations* 2, 404–410. International Atomic Energy Agency, Vienna.

CHAPMAN, J. D., and others (1973) Radiation chemical studies with nitrofurazone as related to its mechanism of radiosensitization. *Radiation Res.* 53, 190–203.

FLUKE, D. J. (1966) Temperature dependence of ionizing radiation effect on dry lysozyme and ribonuclease. *Radiation Res.* 28, 677–696.

GRAY, L. H., and others (1953) The concentration of oxygen dissolved in tissues at the time of irradiation as a factor in radiotherapy. *British J. Radiol.* 26, 638–648.

HOWARD-FLANDERS, P., and ALPER, T. (1957) The sensitivity of microorganisms to irradiation under controlled gas conditions. *Radiation Res.* 7, 518–540.

MICHAEL, B. D., and others (1973) A posteffect of oxygen in irradiated bacteria: a submillisecond fast mixing study. *Radiation Res.* **54**, 239–251.

ORMEROD, M. G., and ALEXANDER, P. (1963) On the mechanism of radiation protection by cysteamine: an investigation by means of electron spin resonance. *Radiation Res.* **18**, 495–509.

PATT, H. M., and others (1949) Cysteine protection against x-irradiation. *Science* **110**, 213–214.

RAAPHORST, G. P., and others (1985) In vitro transformation by bromodeoxyuridine and x-irradiation in C3H-10 T ½ cells. *Radiation Res.* **101**, 279–291.

REVESZ, L. (1983) Studies with glutathione deficient human cells. In *Radioprotectors and Anticarcinogens*, O. F. Nygaard and M. G. Simic, eds. Academic Press, Inc., New York, pp. 237–274.

SAMUNI, A., and CZAPSKI, G. (1978) Radiation induced damage in *Escherichia coli* B: the effect of superoxide radicals and molecular oxygen. *Radiation Res.* **76**, 624–632.

SCHWARZ, G. (1909) Über Desensibilisierung gegen Röntgen- und Radiumstrahlen. *Münchner Medizinische Wochenschrift* **56**, 1217–1218.

THOMLINSON, R. H., and GRAY, L. H. (1955) The histological structure of some human lung cancers and the possible implications for radiotherapy. *British J. Cancer* **9**, 539–549.

WHILLANS, D. W., and HUNT, J. W. (1982) A rapid mixing comparison of the mechanisms of radiosensitization by oxygen and mizonidazole in CHO cells. *Radiation Res.* **90**, 126–141.

YUHAS, J. M., and PHILLIPS, T. L. (1983) Pharmacokinetics and mechanisms of action of WR-2721 and other protective agents. In *Radioprotectors and Anticarcinogens*, O. F. Nygaard and M. G. Simic, eds. Academic Press, Inc., New York, pp. 639–653.

SUGGESTED ADDITIONAL READING

DERTINGER, H., and JUNG, H. (1970) *Molecular Radiation Biology.* Springer Verlag, New York.

VON SONNTAG, C. (1987) *The Chemical Basis of Radiation Biology.* Taylor and Francis, London (particularly Chapters 10 and 11).

PROBLEMS

1. A mammalian cell line has a D_0 of 120 cGy and an extrapolation number of 12. The OER is found to be of a dose modifying character with a value of 2.81. What is the value of m in the Alper/Howard-Flanders equation? The following data have been found for this cell line for OER as a function of oxygen partial pressure.

$[O_2]$ (%)	OER
0.1353	1.10
0.223	1.20
0.2865	1.33
0.3679	1.64
0.6065	1.88
0.7788	2.19
1.00	2.59
1.75	2.77
2.50	2.81

What is the value of K in the same expression? What is the ratio of type 1 to type 2 damage in this cell line?

10

Radiation Biology
of Normal
and Neoplastic
Tissue Systems

INTRODUCTION

On the whole, we would expect that organized tissue systems of the body would respond substantially as predicted from cellular studies of the type discussed in the foregoing chapters. In principle, there are very few, if any, exceptions to this general observation; but often the modifying effects of the tissue environment, under physiological, hormonal, and homeostatic controls, are not wholly understood or even appreciated. Many aspects of the response of tissue systems are strongly affected by the state of the cell in its cycle, for example, the state of oxygenation of the cell. The supply of metabolic substrates and the removal of metabolic products also play a role in modifying the response of tissue systems. Some controls external to the tissue system being observed will also come into play; for example, those hormonal substances responsible for regulatory performance of the tissue will have an important modifying role.

The most significant aspect of the radiosensitivity of a tissue or organ system centers about the state of reproductive activity normally associated with that cellular system. This proliferative state varies widely among the tissues of any mammalian species. At one extreme are the tissues of the central nervous system, some of which never undergo division during the organism's adult life and for which loss of clonogenic ability is an irrelevant end point. At the other extreme are the blood forming organs, which are proliferating at a rate approaching that of an exponentially growing in vitro culture.

NATURE OF CELL POPULATIONS IN TISSUE

One of the earlier systematic overviews of the nature of cell population kinetics in normal and malignant tissues was that of Gilbert and Lajtha (1965). Their classification of the various kinetic systems found in mammalian (and, incidentally, in other organisms) organs and tissues is shown in Figure 10.1. From the figure the definitions of each of the systems are the following. (The double arrows in classifications D, E, and F are meant to signify the mitotic division of one of the cells of the compartment, giving rise to two daughter cells.)

A. *Simple transit population*: Fully functional cells are added to the compartment while a population of either aging or randomly destroyed cells are disappearing from the pool. There are many examples of functional "end-cells" in this category. Examples would be spermatozoa, which are constantly being replaced, as well as red cells or other end cells of the blood.

B. *Decaying population*: Cell numbers are decreasing with time without replacement. The population of oocytes in the mammalian female is often quoted as an example. Populations of this classification are rare in mammalian systems, but not in insects.

C. *Closed, static population*: There is neither decrease nor increase in cell numbers during life. It is unlikely that such a population truly exists. The differentiated neurons of the central nervous system are quoted as an example of a static population, but there is probably a decline in cell numbers even in this population.

D. *Dividing, transit population*: In addition to the transiting cells, division of the cells in the compartment occurs, leading to a larger number leaving than entering. It is assumed in this model that the number of cells within the compartment remains more or less static.

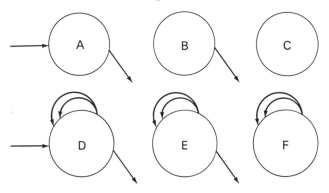

Figure 10.1 Classification of cell kinetic types in the system of Gilbert and Lajtha (1965). The descriptions of each of these models are given in the text.

The differentiating and proliferating blood cell types (for example, the proerythroblast of the bone marrow) that follow the stem cell would be examples of this type of population.

E. *Stem cell population*: A self-sustaining population, which relies on its own self-maintenance for its continued existence. All the progeny of this type of cell line depend on the continued existence of the stem cell pool. Every self-maintaining, dividing cell population must have such a precursor pool. An example is the stem cells responsible for sustained spermatogenesis.

F. *Closed, dividing population*: Such a population is best represented in neoplastic growth. No cells enter or leave the compartment in the early stages of growth of the tumor. In the long run, neoplastic growth is probably best represented by a stem cell population, since as the tumor enlarges there is cell death, suppression of growth by metabolic and other nutrient shortages, and a highly variable rate of division. The epithelial cells responsible for cell renewal in the lens of the eye are another example of this type of population.

Cell Kinetics and Radiation Damage

It should be almost self-evident from the descriptions in the previous chapters that the kinetic types represented by D, E, and F will be most vulnerable to radiation damage. It has been established that the principal target of ionizing radiation is the cellular genome, and the principal outcome of disturbances to the dynamic replicative activity of the genome is altered clonogenic ability. That is indeed the case, and the most critically sensitive of these systems would be the stem-cell-type tissue, which depends for its continuing function on continued clonogenic potential. The ultimate functional viability of a tissue that is dependent on stem cell activity will be determined by whether, after radiation exposure, there are adequate numbers of surviving and still clonogenic stem cells to repopulate the compartment and finally to produce functionally competent progeny. The most resistant tissues would be those that require neither input of cells from a prior compartment nor division within the compartment. The closed, static model is such a case, and in the case of the central nervous system, its high degree of radioresistance can be attributed to its lack of need for cell replication.

Growth Fraction and Its Significance

The concept of growth fraction as a descriptive parameter for the kinetics of proliferating tissue appears to have been first proposed by Mendelsohn (1962) as the result of his observations that all cells in a growing tumor were not in the active process of proliferation as determined by the cellular incorporation of radioactive labels of DNA synthesis. Lajtha (1963), based on his own studies as well as those of others, proposed the concept of the G_0 phase of the cell cycle, a state of the cell in which the cell was not engaged in active proliferation, but in which the cell could reenter the proliferative state. The G_0 cell was visualized by him as a cell that has been removed from the actively dividing population by regulatory activities rather than as a result of metabolic deprivation. Subsequently, it became apparent that cells could also be removed from active division in a reversible manner by deprivation of oxygen, glucose, or other metabolites. The restoration of the lacking nutrient would lead to the reentry of the cell into active proliferation.

To resolve the confusion as to the nomenclature for nondividing cells in normally dividing populations, Dethlefsen (1980) has urged that the broader terms, quiescent (Q) and proliferating (P) be used to identify the two cell classes functionally, rather than by the regulatory or nutrient state.

The *growth fraction* is defined as the fraction of the total cellular population that is clonogenically competent and is actually in the active process of DNA replication and cell division. The growth fraction may be estimated by any one of several techniques, most of which depend on incorporation of a radioactively labeled DNA precursor into those cells that are actively dividing. For details of computational methods, the reader is referred to Steel (1977). One of the simpler methods for the determination of the growth fraction is the exposure of a growing culture of cells, in vitro or in vivo, to an appropriate radioactive label for the synthesis of DNA. A typical and frequently used label is ^{3}H-thymidine. The cells are exposed for at least the full length of a cell cycle, and usually for half again as long. Under these conditions, all cells that synthesize DNA, thus indicating their passage through the S period of the cell cycle, are labeled and can be identified by autoradiography. The percentage of cells that is labeled constitutes the growth fraction, since every cell in cycle will have passed through the S period at least once during exposure to the radioactive label.

The radiobiological significance of the growth fraction was unclear until the appearance of new data in the late 1980s. As recently as 1980, Dethlefsen indicated that the role of quiescent cells in radiobiological response was not satisfactorily delineated. A number of recent studies have indicated that cells that are out of cycle have the capability for a more significant amount of repair of potentially lethal damage, simply because there is more time before the cell is called on to replicate its DNA. A recent report by Rodriguez and others (1988) has confirmed that a good deal more repair of potentially lethal damage is, indeed, accomplished in cells that are out of cycle. A number of other reports have confirmed this observation. Cells that are out of cycle in their system, the multicellular tumor spheroid, take as long as 6 hours to reenter the cell cycle after the cells of the spheroid are dissociated.

It is possible, but by no means proved, that the concentration of enzymes necessary for repair of DNA damage may be depleted in the noncycling cell, but, in spite of this, the additional time allows effective repair to proceed with the lower concentration of repair enzymes. In summary, the weight of the evidence presently suggests that noncycling cells will generally be more resistant to irradiation than will cycling cells, although several exceptions to this rule have been noted.

Cell Kinetics in Normal Tissues and Tumors

Both normal and neoplastic tissues have a cellular kinetic pattern that follows the accepted model of a G_1–S–G_2M cycle, and the cell cycle parameters are not very different for tumors as compared to other growing tissues. The total cycle time and the time devoted to DNA synthesis in the S period are very much alike for both tissue types. There are, however, significant differences in some of the characteristics of the kinetic pattern as the tumor reaches a size where vascularization is required for continued tumor growth. The orderly vascularization of normal tissues that originates in embryonic life and that is maintained throughout the existence of normal, nonpathological function assures that the supply of oxygen and nutrients is adequate for survival of cells. Most, if not all, tumors, on the other hand, originate as nonvascularized aggregations of cells and develop a vascular supply sometime after the origination of tumor growth. The development of vascular supply in a tumor depends on the activities of angiogenic factors occurring in normal tissues. The newly developing vascular supply is, at best, chaotic and disorganized. Some parts of the tumor tissue will be at such a distance from the source of oxygen and nutrients that cell survival will be impossible. Other parts of the tumor will have nutrient and oxygen supply only adequate for survival of cells without replication. Tannock (1968), and subsequently he and his co-workers, examined the morphological relationship between the cells in a tumor and nearby blood vessels. Tannock reported that dead and dying cells were found at a radius of 85 to 100 μm and greater from the nearest blood vessel. The growth fraction was found to be highest adjacent to the blood vessel and to fall with distance from the nearest vessel. Necrotic regions were found at distances of approximately 100 μm. Tannock and others suggested a strong correlation between oxygen diffusion distances and the decrease in growth fraction and, ultimately, necrosis. More recent findings have demonstrated that not only oxygen lack, but also a lack of glucose can lead to a decrease in the growth fraction, and probably both nutrients play an important role in the determination of quiescent and proliferating cells in tumors (Hlatky, Sachs, and Alpen, 1988).

One important difference between normal tissues and tumor tissues is the determinant of the fraction of quiescent cells in the organ or tumor. Because of the orderly vascular architecture of the normal tissue, the movement of cells from the proliferating to the quiescent compartment is probably not the result of nutrient lack, but rather is the result of the activity of normal humoral factors regulating the growth and development of the tissue. The quiescent cells of normal tissues are the ones for which Lajtha coined the G_0 appellation. Tumor cells become quiescent and even die as the result of the lack of nutrients. The question that cannot be answered now is whether the quiescent cell of normal tissue can be expected to have the same properties of DNA repair and control of movement into proliferative activity as the quiescent cell of tumors. The experimental data are for tumor quiescent cells, and extrapolation would be, at best, hazardous.

MODELS FOR CELL SURVIVAL IN NORMAL TISSUES AND TUMORS

The survival data for cells discussed in earlier chapters applies to cells grown in in vitro tissue culture, and it is clearly not possible to generalize these data to predict the radiation response of normal or neoplastic tissues. Fortunately, a number of novel and innovative systems have been developed that permit an examination of the radiosensitivity of normal and neoplastic tissues in a more or less undisturbed in vivo environment. Because many of these assay systems provide a special insight into the radiosensitivity and repair capacity of normal tissues and tumors, they will be examined in some detail.

MODELS FOR RADIOBIOLOGICAL SENSITIVITY OF NEOPLASTIC TISSUES

The earliest attempts at assaying the sensitivity of organized tissue systems were directed to establishing the radiosensitivity of tumor tissues. This was partly because these tissues offered opportunities for analysis not available for normal tissues. The possibility for syngeneic transplantation of the cell lines from host to recipient animal was the most important characteristic of these in vivo tissue systems. After irradiation of the tumor in the host in which it was growing, it was possible to transplant the tumor cells to an unirradiated recipient animal and to observe the growth response of the irradiated tumor cells.

There was also a strong interest in the understanding of tumor biology arising from the treatment of cancer by radiotherapy. It was important to establish the role of oxygen in the sensitivity of cancer cells, as well as the importance of repair or repopulation in these tissues, in order to maximize the effectiveness of radiotherapy for cancer control in patients.

Hewitt Dilution Assay

Probably the first in vivo assay for mammalian tissues was that developed by Hewitt and Wilson (1959) with a syngeneic mouse tumor system. At that time a number of tumor cell lines had been developed that were grown in the peritoneal cavity of mice. The cells from these ascites tumors can be harvested or, if allowed to continue to grow in the peritoneal cavity of the host, will cause the death of the animal. It occurred to Hewitt and Wilson that this end point of death of the host animal could be used to measure the clonogenic potential of the tumor cells after irradiation. Figure 10.2 shows the essentials of a Hewitt assay for a single

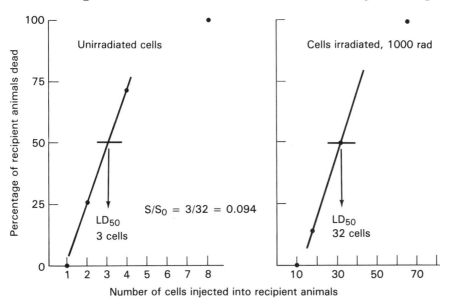

Figure 10.2 Typical data set for a Hewitt dilution assay. The tumor cells are irradiated in the donor animal or are collected without irradiation. A serial dilution provides injectates for the recipients with graded number of cells. By injecting a group of animals at each cell dose level and observing the survival, an LD_{50} for the cells is established. In this example the surviving fraction after 1000 rad (10 Gy) is calculated. To complete a survival curve, it is necessary to repeat the assay at a number of dose levels. The data shown are plotted from Andrews and Berry (1962).

dose point of 1000 rad (10 Gy). In the case shown, cells harvested from the mouse ascites tumor P388 were used. Unirradiated cells were collected from the donor and a series of dilutions prepared from a stock suspension of the tumor cells. A typical microbiological-type binary dilution is carried out to produce cell suspensions with low concentrations of cells that will allow the injection of the recipient animal with cell numbers that are correct for killing about half the animals. For the tumor line used, the usual cell dose required to kill half the animals is about two to three cells. A small number of animals, 5 to 10, is injected with the same cell dose and the survival followed. The same procedure is used for several additional cell doses. The resulting data on percent survival at each of the cell doses are plotted as shown in Figure 10.2, and the LD_{50} (lethal dose for 50% of the animals) is determined by graphical or analytical means. The procedure is repeated, but with the cell suspension prepared from animals that have been irradiated before cell collection. Animals are irradiated at several doses, and the LD_{50} values can be used to construct a survival curve. Figure 10.2 shows an example for only one dose, with the calculated surviving fraction. The surviving fraction is estimated for

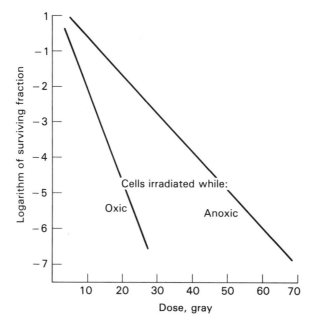

Figure 10.3 Survival curve obtained by the Hewitt assay method for two mouse leukemias and a sarcoma by Berry (1964). The data are not shown in the figure, but they are tightly scattered about the drawn lines. Cells were either collected from the donor normally, or they were made anoxic by killing the donor a few minutes before collecting the cells.

each of the other doses, and a survival curve of surviving fraction against dose is plotted in the usual way.

The Hewitt assay has been the tool used for a number of significant studies of tumor cell sensitivity to radiation. Figure 10.3 is a very good example of such studies. Berry (1964) developed survival curves for three mouse tumors, two leukemias and a sarcoma. Some of the data were his own unpublished observations, and some were provided by Hewitt. The clonogenic survival curves were developed for both anoxic and oxic conditions. All three cell lines could be plotted on the same curve for oxic cells or for anoxic cells as appropriate, and the line produced was a good fit for the appropriate condition of oxygenation. The OER for these cells was about 2.4, not far from the 2.8 or so for cell lines that are irradiated in vitro and analyzed for clonogenic survival in vitro. The D_0 for the cells irradiated under oxic conditions was about 150 rad (150 cGy), and the extrapolation number was about 3 to 4 for this set of data.

A significant shortcoming of the dilution assay system is that donor cells that are grown in ascites fluid are usually irradiated when the cell number in the peritoneal cavity is very large. Under these conditions it is not always clear that the cells are fully oxygenated at the time of irradiation. If that is indeed the case, there is the possibility of significant anoxic protection of the cells and, subsequently, an overestimation of the resistance of the cells to the irradiation. The data reported in the Berry study do not seem to be affected by such hypoxia. The D_0 (oxic) is about 150 rad (150 cGy), a number quite like that found for many cell systems in vitro. The OER of 2.4 or so is, again, not very different from the 2.5 to 2.8 seen for in vitro systems. We must conclude, at least for the cell lines reported on in this study, that adequate oxygenation probably existed at the time of irradiation.

Another shortcoming of the method is that the irradiated tumor cells must be capable of expressing clonogenic potential while growing in the ascites medium. For example, most leukemias grow readily in this environment, and it usually only requires an inoculum of two to three cells to cause the death of 50% of the recipient animals. In the Berry data just described, the sarcoma cells required an inoculum of more than 80 cells to kill 50% of the recipients. In many cases no cell growth is seen, and no assay is possible. To avoid this shortcoming, other assays have been developed.

Lung Colony Assay System

A modification of the Hewitt dilution assay is that developed by Hill and Bush (1969) for the measurement of clonogenic survival of cells derived from solid tumors. In principle, the assay measures the clonogenic survival of tumor cells by determining their ability to form colonies in the lung of recipient syngeneic mice. The cells from a tumor, irradiated either in vivo or, after dissection and cell dissociation, in vitro, are injected into a recipient mouse, and after 18 to 20 days the animals are killed, the lungs dissected, and the number of tumor colonies in the lung counted. The authors were able to demonstrate a linear relationship between cell number injected and the number of colonies formed in the lung. A very large enhancement of the number of colonies in the lung was found if, along with the experimentally irradiated cells, a large number of heavily irradiated, nonclonogenic cells were injected. Typically, such a procedure produced a 10 to 50 fold increase in the number of colonies formed from the clonogenic survivors. The authors were not able to establish the mechanism of this enhancement, but it was not due to an immune response on the part of the recipient.

Very consistent survival curves were obtained, and the authors reported for the KHT transplantable sarcoma that the D_0 was 134 rad (134 cGy), with an extrapolation number of about 9.5. Again, these data were found to be quite consistent with the values found for the same tumor with the Hewitt assay. Such an agreement not only validates the lung colony assay; it also demonstrates that there was little protection of partial hypoxia for the KHT cells tested by the dilution assay.

A significant limitation of the lung colony assay is that cells must be injected into syngeneic recipient mice, that is, inbred mouse lines of the same genotype as that from which the tumor is derived. There is no published evidence that anyone has yet attempted the experiment, but it should be possible to carry out the lung colony assay not only with cells of different mouse strains, but also even from different species, such as the human being, if nude mice are used as recipients. The special nude mouse line has little immunologic reactivity, and nonsyngeneic cell lines grow without rejection.

Tumor Growth and Tumor "Cure" Models

Since there is a very limited set of models for examining the clonogenic potential of tumor cells, much of the radiation biology of tumors has been developed using a set of tools that was developed for general use in tumor biology. Some of these have been more valuable than others for radiation effects studies because of the inherent lack of the ability for precise quantitation.

Tumor volume versus time. A widely used and relatively powerful tool in tumor radiobiology has been the tumor growth curve after implantation of an inoculum of cells, usually in the flank region of recipient syngeneic mice or rats. The simplest application of the growth curve for implanted tumors is the analysis of the rate of increase of the tumor volume. For analysis of the radiation effect, we can measure the time for the tumor to reach a preselected volume. The measurements of tumor volume are at best imprecise. The volume is usually determined from a caliper measurement of two or more diameters of the growing tumor and calculation of the volume from the average diameter.

After the tumor has been irradiated, the time course of volume change is shown in Figure 10.4. There may be a slowing of growth for a brief time, followed by a period of decreasing tumor volume. This decrease is due to lack of replacement of the normal cell loss from tumors, associated with local necrosis, nutrient lack, or other causes unrelated to the radiation exposure. It is not due to the interphase death of cells as the result of irradiation. As the surviving clonogenic cells repopulate the tumor, regrowth will be observed; the surviving clonogenic cells will ultimately produce progeny exceeding the cell-loss factor. The criterion for measurement of the radiation-dependent response is taken as the time for the cell volume to reach again the value observed at the time of irradiation. This time is shown on Figure 10.4, and it is measured as shown, as the time from irradiation until the tumor volume achieves the value existing at the time irradiation occurred. This time value is called the *growth delay*. The important limitation of the growth delay model for testing the radiobiological response of tumors is that a significant number of trans-

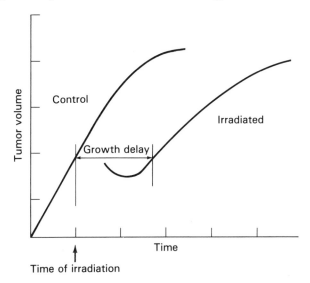

Figure 10.4 The ordinate is the tumor volume as a function of time after implantation of an inoculum in the flank of a recipient mouse. The abscissa is time after implant and is usually no more than about two to three weeks. The time of irradiation is shown by the vertical arrow. With many tumor lines there is a brief lag before exponential growth commences, but as long as the irradiation occurs at a time of exponential growth, it is of little importance.

plantable tumors do not show any decrease in the volume of tumor after irradiation. Presumably, this failure to decrease in volume is the result of a small cell-loss fraction in the growing tumor. When irradiation takes place, clonogenic activity is reduced until repopulation from competent clonogenic cells occurs. During the period before regrowth commences as the result of repopulation, the normally small cell-loss fraction of the tumor does not lead to reduction in tumor volume. In these cases it is necessary to revert to the simpler measure of tumor volume versus time and the use of the time to reach a preset volume, or the differences in this time for control and irradiated tumors may be taken as the end point.

An alternative analysis of the tumor regrowth curve uses the logarithmic plot of tumor volume against a linear time scale. The relative volumes of the growing tumor are used, against a value of 1.0 for the starting volume of the nonirradiated tumor. Assuming that the tumor regrowth is a function of the number of surviving clonogenic cells, the linear portion of the regrowth curve for each of several irradiation doses is extrapolated back to the zero time ordinate, and the relative tumor volume for this back-extrapolate is taken as the surviving fraction of clonogenic cells in the irradiated tumors. From these data, a survival curve for in vivo irradiation may be developed. The survival curve will necessarily be limited to a rather small part of the normal range of surviving fractions. It is difficult to carry out these studies at estimated surviving fractions of less than about 0.05.

A significant constraint on the reliability of the analysis by the back-extrapolation just discussed is the possibility that tumor regrowth rate is not a simple function of the number of surviving clonogens. This is almost certainly the case, since a number of workers have shown a *tumor bed effect*. This effect is most probably due to the results of the irradiation on the development of the vascular tissue, the development of which determines the growth rate of the tumor after it has exceeded a very small size.

TCD$_{50}$, tumor cure.   Another end point widely used in tumor biology is the dose required to "cure" an implanted tumor. For this model a large number of implanted tumors is irradiated with graded doses at the same time period after implantation of the tumor inoculum. The end point is the fraction of animals that has received a given dose in which the growth of the tumor is controlled. This *local control* index can then be plotted for each of the doses, and the dose required to control tumor growth in 50% of the animals is estimated by a variety of statistical techniques. This value is usually called the *50% tumor cure dose* (TCD$_{50}$).

Radiobiological Response of Tumors

Using a number of end points, including dilution assay, lung colony assay, primary cell cultures, and tissue derived in vitro cultures, it has been possible to define rather clearly the radiobiological responsiveness of various tumor lines, both animal and human. With only a few important exceptions, the various tumor cell lines in wide and long-term use have been found to have clonogenic survival characteristics that are generally stable and for which the relevant survival paraments are not very variable, considering the range of cell types and tissues from which these transformed and immortal cell lines have been derived. In Chapter 8, the range of variability for these lines was shown by comparison among the HeLa, T1, and CHO lines. The first two of these are of human derivation, and the CHO is from the Chinese hamster. Typically, we find a range of D_0's covered roughly by a factor of 2 to 3 around the mean value and a range of n values (extrapolation numbers) from 5 to 20 for all of the transformed cell lines in common laboratory use. We must exclude from this generalization those cell lines that are derived from sources characterized as having genetic defects in the DNA damage repair mechanisms. An important example of such a line is of human derivation from patients who have been diagnosed to have the disease *xeroderma pigmentosum*. In these cells the ability to repair DNA damage induced by ionizing or ultraviolet radiation is very much reduced, if not completely absent. The survival curves for these cells are characterized by an extrapolation number of 1 and a D_0 that is a small fraction of that for other cell lines.

Rather different findings have been reported for the survival curve parameters of freshly derived culture systems grown from naturally occurring malignant tumors. Extensive efforts have been devoted to the characterization of the radiosensitivity of cell lines from human tumors, especially by Fertil and Malaise (1981, 1985). They point out that the best fit to the data for a large number of human cell lines, both nontransformed fibroblasts and tumors, was with the linear–quadratic model. It should be pointed out that they did not attempt to fit their data with any of the more advanced models for cell killing mentioned in Chapter 7.

The analysis of Fertil and Malaise of the survival curves fitted to the L–Q model revealed extreme cases. Untransformed fibroblasts were the most radiosensitive, and, in general, they were best described as having exponential survival curves, that is, in MTSH terms, single-hit characteristics, or in L–Q terms, a predominant α term and a negligible or zero β term.

In the second of these reports, the same authors analyzed not only their own data, but that of several other workers who had collected data on the radiosensitivity of newly derived cell lines from human tumors. They were able to correlate the radiosensitivity of the tumor, as defined by the parameters of the survival curve, and the survival of the cells after a single dose of 2 Gy. This dose is typically the one used for the daily irradiation fractions in the treatment of human malignancy, and it should be a good indicator of tumor curability. It was, however, difficult for them to unequivocally demonstrate from the survival curve parameters that a relationship existed. In the 1985 paper, they showed that the most reliable indicator for radiosensitivity and tumor curability, as reflected in survival after 2 Gy, was the magnitude of α, the coefficient for the first-order term in dose of the linear–quadratic model. This is not necessarily surprising, since the first-order dose term will predominate at doses of 2 Gy and below.

These authors classified the radiosensitivity of the various cell lines into three groups. There is a very good correlation with the known responsiveness of the tumors to radiotherapy. Lymphomata, known to be highly curable, were the most radiosensitive of the derived cell lines, and melanomata were both the most resistant for tumor curability and the most radioresistant in the survival of the cell lines in culture.

It is important to realize that the immediate responsiveness of a tumor to radiation, as determined by reduction in the tumor volume, does not necessarily predict with high efficiency the curability of the tumor. The degree of responsiveness will be determined by many of the cell kinetic parameters of the tumor system. A high cell-loss factor and a high growth factor associated with a small fraction of cells out of cycle, and associated with inherent cellular radiosensitivity, will assure a high degree of responsiveness of the tumor, as measured by volume changes. Curability, on the other hand, will depend in a complex way on the ability of the clonogenic cells to repopulate the tumor after irradiation is over.

Hypoxia and Radiosensitivity in Tumor Cells

The extensive review of the modifying action of oxygen on the radiosensitivity of cells that was presented in Chapter 9 emphasized repeatedly that, under circumstances where severe anoxia could occur in tissues or cellular preparations, we would expect to see significant protection from the effects of ionizing radiation. References were also provided in that chapter to show that, indeed, we would expect to find conditions of moderate to severe anoxia in growing tumors in vivo. For cells grown in suspension, careful attention to culture conditions will usually prevent the development of such anoxic conditions with concomitant radioprotection. For the tissue assay systems, such as the Hewitt dilution assay and others, there is clearly a protective effect of oxygen lack under the correct conditions. Figure 10.3 shows such radioprotection for cells deliberately

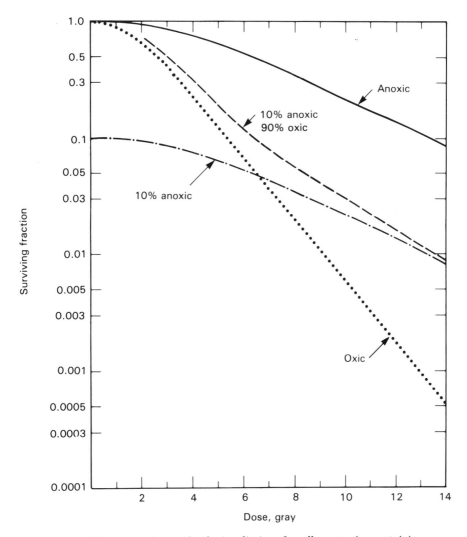

Figure 10.5 Survival curve for the irradiation of a cell suspension containing a fraction of hypoxic cells. Shown are the survival curves for the populations of anoxic cells and for fully oxic cells. If the irradiation mixture contains 10% hypoxic cells, the expected result for the survival curve is shown. The analysis of this curve, given the survival data for oxic and anoxic cells, will provide an estimate of the fraction of hypoxic cells in the mixture.

made anoxic by killing the host animal or by allowing the cell number for cells growing in the peritoneal cavity to reach very high levels. Figure 10.5 demonstrates methods by which the fraction of hypoxic cells in a mixture with fully oxygenated cells can be detected and measured quantitatively.

The radioresistant tail for the survival curve shown in Figure 10.5 is a common observation for cells from tumors and indicates the presence of a mixed population of cells, part of which have a radioresistance relative to the remainder of the population. This resistant fraction is due to hypoxia and the radioprotection that this state affords.

The classic work of Thomlinson and Gray (1955) laid the foundations for our understanding of both hypoxia as well as reoxygenation in tumors during growth and regrowth. Figure 10.6 (from Thomlinson, 1967) illustrates the processes proposed by these authors. The very young tumor is well oxygenated, since it is so small that no cells are beyond the useful diffusion distance of oxygen from nearby capillaries. As the tumor continues to grow, portions of the tumor volume may be beyond easy access to diffusing oxygen. The tumor must depend for its supply of oxygen on the development of newly formed vessels arising from the adjacent normal tissue and penetrating the tumor volume. This neovascularization of the

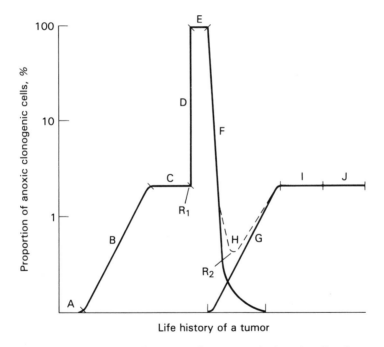

Life history of a tumor

Figure 10.6 Development of hypoxia and reoxygenation in an irradiated tumor (from Thomlinson, 1967; reproduced by permission of Butterworth Scientific Ltd.). Segment AB is the development of an hypoxic fraction. Segment C is the steady-state hypoxic fraction characteristic of the tumor. D is the rapid rise after irradiation at R_1 in the hypoxic fraction as the result of death of the sensitive oxygenated cells. E is the maximum for the hypoxic fraction, followed by a rapid fall shown as F, the reoxygenation phase. GIJ is the repeat of the process.

tumor is not as well organized as is the blood supply in normal tissues, and the expanding volume of tumor will contain regions in which oxygen is inadequate for the maintenance of metabolism, and some fraction of the cells will be anoxic. In Figure 10.6 the authors show that the fraction of anoxic cells in the growing tumor may rise to several percent and, in some tumor types, to as much as 10%. When the tumor is irradiated (position R_1 in the figure), the more radiosensitive, fully oxygenated cells are killed, and the remaining hypoxic cells are in an environment of dead and dying cells with lesser demand for metabolic oxygen. The shrinking of the tumor volume and the lowered oxygen demand allow for reoxygenation of the hypoxic cells, indicated by a rapid fall to near zero for the anoxic fraction. After this period of reoxygenation, tumor regrowth commences and the complete cycle is repeated.

The significance of the reoxygenation phase in fractionated radiotherapy of human tumors is undergoing careful reexamination, partly since treatment modalities designed to optimize the kill of anoxic cells (high LET radiation, radiation under hyperbaric oxygen conditions, and so on) have not been particularly successful. According to Figure 10.6, the optimum time for a second irradiation of a fractionated scheme would be at point H in the curve, when the population of hypoxic clonogenic cells is at a minimum. Recent data would suggest that the reoxygenation phenomenon actually occurs very soon after irradiation, and indeed may take place while the irradiation is in progress. Kolstad (1964) has actually measured with an oxygen electrode that reoxygenation progresses significantly during the treatment interval.

Only time and additional experimental and clinical investigations will determine the role of reoxygenation in radiotherapy of human tumors. Recent emphasis on the role of quiescent cells in the radiosensitivity of tumors is now a major topic for consideration.

ASSAY MODELS FOR NORMAL TISSUES IN VIVO

Acute Response of Normal Tissue

To determine the radiosensitivity of normal tissues in vivo, it has been necessary to develop a range of assays that allow either direct or indirect measurement of the surviving fraction of clonogenic cells in the tissue. Such assays are of importance only for those tissues of the living mammal that are ordinarily cell renewal systems, that is, type D of the Gilbert and Lajtha scheme described at the beginning of this chapter. From the clonogenic survival curves developed from these assays, the radiosensitivity of the clonogenic cells of the tissue may be described. The Hewitt dilution assay would seem to be a useful tool for such analysis, but the production of single-cell suspensions of normal tissues that would grow in the Hewitt assay has generally not been possible.

The principal cell-renewal systems in mammalian organisms are the following.

1. Hematopoietic system (the bone marrow)
2. Skin
3. Gastrointestinal tract
4. Testes (regeneration of spermatozoa)
5. Lens of the eye

There are other cell-renewal systems in the mammal, such as hair and some specialized parts of the central nervous system, but they play no significant role in the response of the living organism to irradiation.

Useful assays have been developed for each of these principal cell-renewal systems. These will be described and the deductions from the measurements as to the radiosensitivity of the tissues will be outlined.

Hematopoietic System: The CFU-S Assay

The CFU assay system (colony forming unit) was developed by Till and McCulloch (1960). Since this assay is carried out in the spleen of the recipient animal, it has come to be known as the CFU-S assay in more recent time. The assay grew out of the observation that, in irradiated animals that received bone marrow transplants from an unirradiated host, nodules developed in the spleen of the recipient. These nodules were observed by a number of workers in the field of bone marrow transplantation, but Till and McCulloch were able to appreciate the quantitative utility of the nodules for an assay system for hematopoietic cells. Because the assay depends on the counting of nodules or colonies in the spleen, it has also been called the *spleen colony assay*. The mechanics of one approach to the assay are shown in Figure 10.7.

The colony forming unit assay (CFU-S) depends on the nature of the bone marrow regenerating cellular system. At the time that Till and McCulloch first developed their assay, the cell-renewal system was visualized as a single stem cell compartment that was pluripotential in its function. That is, a single stem cell type was thought to be the precursor of all the functional end cells of the bone marrow: erythrocytes, granulocytes, mononuclear cells, platelets, and other less common cell types. That view has now been shown to be less than complete. It is now known that there are several hierarchical levels within the stem cell compartment, and at each level the degree of commitment of the stem cell to the final functional cell lines becomes more dedicated and less versatile.

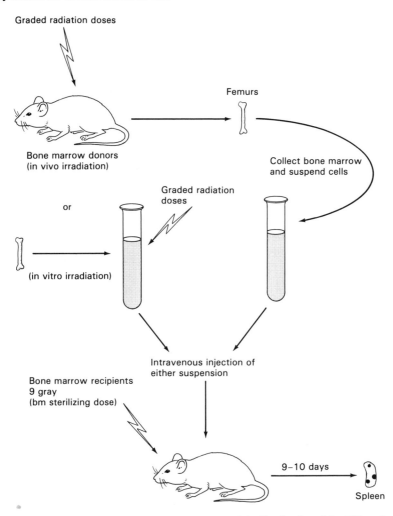

Figure 10.7 Colony forming unit assay as originally developed by Till and McCulloch. The assay may be accomplished as shown either with the irradiation of the donor bone marrow in vivo or in vitro. Either the donor intact mouse or bone marrow collected from an unirradiated mouse and placed in suspension is irradiated with graded doses. The collected and suspended marrow is then injected into recipient mice that have received a radiation dose that otherwise would be lethal.

When a cellular suspension of bone marrow is injected into lethally irradiated recipient mice of the same strain, a few of the cells circulating in the animal lodge in the spleen, and if these cells are clonogenic stem cells, they will give rise to colonies of bone marrow functional end cells. It has been observed by a number of investigators that, when the spleen colonies are examined for morphological characteristics and the functional end cell identified, the cells found in the nodules are more often of a single type. That is, the progeny of the transplanted cell are, for example, all erythrocytes, all leucocytes, or one other characteristic cell. The deduction is that the colonies are monoclonal, that is, arising from a single clonogenic stem cell. Occasionally, mixed cultures are observed in the nodules, suggesting that once in a while the implanted cell is early enough in the stem cell hierarchy to provide multifunctional development pathways. Becker, McCulloch, and Till (1963) were able to demonstrate that all the colonies formed, mixed or monofunctional, were all monoclonal. Since the cells that are capable of polyfunctional expression are earlier in the developmental pathway, they should take somewhat longer to develop end cell progeny. This is the case, and if the time of sacrifice and collection of the spleens is extended for up to two weeks, more multifunctional colonies are found. CFU-S subpopulations are scored at 7, 10, and 13 days after injection into the recipient mice.

Radiation Sensitivity of CFU

The CFU assay has been used to develop complete survival curves for this class of cells. The cells are irradiated at several doses, and appropriate dilutions based on expected survival rates are injected into the lethally irradiated recipient animals. If the number of nodules per spleen at 13 days exceeds 15 or so, the counting of discrete colonies is difficult; therefore, the dilutions have to be made to keep the nodule number between a few to a maximum of 15 or 20. If earlier sacrifice times and resultant smaller colonies are scored, up to 25 colonies per spleen can be counted accurately.

The irradiations may be performed on the donor cells by irradiating the intact mouse, or the suspension of cells collected from an unirradiated mouse may be used as a cell pool. Only minor and generally inconsistent differences in radiosensitivity result from irradiation by the two alternative methods. The multitarget single-hit parameters for mouse marrow cells irradiated in vitro by Till and McCulloch were extrapolation number, n, 2.5; D_0, 105 rad (105 cGy). The parameters for the same cells irradiated in vivo were extrapolation number, 1.5; D_0, 95 rad (95 cGy). Many more investigators have made these determinations with approximately the same results.

The general characteristics of the radiation sensitivity of bone marrow cells is that there is nearly exponential cell killing, with little or no shoulder, and a D_0 of about 100 rad (1 Gy). There is no doubt that the difference in extrapolation number between in vivo and in vitro irradiation is real, since it has been observed by a number of investigators. The basis for this difference is not known with certainty, but it has been suggested that for the in vivo irradiations, during the two hours or so required for cell collection and injection into the recipient, there is movement of cells from another, probably resting or G_0 phase, cell compartment into the CFU compartment.

As we would predict from the low extrapolation number, there is limited protection provided to bone marrow cells by protraction or fractionation of the radiation exposure. Sublethal damage repair, as measured from the standard Elkind–Sutton divided dose experiments, is rarely able to increase the D_0 by more than a factor of about 1.5. The extent of potentially lethal damage repair that can be observed from holding irradiated cells after exposure is inconsistent, because the holding of CFU-S before injection results in a significant drop in plating efficiency. The repair of PLD is likely, in any case, to be small and insignificant. The other characteristics of the radiation responsiveness of CFU-S are much as we would predict. All the response modifiers work as predicted. Chemical protectors are active, and anoxia protects the cells.

Independent verifications of the radiation sensitivity of bone marrow cells lead us to put a good deal of faith in the CFU assay as a real measure of radiosensitivity of bone marrow cells. Isogeneic bone marrow cells can rescue otherwise lethally irradiated mice if an adequate cell number is provided. Using this end point, the same survival parameters have been found for bone marrow cells from irradiated donors. However, it must be realized that the CFU-S is not identical with the primitive bone marrow stem cell, and it represents subpopulations from the maturing and differentiating bone marrow stem cell compartment. The radiosensitivity of 10-day CFU-S does not correlate with the LD_{50} for various mouse strains, and divided dose experiments with CFU-S are not predictive of divided dose LD_{50}'s.

More recently, a wide range of techniques has become available for growing monoclonal cultures of bone marrow stem cells in vitro. All these procedures provide survival parameters about like those found by the CFU assay. Frequently, the monoclonal cultures are seen to have nearly exponential killing, and the extrapolation numbers are often not different from 1.

An interesting but not very widely used version of the CFU assay is the measurement of the radiosensitivity of leukemic- or lymphoma-type cells that will grow as colonies in the spleen after intravenous administration. Bush and Bruce (1964) measured the radiosensitivity of a lymphoma line by this method and found a result that was not different from that found with the Hewitt dilution assay.

As will be seen shortly, the low extrapolation number for bone marrow cells makes this population one of the more radiosensitive of the cell-renewal populations of the mammal. As a result, this organ becomes dose limiting for survival at a much lower dose than for other of the cell-renewal systems to be described that affect survival. Only the spermatogonia are

more sensitive than bone marrow clonogens, and the former are not required for survival.

Gastrointestinal Crypt Cell Assay

The gastrointestinal tract is lined by villi, which are small papillary
processes that project from the base of the intimal layer of the gut. These
villi, which give a vastly expanded surface area to the gut to carry out
its absorptive and secretory processes, are constituted on their outer surface by a classical cell-renewal system that also happens to be geograph-
ically organized. At the base of the villi, in the intimal wall, are the
gastrointestinal crypts (the crypts of Lieberkuhn) in which the stem cells
for this population are situated. The lowest portion of the surface of the
villus is a population of mostly dividing cells. Above it, toward the tip of
the villus, are found regions of dividing and differentiating cells, followed
by a region of fully differentiated and functional cells. Most of the metabolic activity of the villus is found in the distal third of the structure.
In mammals in general, the time for complete turnover of this population,
from crypt cell to tip, is about four days.

It had been known for some time that the gastrointestinal lining
was sensitive to irradiation and that this sensitivity was seen as a massive
loss of fluid and electrolytes from the irradiated intestine. It remained,
however, for Withers and Elkind (1970) to develop a quantitative assay
for the stem cells of this cell-renewal system. If we examine a histologically
prepared cross section of the small intestine (usually the jejunum of the
mouse is the test system), we find various degrees of depletion of the villi
in that section after irradiation. At very high doses, the villi are completely
abolished at about four days after irradiation. Withers and Elkind were
able to detect visible signs of clonogenic activity in the intestinal crypts
if the radiation dose was not so high that the stem cells of the crypts were
completely sterilized. They found that they were able to develop a dose–
response curve for the survival of the clonogenic stem cells of the intestinal
crypts, assuming that repopulation of the crypt stem cells occurred as the
result of remaining clonogenic activity in one or a small number of these
clonogenic cells. The data are amenable to a simple statistical analysis
to correct for the possibility that regeneration of a crypt could have occurred from more than one remaining survivor stem cell.

The criteria for determination of the survivability of clonogenic function in a single crypt are not entirely objective. About $3\frac{1}{2}$ to 4 days after
irradiation, the mice are sacrificed and a section of the jejunum is prepared
by usual histological methods, being sure that sections are as orthogonal
as possible to the long axis of the gut. Examination of these sections
reveals the presence of darkly staining, mitotically active centers where
the crypts are expected. The villi at this time are flattened and denuded

of functional absorptive cells if the radiation dose was large enough. The number of regenerating crypts in a transverse section is recorded, and a correction is made for the possibility that more than a single clonogenic stem cell survived in a single crypt.

The assumption is made in this analysis that cells survive independently and that a single clonogenic survivor cell in the crypt is adequate for its repopulation. If the number of regenerating crypts seen at time of sacrifice is x, and the number of crypts counted per circumference in unirradiated animals is N, then the fraction of crypts sterilized by irradiation, f, is

$$f = \frac{N - x}{N} \tag{10-1}$$

This fraction f represents the fraction of crypts in which there are no surviving clonogens. For the surviving crypts, the number of surviving clonogens could be 1, 2, . . . , n, and the number of surviving clonogens per crypt would be described by the Poisson distribution for which the average number of surviving clonogens per crypt is $-\ln f$. The corrected cell survival per circumference is then

$$f' = N(-\ln f) \tag{10-2a}$$

or

$$f' = N\left(-\ln \frac{N - x}{N}\right) \tag{10-2b}$$

Some are confused by the fact that f' is a number larger than the number of crypts per circumference in a normal, unirradiated jejunal section. The reader is reminded that f' is the number of surviving clonogenic *cells* per circumference, not the number of surviving crypts per circumference.

Radiation Sensitivity of Gastrointestinal Crypt Cells

The gastrointestinal crypt cells are remarkably different from other mammalian cell-renewal systems in their response to irradiation. The original Withers and Elkind report (1970) gave the D_0 for these cells to be approximately 100 rad (1 Gy), not significantly different from that for other cell lines; but the clonogenic cells of the gastrointestinal crypt are unusual in that little effect of irradiation is seen until the dose has exceeded about 900 to 1000 rad (9 to 10 Gy). The extrapolation number has been given by various authors as ranging from 50 to several hundred. The determination of the extrapolation number from a survival curve for acute radiation exposures is not particularly satisfactory. Very small changes in the D_0 will be reflected in very large changes in the extrapolation number, since the shoulder (or D_q) is so very large. The best means to determine the extrapolation number for the gastrointestinal crypt cells is to utilize the Elkind–Sutton repair of sublethal injury method described in Chapter 8. After a conditioning dose and an appropriate time delay for construction of the survival curve after the conditioning dose, the shift in the survival curve can be taken to be a measure of D_q, and the extrapolation number can be estimated from the usual expression: $D_q = D_0 \ln n$. The reader is referred to Chapter 8 for the details. With the method just described, the extrapolation number has been variously estimated to be from 100 to 400. Using an older "macrocolony assay," Withers reported an extrapolation number of 300 for gastrointestinal crypt cells.

The significance of the broad shoulder for survival curves for GI crypt cells is that these cells will have a very large capacity for the repair of sublethal injury. Repair of SLD in this tissue has been variously shown to be nearly complete in 3 to 6 hours after irradiation. Because of the high repair rate, we can see very significant effects of dose–rate on GI injury, since substantial repair may occur during irradiation. The repair of potentially lethal injury (PLDR) in GI crypt cells is difficult to demonstrate in the presence of the high SLDR rates.

The dose modifying effect of oxygen for the GI crypt cells has been difficult to demonstrate. Withers and Elkind's (1970) original paper found no difference for the cell survival parameters for animals breathing air at the time of irradiation and for animals breathing oxygen. The anoxia experiment is difficult to do, since the animals must survive for $3\frac{1}{2}$ days after irradiation. There is little doubt that a dose modifying effect of anoxia is present, but it cannot be easily shown.

The radiobiological characteristics of the GI crypt cell of the mouse have been established by other means than the crypt cell assay by a number of workers. The most commonly used method to assess the response of the gut to irradiation has been to evaluate the doses required to cause death of the animals in 4 to 6 days after irradiation. These deaths are taken to be entirely due to denudation of the gut and, indirectly, to sterilization of the clonogenic cells of the crypts. For 4-day lethality, there is a very large shoulder region, up to about 9 Gy, before any deaths occur, and beyond that dose there is a rapid increase in the death rate, indicating a D_0 about like that found from the crypt cell assay.

We would be mistaken to draw one-to-one correlations between the assay of 4-day lethality and GI crypt cell survival characteristics. Repopulation of the villi from stem cells is by no means complete at 4 days, and the physiological mechanics of the 4-day death are not entirely documented.

Spermatogenesis and in Vivo Assays

The production of mature sperm is the result of cell division and functional maturation in a classical hierarchical cell-renewal system. The transit time from the earliest spermatogonial stem cell to mature sper-

TABLE 10.1   Developmental Stages in Spermatogenesis in the Mouse

Cell Type	Time to Mature Spermatozoon (days)	LD_{50} (rad; cGy)
Type A spermatogonia (type A_s, A_1, A_2, A_3, A_4)	35–45	200 +
Intermediate spermatogonia	32–35	20
Type B spermatogonia	30–35	100
Primary spermatocytes[a]	20–35	
Resting (preleptotene)		200
Leptotene		500
Zygotene		500
Pachytene		
Diplotene		800
Diakinesis		900
Secondary spermatocytes	20–22	1,000
Spermatids	7–20	1,500
Spermatozoa	0–7	50,000

[a] Meiotic stages.

matozoa in man is about 60 days, and in the mouse it is about 40 days. Reduction of the numbers of mature spermatozoa in the ejaculate will not be significant until the number of hierarchical predecessors of the spermatozoon begin to be depleted. The radiosensitivity of the various stages in the stem cell development chain varies widely. Table 10.1 lists the stages of spermatogonial development for the mouse.

The multiplicity of the hierarchical model in Table 10.1 is the source of the complicated nature of the radiation sensitivity of the process of spermatogenesis. Several in vivo assays have been developed, but depending on the stage of advancement of the developmental chain that is most involved in the assay, different results are obtained.

Testes weight loss assay.   Very early in the development of radiation biology, it was observed by several investigators that there was a dramatic weight loss in the testes after irradiation. A number of approaches to developing a quantitative tool for the use of this observation are briefly reviewed by Alpen and Powers-Risius (1981). Kohn and Kallman (1954) were one of the pioneers in this effort, but it remained for J. S. Krebs to appreciate that a true survival curve could be developed from this weight loss data. Unfortunately, Krebs died before publishing this method, but it has been documented by Alpen and Powers-Risius (1981).

This version of the testes weight loss assay is based on the assumption that the loss of weight in this organ after irradiation is made up of two parts. There is a radiosensitive portion of the mass of the testes that is rapidly reduced after irradiation, and the logarithm of this weight loss is related linearly to dose. The remaining fraction of the weight of the testes (somewhat less than half) is taken to be radioresistant, but again it falls with dose as an exponential function.

The change in weight is described by the following function:

$$W_D = W_S e^{-k_s D} + W_I e^{-k_i D} \tag{10-3}$$

where W_D = total weight of the testes after dose, D

W_S = weight of the radiosensitive fraction of the testes at zero dose

W_I = weight of the radioinsensitive fraction of the testes at zero dose

k_s, k_i = inactivation constants for weight loss of the sensitive fraction (k_s) and the insensitive fraction (k_i) after irradiation

The data in Figure 10.8 are calculated for typical data published by Alpen and Powers-Risius using this method. The D_0 is taken as 90 rad (90 cGy) for the radiosensitive fraction ($k_s = 1/D_0 = 0.01111$), and W_S is taken as 150 mg. The D_0 for the insensitive fraction is taken as 1800 rad (18 Gy) ($k_i = 1/D_0 = 0.000556$), and W_I is 110 mg.

Testes radiosensitivity by weight loss methods.   Whether the more sophisticated methods just described or the simple weight loss methods of Kohn and Kallman (1954) are used, the conclusions are much the same. The insensitive fraction of the mass of the testes is made up of the Sertoli and connective tissue cells surrounding the tubules. These cell types are radioresistant, with a D_0 of 1800 rad (18 Gy) or more, and this observation is consistent with the radiosensitivity of a generally nondividing cell population.

The weight loss described as the radiosensitive fraction represents the loss of mature spermatozoa and some of their predecessors, but these cells are nondividing and maturing forms that are radioresistant. The weight loss represents the cessation of formation of primary and secondary spermatocytes and spermatozoa from, probably, type B spermatogonia. This latter conclusion may be drawn from the fact that the maximum weight loss occurs at about 28 days after irradiation, representing the transit time from type B spermatogonia to spermatozoa in the mouse. Supporting this conclusion would be the fact that the type B cells account for most of the multiplication in the production pathway and, therefore, would represent the bulk of the weight loss.

The D_0 for type B spermatogonia can then be deduced from the weight loss curves to be on the order of 80 or 90 rad (80 to 90 cGy). There is no shoulder to the survival curve derived by this method, and this is con-

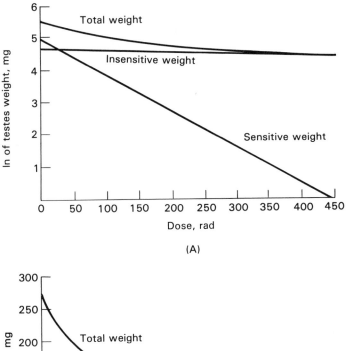

Figure 10.8 Plot of the weight of the mouse testes 28 days after irradiation according to the method of Krebs as described by Alpen and Powers-Risius (1981). Panel A is a plot of the natural logarithm of the total testes weight and the derived lines for the insensitive weight and the sensitive weight components obtained by fitting a biexponential expression and assuming that the dose dependences of both the sensitive and insensitive weights are exponential. Panel B is the same data plotted in arithmetic form after derivation from Panel A.

sistent with the observation that protraction of the exposure or fractionation has no effect on the outcome if the period of exposure is short enough
to preclude cellular proliferation during exposure. The conclusion, then,
is that this cell type is not capable of significant postirradiation repair.
Many different studies, including the early studies of Kohn and Kallman,
have come to the same conclusion that dose rate or fractionation does not
modify the response to irradiation. There is, however, an important exception to this generalization, as follows.

Tubular regeneration clonogenic assay.   The general theme of the
crypt cell clonogenic assay was utilized by Withers and his colleagues
(Withers and others, 1974). The survival curve for clonogenic cells in the
tubules was estimated by examining microscopically the tubules of
the irradiated testes at 35 to 40 days after regeneration and scoring the
percentage of the tubules that showed evidence of repopulation with spermatogonial elements and their postcursors. Using this end point, these
authors found that there was little effect on the tubular regenerative
activity from doses up to 800 rad (8 Gy), and at doses above this there
was an exponential survival curve with a D_0 of about 180 rad (1.8 Gy).
There was still significant tubular regeneration with doses as high as
1300 rad (13 Gy). Also, these studies showed a significant repair capacity
for whatever stem cell system of the spermatogonial hierarchy was being
tested.

Summary of the Radiosensitivity of the Testes

Much of the conflicting data regarding the cellular radiosensitivity
of the spermatogonial hierarchy can be understood in light of the data
presented in the last column of Table 10.1. These data are reported LD_{50}'s
for various elements of this hierarchy as measured by Oakberg and Clark
(1961), using cytological criteria. These data are probably as reliable as
any available today. As expected, the nondividing, differentiating late
forms, such as the spermatids and the spermatozoa, are extremely resistant to cell killing by irradiation. With the exception of the type A_s, the
most primordial stem cell, which has an LD_{50} of about 200 rad (2 Gy),
the stem cells of the spermatogonial series tend to be very radiosensitive.
The intermediate spermatogonia are the most radiosensitive, with an LD_{50}
of about 20 rad (20 cGy). It is not always clear which of these cell types
are involved in any given assay. The late spermatogonia, as mentioned
earlier, are probably the ones most represented in the survival curves
determined by the weight loss method. If so, these are the stem cells with
little or no repair capacity, as determined by fractionation studies. Withers
and others (1974) suggest that the tubular regeneration assay represents
the survival of the earliest and most primordial of the stem cells. We are

tempted to suggest, however, that because of the very high radio-resistance found by them, some of the later, very resistant forms might be involved. In any case, there is evidence of repair capacity for whatever cell line is represented in their studies. It is possible that the radiosensitivity of the earliest cell is being measured by this latter method, but that a very small growth fraction for this population provides extra radio-resistance.

In the human male, sterility lasting for several years may be induced by doses on the order of 250 rad (2.5 Gy), and permanent sterility will result after doses of about 600 rad (6 Gy) to the testes.

Assays for the Radiosensitivity of Skin

The remaining cell-renewal system of the mammal to be discussed is the skin. The epidermis of all mammals is a cell-renewal system for which the stem cell is the cell found in the basal layer of the epidermis. As for all hierarchical systems, the stem cell provides a population of dividing and differentiating cells, and in the case of skin, these cells move outward and ultimately undergo keratinization and are lost from the scaly outer layer of the epidermis.

Withers (1967) was again responsible for developing a clonogenic assay for in situ epidermis. This assay is the first of his several clonogenic in vivo assays. The technique for the assay is carried out by isolating a small island of epidermis by means of a heavily irradiated (3000 rad; 30 Gy) moat of epidermis surrounding a central island that is shielded while the moat is irradiated. The result is an annular region, in which skin destruction is complete, surrounding a central island to which various radiation doses can be given. If graded doses are used, it is possible to determine the dose for which nodular regrowth occurred in 50% of the islands irradiated. This is taken to be the dose associated with the survival of one clonogen. Withers' method finally plots the number of surviving cells per square centimeter of skin, and a survival curve is constructed.

Since the resulting curve is not truly a survival curve, we cannot directly measure the shoulder, or D_q, but it can be measured indirectly by the divided dose technique. For this method, split doses with a separation of 24 hours are used. The D_q is the separation between the survival curves for a single dose and that for a split dose regimen. Withers found the D_q to be about 350 rad (3 Gy) and the D_0 to be about 135 rad (1.35 Gy). Using the expression, $D_q = D_0 \ln n$, it is possible to estimate the extrapolation number n. For his original data, Withers found an extrapolation number of 12.

Radiosensitivity of skin.   Basal cells of the epidermis have a D_0 that is not very different from other mammalian cell lines, in vivo and in vitro.

Associated with the extrapolation number of 12 is a good deal of radio-protection with fractionation or low dose rate.

Because it is relatively easy to develop a scoring method for the destruction of the epidermis and since the tissue is conveniently available, skin has become a widely used tool for radiobiologists to look at all manner of dose modifying end points.

RADIATION EFFECTS ON THE EMBRYO AND FETUS

The next normal tissue system to be considered is the growing embryo and fetus. Growth is associated with continuing cell division, and presumably the embryo and fetus would have a radiation sensitivity much like that of other dividing tissues. The important difference is that these are not steady-state dividing tissues, with a resting or G_0 cell population that can be called on for restitution of a cell population depleted by irradiation; thus there is no possibility for compensatory cell replacement, and fetal death or developmental abnormalities may result.

A clear distinction must always be made between those alterations from normal seen in the term fetus that can be attributed to genetic change in the gametes prior to fertilization and changes that occur in the growing new organism. The mutational effects are heritable in the offspring of the affected individual and are stochastic in nature. That is, there is presumably no dose threshold below which mutational effects will not occur, even though the probability is low at low doses. For the nongenetic changes seen in the term fetus, they are the result of cell killing in the primitive anlage, are not heritable, and have a dose threshold related to the usual survival curve of mammalian cells.

In human populations, major congenital anomalies that have no heritable genetic basis are found in 5% to 8% of the population of newborn children. The radiation induced anomalies must be recognized against this background of naturally occurring defects. For both naturally occurring and radiation induced birth abnormalities, it is a possibility that they may be the result of somatic mutations, leading to a defective developmental cell line.

Since congenital abnormalities have been reported for human beings and rodents at doses as low as several rad or centigray, the question must be posed as to whether there is a peculiar radiosensitivity of the dividing cells making up the growing organism in utero. The answer is that there is no need for a special radiosensitivity of these cell lines to understand the observed results. If the population of cells responsible for embryonic and fetal development is small, for example, two or four cells as in the newly fertilized ovum, or if only a few dozen cells must be the origin of complicated fetal development, then the death of only a small fraction of these cells will dramatically influence the developmental outcome.

Consider the mathematics of cell survival if the anlage cells are typical of other mammalian cell populations. Assume, for example, an extrapolation number of 4 and a D_0 of 100 rad. What would be the probability of killing one cell responsible for a developmental program if it were irradiated with a dose of 50 rad (50 cGy)?

$$\frac{S}{S_0} = 1 - (1 - e^{-D/D_0})^n = 1 - (1 - e^{-50/100})^4 = 0.976$$

The chance then of killing this single cell is 0.024; or, to state it in another way, if this cell were responsible for a given developmental process, on average slightly more than 2 out of every 100 individuals irradiated in this way would have the given developmental abnormality. With the same parameters and a dose of 10 rad (10 cGy), the probability would be 1 in 10,000, still an outcome for a human population that is probably of some concern.

These calculations give some insight into the mechanisms and the timing of the production of developmental abnormalities as the result of radiation exposure. The timetable of embryonic and fetal development is complicated and detailed. At various times during prenatal development there will be crucial times when only a few cells will be responsible for the ultimate development of a complete organ system or phenotypic trait. The destruction of a few cells at a critical time may lead to partial or total disruption of the developmental calendar.

Developmental Sequence and Radiation Effects on Prenatal Development

The most detailed studies of the effects of irradiation of the fetus and embryo are those of Rugh (1964). These data are old, but they are still of great significance, and little except detail has been added by recent work. His model for study was the mouse, and Figure 10.9 is the schematic diagram offered by him in summary of the results. The stages of intrauterine development are conveniently described as (1) preimplantation, the period before the fertilized ovum has attached to the uterine wall; (2) embryogenesis, the period of organ differentiation and development, sometimes referred to as organogenesis; and (3) the period of fetal development, mostly a time of growth and elongation, but also a time in which some late organogenesis is taking place. Each of these periods is characterized by a somewhat different evident response to irradiation.

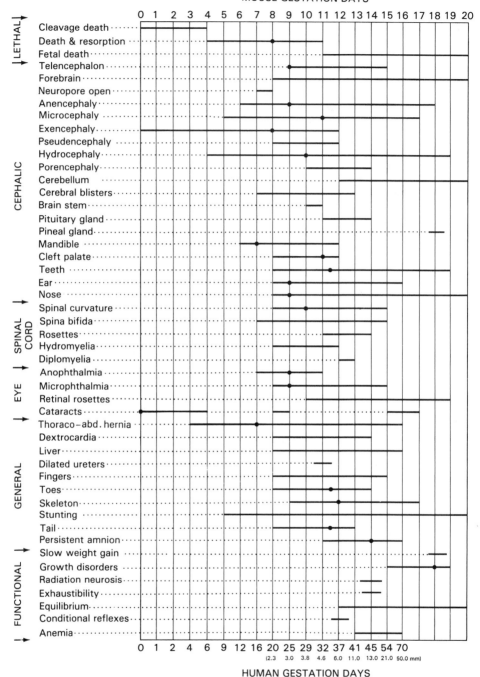

Figure 10.9 The periods of maximum radiation sensitivity of each of the developmental episodes from conception to term for the mouse and the type of congenital abnormality resulting. The large dot on the line represents the period of maximum sensitivity. At the bottom is shown a scale of the relative times during development of the human embryo and fetus. (From Rugh, 1964; reprinted from *Radiology* by permission of the publisher)

Preimplantation period. During this period the fertilized ovum completes its descent through the fallopian tube to arrive in the uterus. Prior to implantation in the uterine wall, with concomitant formation of the placenta, the fertilized ovum will undergo several divisions, but the loss of only one or two of these cells to irradiation will cause the loss of viability of the product of the conception. The detectable result of irradiation of the early, preimplantation embryo will be a reduction in the litter size for animals usually having multiple births. Careful study of the dissected uterus can lead to detection of early implantation sites where the embryo failed to continue its growth after implantation. If the embryo fails to implant before its death, there is no visible sign of the loss. In the human being, these losses in the preimplantation period could not reasonably be expected to be detected.

Doses on the order of 10 rad (10 cGy), delivered to mice on the first day of gestation, can lead to death of the fertilized ovum, and an increased number of vacant resorption sites has been seen in the rat for doses in the same range. Russell and Russell (1954) showed that a dose of about 200 rad delivered in the first day or two after conception in mice could lead to a prenatal (preimplantation) death rate on the order of 80%.

Congenital malformations only rarely result from irradiation of the embryo at these early preimplantation times. Either the embryo survives or it dies. The rare exception to this rule is that occasionally Rugh observed exencephaly resulting from irradiation in the first several days after conception.

Period of organogenesis. This period of embryonic life, which in the mouse extends roughly from the 6th to the 13th postconception day and in the human being extends from about the 9th or 10th to about the 40th day postconception, is the period when development and differentiation of all the organ systems that will be present in the term fetus commence. Following the reasoning given at the start of this section, since each of the organogenetic phases commences with a primitive anlage of only a few cells, and with no backup, replacement stem cell population, loss of only a few cells at critical times will lead to congenital abnormalities.

Following the schema laid down by Rugh as shown in Figure 10.9, each organ system, in turn, becomes susceptible to radiation induced abnormalities as that system enters into its development phase.

The earliest developing elements that are susceptible to induction of abnormalities by irradiation are the elements of the central nervous system as shown in Figure 10.9. As can be seen, essentially all developmental characteristics can be altered by irradiation at appropriate times. The later stages of organogenesis involve mostly skeletal elements, as well as the continuing development of some of the elements with more protracted periods of growth and development, such as the eye, the teeth, and some external features. Russell and Russell (1954) state that a dose of 100 rad (1 Gy) can induce abnormalities in essentially all embryos irradiated during organogenesis. It has been variously reported that doses in the range of 15 to 20 rad (15 to 20 cGy) can be seen to produce significant numbers of abnormalities. In the mouse the period of greatest sensitivity is during days 9 to 12 after conception. For the human being, from a comparison of the organogenetic calendar, we might predict that the comparable time of maximum sensitivity might be during the fourth or fifth week of pregnancy. It is worth noting that, for human pregnancy, detection of pregnancy at this early time requires an alert and observant female.

Prenatal or neonatal death can result from malformations that are not consistent with continued life, and these events can occur with a high frequency with doses on the order of 100 to 200 rad (1 to 2 Gy) during organogenesis. Another sequela of the developmental defects is low birth weight resulting from cell loss during development.

Stage of fetal growth. During this stage of pregnancy, growth and elongation are the principal activities. Some organs complete their development late in the fetal stage, and it is possible that some developmental abnormalities might result from irradiation at this time. Only large doses can cause neonatal death, and a more likely outcome of irradiation during the fetal stage is functional disorders that are not easily detected. Associated with fetal irradiation are some retardation of growth, and a particular trait found in human babies irradiated in utero is a smaller head size. This effect is detectable for doses of 100 rad (1 Gy) or less.

REFERENCES

ALPEN, E. L., and POWERS-RISIUS, P. (1981) The relative biological effect of high-Z, high-LET charged particles for spermatogonial killing. *Radiation Res.* **88,** 132–143.

ANDREWS, J. R., and BERRY, R. J. (1962) Fast neutron irradiation and the relationship of radiation dose and mammalian cell reproductive capacity. *Radiation Res.* **16,** 76–81.

BECKER, A. J., McCULLOCH, E. A., and TILL, J. E. (1963) Cytological demonstration of the clonal nature of spleen colonies derived from transplanted mouse marrow cells. *Nature* **197,** 452–454.

BERRY, R. J. (1964) On the shape of x-ray dose–response curves for the reproductive survival of mammalian cells. *British J. Radiol.* **37,** 948–951.

BUSH, R. S., and BRUCE, W. R. (1964) The radiation sensitivity of transplanted lymphoma cells as determined by the spleen colony assay method. *Radiation Res.* **21,** 612–621.

DETHLEFSEN, L. A. (1980) In quest of the quaint quiescent cell. In *Radiation Biology in Cancer Research*, R. E. Meyn and H. R. Withers, eds. Raven Press, New York, pp. 415–435.

FERTIL, B., and MALAISE, E. P. (1981) Inherent cellular radiosensitivity as a basic concept for human tumor radiotherapy. *Intl. J. Radiation Oncol. Biol. Phys.* **7**, 621–629.

———. (1985) Intrinsic radiosensitivity of human cell lines is correlated with radioresponsiveness of human tumors: analysis of 101 published survival curves. *Intl. J. Radiation Oncol. Biol. Phys.* **11**, 1699–1707.

GILBERT, C. W., and LAJTHA, L. G. (1965) The importance of cell population kinetics in determining the response to irradiation of normal and malignant tissue. In *Cellular Radiation Biology*. Williams and Wilkins, Baltimore, pp. 474–497.

HEWITT, H. B., and WILSON, C. W. (1959) A survival curve for mammalian cells irradiated in vivo. *Nature* **183**, 1060–1061.

HILL, R. P., and BUSH, R. S. (1969) A lung colony assay to determine the radiosensitivity of the cells of a solid tumor. *Intl. J. Radiation Biol.* **15**, 435–444.

HLATKY, L., SACHS, R. K., and ALPEN, E. L. (1988) Joint oxygen–glucose deprivation as the cause of necrosis in a tumor analog. *J. Cell Physiol.* **134**, 167–178.

KOHN, H. I., and KALLMAN, R. F. (1954) Testes weight loss as a quantitative measure of x-ray injury in the mouse, hamster and rat. *British J. Radiol.* **27**, 586–591.

KOLSTAD, L. (1964) *Vascularization, Oxygen Tensions and Radiocurability in Cancer of the Cervix*. Scandinavian University Books, Stockholm.

LAJTHA, L. G. (1963) On the concepts of the cell cycle. *Cell Comp. Physiol.* **62**, 143–145.

MENDELSOHN, M. L. (1962) Autoradiographic analysis of cell proliferation in spontaneous breast cancer of C3H mouse. III. The growth fraction. *J. Natl. Cancer Instit.* **28**, 1015–1029.

OAKBERG, E. F., and CLARK, E. (1961) Effect of dose and dose-rate on radiation damage to mouse spermatogonia and oocytes as measured by cell survival. *J. Cell Comp. Physiol.* (Suppl. 1) **58**, 173–182.

RODRIGUEZ, A., and others (1988) Recovery from potentially lethal damage and recruitment time of noncycling clonogenic cells in 9L confluent monolayers and spheroids. *Radiation Res.* **114**, 515–527.

RUGH, R. (1964) Why radiobiology? *Radiology* **82**, 917–920.

RUSSELL, L. B., and RUSSELL, W. J. (1954) An analysis of the changing radiation response of the developing mouse embryo. *J. Cell Physiol.* (Suppl. 1) **43**, 103–149.

STEEL, G. G. (1977) *Growth Kinetics of Tumors*, Chapter Two, Basic theory of growing cell populations. Oxford University Press, New York.

TANNOCK, I. F. (1968) The relation between cell proliferation and the vascular system in a transplanted mouse mammary tumor. *British J. Cancer* **22**, 258–273.

THOMLINSON, R. H. (1967) Oxygen therapy—biological considerations. In *Modern*

Trends in Radiotherapy, T. J. Deeley and C. A. P. Wood, eds. Butterworth, London, pp. 52–72.

———— and GRAY, L. H. (1955) The histological structure of some human lung cancers and the possible implications for radiotherapy. *British J. Cancer* **9,** 539–549.

TILL, J. E., and McCULLOCH, E. A. (1960) The radiation sensitivity of normal mouse bone marrow cells determined by quantitative marrow transplantation into irradiated mice. *Radiation Res.* **13,** 115–125.

WITHERS, H. R. (1967) The dose–survival relationship for irradiation of epithelial cells of mouse skin. *British J. Radiol.* **40,** 187–194.

————, and ELKIND, M. M. (1970) Microcolony survival assay for cells of mouse intestinal mucosa exposed to radiation. *Intl. J. Radiation Biol.* **17,** 261–267.

————, and others (1974) Radiation survival and regeneration characteristics of spermatogenic stem cells of mouse testes. *Radiation Res.* **57,** 88–103.

SUGGESTED ADDITIONAL READING

ELKIND, M. M., and WHITMORE, G. F. (1967) *The Radiobiology of Cultured Mammalian Cells*. Gordon and Breach, New York (in particular, Chapters 4 and 7).

MEHN, R. E., and WITHERS, H. R., eds. (1979) *Radiation Biology in Cancer Research*. Raven Press, New York.

STEEL, G. G. (1977) *Growth Kinetics of Tumors*. Oxford University Press, New York.

THAMES, H. D., and HENDRY, J. H. (1987) *Fractionation in Radiotherapy*. Taylor and Francis, London.

PROBLEMS

1. The Hewitt dilution assay was carried out with a mouse lymphoma line that grows well as a single-cell suspension. The following table lists the mortality rate (percent dead) in groups of 10 mice per data point that received the injected number of lymphoma cells shown.

Dose, rad					
0		100		200	
Cells Inj.	% Dead	Cells Inj.	% Dead	Cells Inj.	% Dead
2	0	6	2.5	8	12
4	14	8	20	12	32
6	42	10	30	14	52
8	75	12	55	16	60
10	100	14	70	18	70
12	100	18	100	—	—

Dose, rad					
400		500		600	
Cells Inj.	% Dead	Cells Inj.	% Dead	Cells Inj.	% Dead
12	0	50	10	130	0
14	6	60	25	140	10
18	40	65	50	150	42
20	49	70	52	160	51
22	70	75	69	170	80
26	100	80	95	180	100

Dose, rad	
800	
Cells Inj.	% Dead
400	0
450	5
500	17
550	30
600	51
650	70

Given these data, construct the curves for animal lethality versus dose of injected cells for each of the irradiation dose levels given. Construct the derived survival versus dose curve for the irradiation of this line of lymphoma cells. Estimate the D_0 and the extrapolation number n for this cell line. Using the linearized form of the linear–quadratic expression, plot the survival curve in L–Q format. In your opinion, which is the better fit for the data?

2. The gastrointestinal crypt assay was carried out on a group of mice with the following results. Five mice were irradiated at each of the dose points. The animals were sacrificed at 4.5 days after irradiation, and the average number of surviving crypt cells per jejunal circumference was calculated from the individual data.

Dose, rad	Surviving Crypts	Dose, rad	Surviving Crypts
900	93	1250	16
950	67	1300	11
1000	61	1400	12
1100	61	1500	3.4
1150	38	1600	3
1200	32		

The average number of crypts per circumference in the unirradiated animals was 120.

Calculate the corrected number of surviving clonogens per circumference and plot the data. Assuming MTSH kinetics of killing, estimate the D_0 for the cells.

To estimate the extrapolation number, another experiment was conducted by irradiating the animals with 1000 rad, and after 12 h a series of additional graded doses was given and the surviving crypts were estimated as before. The data are as follows:

Dose, rad	Surviving Crypts	Dose, rad	Surviving Crypts
900	112	1250	12
950	70	1300	11
1000	62	1400	9
1100	54	1500	4
1150	32	1600	4
1200	32		

The control crypt number was the same as in the earlier part of the problem.

Calculate the corrected number of surviving clonogens per circumference and plot the data in the typical Elkind–Sutton fashion. Assuming MTSH, estimate D_0 for the reirradiated cells and estimate the extrapolation number for the gastrointestinal crypt clonogens.

11

Late Effects
of Radiation
on Normal Tissues:
Nonstochastic Effects

INTRODUCTION

In Chapter 10, all discussions of the effects of ionizing radiation on normal and neoplastic tissues were centered on the prompt responses observed within a few days to a few weeks of irradiation. Generally, all these responses could be related to depletion of clonogenic cellular elements that served as stem cells for the normal cell renewal of the population of cells at risk.

The terms stochastic and nonstochastic will be defined in detail in the next section of this chapter, but for the moment it may be generalized that *nonstochastic effects* are those for which there is a dose threshold, while *stochastic effects* may be taken to be those for which there is no known threshold dose. The early responses described in Chapter 10 are clearly nonstochastic, since they have a dose threshold and they also have describable dose–response relationships related to the underlying survival curve of the clonogenic cells. There is another class of responses of normal tissues that is also nonstochastic. These are responses of the tissues that can begin to be first seen many weeks to many months after the radiation exposure. The doses required to elicit these late responses may frequently be larger than those associated with most of the early, acute responses, and for this reason they are most often encountered with regional radiation of the type administered in radiotherapy for malignant or other disease. The animals or patients would not normally survive total body radiation doses at the dose levels usually associated with the onset of effects in these late responding tissues. Because of the association of late effects in normal tissues with typical radiotherapy regimens, these end points have been a great concern mostly to the radiotherapists and experimentalists supporting this particular clinical application of radiation.

In spite of this specialist approach of the clinical and experimental radiooncologists to the late nonstochastic effects of radiation in normal tissues, there is much of a fundamental nature to be learned from studying these end points. Because of the generally higher doses usually associated with these late changes, there has been much speculation as to their pathogenesis. We must point out, however, that late effects have frequently been studied for their importance in limiting radiotherapeutic regimes, so the emphasis has been on the higher doses associated with these regimes. One central question about late effects is whether simple cellular depletion of clonogenic cell lines could produce a resulting lesion many weeks or months after exposure, but no completely satisfactory models have been formulated to date. Two approaches to explanations of the underlying cause of late effects from ionizing radiation have received a great deal of recent attention. One group emphasizes cellular depletion (target cell depletion) of clonogenic elements or functional elements that require replacement from clonogenic pools. Another group suggests that the central process underlying all late effects in organ systems is damage to the vascular endothelium, leading to vascular insufficiency.

In the *Proceedings of the Twelfth L. H. Gray Conference*, which took place in 1985, Hendry (1986) keynoted the conference by summarizing the questions that needed to be addressed in understanding the pathogenesis of late effects in normal tissue, whatever the mechanism. These are worth repeating here.

1. Can target cell populations be defined in the tissue for specific types of injury?
2. Are there qualitative differences in the target cell populations for different cytotoxic agents, for example, resistant subpopulations?
3. What is the cellular basis to the volume effect (that is, greater injury per unit volume with increasing volume of tissue treated)?
4. What is the contribution of intracellular repair of potentially lethal damage to the sparing of injury in late responding tissues?
5. What are the contributions of repopulation by colony forming cells and cells with very limited division potential to the sparing of injury in the late responding tissues?
6. How well is the functional (effector) cell number correlated with overall *capacity* for function?
7. What is the contribution of stromal (including vascular) injury in the response of the parenchyma?

All these questions have been addressed through experimental studies of the various normal tissue systems, but, as yet, no clear answer has developed to any of them.

STOCHASTIC VERSUS NONSTOCHASTIC EFFECTS

In Chapter 10, it was generally possible to describe the dose–response relationships for these acute effects in the usual pharmacological terms of a sigmoid dose response curve with some dose below that for which we would not expect any effect in any exposed individual (the *threshold*). These same threshold-type dose–response relationships hold for the responses of late responding tissues also; but while the acute effects can almost always be directly related to the survival of clonogenic cells, the underlying mechanism for the delayed and late effects is not at all clear.

There is another group of responses, genetic and oncogenic changes arising as the result of irradiation, for which the concept of a threshold is less applicable. For these latter responses to ionizing radiation, the scientific community has come to believe that changes can occur at any exposure level; even though for low doses the frequency of the change in an exposed population might be thought to be low, the frequency of occurrence is still not zero. There is still a great deal of discussion as to the actual shape of the dose–response relationship for cancer and genetic damage, but there is no longer any argument that these are indeed effects without a dose threshold after exposure to ionizing radiation. We might postulate that there is a fundamental difference between these two types of responses. The nonthreshold responses, such as carcinogenesis or genetic mutations, probably require only the alteration of one or, at most, a few cells to bring about the response. For the threshold responses it is probably necessary to alter the function of many cells, and destruction or alteration of less than this minimum number of cells leads to the threshold nature of the response.

Because of the fundamental difference in the effects of ionizing radiation in the two classes of responses, the radiation protection community has devised the terms *stochastic* and *nonstochastic* to describe each class of effects (International Commission on Radiological Protection, 1977). Effects for which the dose–response relationship is thought to have a dose threshold and for which the expectation is that any subject exceeding this threshold dose is likely to experience the effect, subject to the usual biological variability, are defined as nonstochastic effects. Another characteristic of the nonstochastic class of effects is that there is usually a severity scale associated with dose, and increasing the dose to an individual subject is expected to increase the severity of the outcome for that subject. Stochastic effects, on the other hand, are thought to be effects for which there is no threshold for the response and for which the severity of the response is independent of the dose, that is, "all or nothing." The two effects for which the stochastic definition is appropriate are the production of cancer and the induction of defects in the genetic substance of the germ cells. These two end points will be described in detail in successive chapters.

An important inference is associated with the terms stochastic and nonstochastic. The definition of stochastic is that it describes a process that contains an element of chance in the outcome or, in other words, predictions about the process are developed on a stochastic or probabilistic base. In biological terms, we might best describe the nonstochastic class of effects as those that require the reduction of function in some class of cells that is essential to maintain a certain function, and if such a reduction is indeed achieved, then the expected altered or diminished function will be revealed. The intensity of the response might be expected to be associated with the extent to which the functional population of cells is diminished in an individual as the result of the irradiation. For stochastic effects, we envision an event, or a chain of events, in a single cell that will ultimately lead to an irreversible alteration.

The adjective stochastic can be taken to mean that, among any very large population of cells, there is a probabilistic expression for the likelihood of any single cell being transformed into a potentially clonal line for some new trait. Presumably, this trait may also be passed on to any progeny cells as a permanently heritable trait, while for the nonstochastic processes such a heritability is not a necessary condition even though somatic mutations will exist.

This chapter deals with the nonstochastic late effects, while future chapters will examine the two important stochastic effects, carcinogenesis and heritable changes in the genome of the germ cells.

RADIATION-INDUCED LATE PATHOLOGY IN ORGAN SYSTEMS

A complete and detailed examination of the late delayed functional alterations and histopathology of all organ systems that have been exposed to ionizing radiation is beyond the scope of a text emphasizing the biophysical aspects of radiation effects. The purpose of the descriptive detail that is provided is to lay the groundwork for the examination of the extensive literature on the modeling of dose and dose-rate effects on these late changes. This will be the principal goal. For an extensive documentation of the late effects of ionizing radiation, the readers are referred to the definitive texts by Rubin and Casarett (1968) and Casarett (1980). See Suggested Additional Reading.

Before beginning an organ by organ discussion of the late effects of radiation, it is worth a moment to outline the mechanistic model first proposed in a formal way by Casarett (in Creasey, Withers, and Casarett, 1976). It should be pointed out that the model proposed by Casarett deals almost entirely with one of the two possible bases for late effects in normal tissue. This model suggests that the larger part of the late effects syndrome may be caused by an alteration of the microvasculature of individual organs, which includes a powerful feedback loop promoting fibrosis and vascular insufficiency. This model is described in Figure 11.1. The cellular

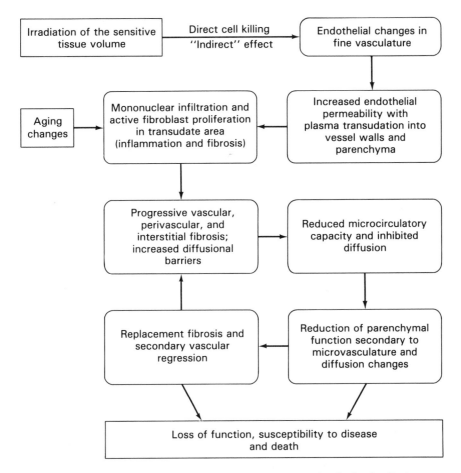

Figure 11.1 The model of Casarett proposing the generalized role of radiation damage to the microvasculature as the underlying cause of late tissue damage. Note the cyclic effect of fibrosis and secondary vascular regression of radiation induced early fibrosis and vascular change. (Modified from W. Creasey, H. R. Withers, and G. Casarett, 1976, *Cancer* **37,** 999–1013, with permission of the author, the American Cancer Society, and J. B. Lippincott Co.)

depletion (or target cell depletion) model will be discussed in upcoming sections of this chapter.

Since, as alluded to earlier, late effects are seen in organs in which there is little or no normal proliferative activity, we must search for a mechanism that does not depend on the killing of mitotically active cells or one in which some indirect effector mechanism is responsible. The model of Casarett, which has wide acceptance, particularly among pathologists in the field, suggests that the fundamental mechanism for late damage centers around changes to the microvasculature within the organ. The endothelial cells lining the microvasculature are indeed part of a cell-renewal system in which depletion of clonogenic potential might lead to the diminished function, although Casarett does not explicitly deal with the importance or lack of importance of the destruction of cells within a cell-renewal system of the vascular endothelium. As shown in Figure 11.1, the first result of irradiation is damage to the endothelium of the microvasculature of the organ (principally the capillary network and the smallest vessels feeding them). This damage can either be direct, in which case it may be presumed that the radiation is causing interphase death of essential cells making up the endothelial wall, or indirect. The indirect mechanisms suggested by Casarett and others is via obstruction of the endothelial lumen by radiation induced cellular proliferation or the formation of thrombi, or possibly by reduced functional integrity resulting from cell loss. The root cause of the cellular proliferation can be presumed to be due to either interphase death of functional cells (less likely) or to loss of clonogenic cells responsible for maintenance of the functional population.

The model proposes a series of steps, starting with increased vascular permeability of the microvessels with leakage (transudation), followed by classic inflammatory changes leading to progressive permanent changes in the endothelial tissue. These changes include fibrosis and wall thickening that cause reduced blood perfusion and increased barriers to trans-vessel wall diffusion and osmotic activity. Once this process starts, the degeneration of the vascular system becomes cyclically worse. Fibrosis and lack of perfusion lead to further fibrosis, functional loss, and ultimately severe functional impairment.

To what extent is this model universally applicable? The question remains open at this time, although it will be seen that, at least in some organs, microvascular degeneration can be only partly responsible for the functional defect seen. Examples of these might be the spinal cord, in which demyelination resulting from the killing of stem cells might account for most of the functional loss, and the kidney, where early functional changes in the tubular epithelium appear to precede any vascular change. The exceptions do not, however, invalidate the model, which seems to be entirely adequate to explain many of the late postirradiation findings in mammalian organ systems. It is also equally likely that cellular depletion of important functional elements also plays a role in this complex picture of cyclical loss of perfusion and fibrosis.

It might be said that the changes in the vascular endothelium are a special case of cellular depletion related to killing of clonogenic progenitors, and that this depletion and resulting change in the vessels simply happens to be the most sensitive or most critical change in many organs.

LATE EFFECTS IN NORMAL TISSUE SYSTEMS AND ORGANS

Gastrointestinal Tract

The late radiation pathology of the gastrointestinal system can best be considered by its separate parts. These are esophagus, stomach, small intestine, large intestine, and rectum.

Esophagus. The esophagus is a tubular structure of striated muscle, a many layered, interior squamous epithelium, and extensive loose connective tissue. It has contractile activity that arises through action of the striated muscle components of the tubular structure. The squamous epithelial lining is a relatively rapidly turning over tissue that is generally radiosensitive, with the outcome after high doses being early denudation of the interior surface of the structure. This relatively prompt response of the epithelial lining might properly be listed as an acute response of the tissue, since the resulting ulcerative esophagitis arises in the second to the fourth week after irradiation in mice and in the first postirradiation month for man. The dose required for the onset of the esophagitis has been shown to be about 2400 rad (24 Gy), a very large dose compared to that necessary to bring about depletion of the cell-renewal potential in the tissue systems described in Chapter 10. This dose estimate for esophagitis was developed by Michalowski and others (1983), who also demonstrated that a tissue volume effect was detectable. Doubling the size of the section of esophagus irradiated reduced the effective dose required for ulceration.

Significant late injury to the esophagus is an important obstacle to radiotherapeutic treatment of malignant disease in this organ and in nearby structures. The important pathology arising some months after irradiation of the esophagus is stricture of the organ (stenosis), and this can be so severe as to cause complete loss of the lumen, with concomitant inability to pass food to the stomach. The change in the organ, as far as can be determined experimentally and from human pathology, is not the result of altered function in the epithelial layer, but is rather due to impaired function in the striated muscle and connective tissue, with a probable significant source of the difficulty being due to impaired function in the vascular supply to the tissue.

Stomach. The interior surface of the stomach, in its transition from the stratified epithelium of the esophagus, becomes a simple columnar epithelium richly endowed with secretory elements at various portions of its surface. The postmitotic cells of the functional columnar epithelium are relatively radioresistant, as one would predict, but the stem cells located deeper in the gastric pits are actively mitotic, replacing the surface columnar epithelium in a regular way. These progenitor cells are indeed radiosensitive. An analogy can be drawn to the crypts of Lieberkuhn of the small intestine, which were described in some detail in Chapter 10. The stem cells of the stomach appear to have a similar dose–response characteristic to the crypt cells of the small intestine. Early responses of the stomach lining are much governed by the changes occurring as the result of cytocidal effects of radiation on the mitotically active precursor cells. The stomach returns quickly to normal after moderate doses of radiation, although secretory depressions may be detected for months to years after irradiation.

The late effects in the stomach are predominantly those that might arise as the result of vascular insufficiency. There is a progression from interstitial fibrosis with arteriocapillary fibrosis to extensive and widespread fibroatrophy. The late syndrome for the stomach is described as fibroatrophy of the gastric mucosa, impairment of motility, loss of distensive ability, and outright stenosis.

Small and large intestines. The small and large intestines are anatomically similar except for the extensive proliferation of the microvilli of the small intestine. The large intestine is without villi, and, except for the goblet cells that are the mucus secreting cells of the large intestine, it is a smooth surface. There is an extensive submucosal connective tissue and smooth muscle development in both small and large intestines, which is richly supplied with vascular elements.

The early effects of irradiation of the small intestine have been examined in great detail in Chapter 10. These acute effects will not be reexamined here, except to report that, as we might expect, the large intestine is not involved in the acute changes induced by irradiation because of the lack of the rapidly turning over crypt cells found in the small intestine.

Vascular and connective tissue changes are predominant in the late responses of all portions of the intestine. The earliest visible precursor of the late effects in the intestine is the alteration of endothelial cells in small vessels, accompanied by extensive thrombosis. All these changes in the microvasculature are precursors to the familiar late syndrome outlined for other parts of the gastrointestinal tract. These changes are fibrotic thickening of the submucosal tissues, vascular insufficiency, and generalized fibroatrophy. Again, stenosis or complete intestinal blockage can occur (see Geraci and others, 1977).

Rectum. The rectum is a specialized extension of the large intestine, the function of which is the storage and elimination of the formed feces. The organ is similar in many respects to the large intestine, and the early radiation response as well as the late outcome of radiation to the rectum is essentially the same as for the large intestine. This organ is of particular importance in radiotherapy since it is frequently in the field of irradiation for tumors in the lower abdomen, for example, the bladder, uterus, prostate, and cervix. Severe late radiation damage to the rectum is seen as the same sort of extensive fibrotic alteration, leading to stenosis or complete blockage. In addition, because of the structural characteristics of the submucosal portion of the organ, thinning and ultimately perforation of the rectum are possible outcomes.

Skin

Skin is an organ that has been considered before as a cell-renewal system for which convenient bioassays have been developed. These are described in Chapter 10. The rapidly turning over cell-renewal system of the skin is the epidermis, made up of a basal cell layer that acts as the stem cell reservoir for the constantly aging cells of the epidermis. The ultimate fate of these cells is to become keratinized to form the outer scaly layer of the skin and then to fall away. The remainder of the skin as an organ is the dermis, a layer of loose connective tissues richly supplied with blood vessels. Hair follicles and sebaceous glands penetrate the dermis. These latter elements are rapidly renewing cell populations, but most of the dermis is a nonrenewing cellular system.

The immediate reactions of skin to irradiation are of both historical and practical importance. The changes in the skin occur over a period of several weeks to two or three months and are related to the cytocidal effects of radiation on the cell-renewal systems. With moderate doses of radiation, after the first two to three weeks an erythema of the skin is seen, and the time for appearance of the erythema can be as short as two or three days as the dose is increased. If the dose is high enough, there is progressive change to dry desquamation and finally wet desquamation of the epidermis. The latter condition is characterized by the loss of intact epidermis and exudation of fluid from the underlying dermis. These changes are generally reversible, although, in the case of wet desquamation, there is residual thinning of the epidermis and alterations of the underlying capillary bed (telangiectasia). The rapidly turning over cells of the hair follicles and sweat glands are also affected, and the result is loss of hair and dysfunction of the sweat glands.

Before physical dosimetry was developed for ionizing radiations used in medical practice, one of the most commonly used dosimetric methods was based on the threshold for appearance of the erythema following exposure, the *erythema threshold dose*. The skin continues to be a very practical limiting organ for the application of radiotherapy in many areas of the body, although modern methods for reducing the dose to the skin while achieving the dose in the tumor have reduced the importance of skin sensitivity.

The late effects of irradiation of skin appear to be undoubtedly due to changes in the microvascular architecture of the dermal layer beneath the dermis. The point has been made forcefully and clearly, particularly by Hopewell (1986).

The late lesion is described as being due to extensive alteration of the vascular network of the dermis, leading to dermal thinning and even dermal necrosis at the higher doses. In some cases, extensive fibrotic alteration of the dermal layer leads to cicatrix formation in the radiation field. These changes are all in accord with the model described earlier, which identifies a cyclic inflammation, fibrosis, and vascular insufficiency leading to the final irreversible changes.

Because of its easy accessibility for damage scoring and the long history of erythema as a dosimetric tool, the organ has been used for a number of studies of the effects of fractionation and protraction on biological effectiveness. These fractionation models will be discussed in detail in the next section, but it is worth citing the landmark work of Strandqvist (1944).

The findings of late changes in the skin are shared by other epidermoid mucosal organs of the body. One which already has been mentioned is the *esophagus*, with its epidermislike structure. In addition, the *bladder*, the mucosa of the *oropharynx*, and lower *nasopharynx*, the cardiac portion of the *stomach*, the *ureters*, *anus*, *vagina*, and *cervix* of the uterus all share a very similar radiation response with skin because of their histopathological similarity.

Liver

The liver has been long considered to be a radioresistant organ. This is not surprising since it is, in general, a postmitotic, highly differentiated organ. An interesting early observation of Leong, Pessotti, and Krebs (1961) was that, although the irradiated liver showed no detectable acute responses to moderately high doses, if the liver of the rat is resected sufficiently to initiate the typical proliferative response, severely disorganized mitotic cells can be seen. This simple experiment demonstrates that DNA damage incurred by an intermitotic cell will be expressed by causing the cell to divide.

Observations on irradiated patients and on animals have shown that late sequelae of irradiation of the liver can be significant and even life threatening. Again, the response appears to be mediated via damage to the microvasculature of the organ. Whether this microvascular damage is the precipitating cause or whether it, in turn, is the result of damage at some other level cannot be shown; but it is reasonable to assume that the general pattern of destruction of the microvascular endothelial cells is again at work here.

Radiation induced hepatitis is a distinct entity resulting usually from the inevitable inclusion of the liver in the radiation field during many types of radiotherapy. The histopathological changes seen are occlusion of the small hepatic veins, leading to sinus congestion, hyperemia, and hemorrhage. At quite late times after irradiation, severe fibrotic alteration of the liver can be seen with clinically important doses. Liver failure, hepatitis, and ascites have been reported for radiotherapy patients who have received high doses to the liver.

Kidneys

The radiosensitivity of the kidney to late radiation damage has been well known and well documented. Its causes are less clear. In patients with advanced degenerative changes in the kidney resulting from irradiation, the disease is described as arteriolar nephrosclerosis. This disease state is an advanced fibrotic and sclerotic alteration of the arteriolar and capillary blood supply to the renal glomerulus, which leads to sclerotic alteration of the glomerulus and loss of its functional capacity to produce the plasma ultrafiltrate necessary as the first step in the process of urine formation. The kidneys have certain other hormonal functions unrelated to the production of urine. The sclerotic changes lead to expressions of dysfunction in these special characteristics in the form of hypertension and the anemia of renal insufficiency.

Although it is clear that the ultimate outcome of the irradiation of the kidney at high doses is as just described, recent animal studies, particularly those of Stewart and others (1984), have shown that the picture is much more complicated. These workers have shown that even before significant sclerotic changes have occurred, as judged by microscopic pathology, there are very large changes in both glomerular function and tubular reabsorptive function. It cannot be ruled out that these might be secondary to vascular alterations, particularly because the regulatory mechanisms of the kidney are so complex. Associated with each glomerular–tubular unit are effective baro- and osmoreceptors that can regulate blood flow to the glomerulus in complicated ways. It does appear that the polyuria reported by Stewart and others (1984) would indicate a significant tubular dysfunction unrelated to vascular disorders. This polyuria occurs at the same time that the glomerular filtration rate is sharply reduced; so in spite of the smaller volume of plasma ultrafiltrate presented for processing in the tubule, the final daily urinary output is increased several fold. A recent study by Moran, Davis, and Hagan (1986), extended by Parkins, Rodriguez, and Alpen (1989), has examined the effects of irradiation on the glucose reabsorbing function of pig kidney tubule cells grown in culture. These cells grown in culture retain at least some of the functional activity known to exist in the tubule of the intact kidney. When the cultured cells are exposed to an increased external concentration of glucose, they respond by generating increased glucose transporting function on the cellular surface. When the cells are irradiated (by doses of 10 Gy and higher), their ability to respond to a regulatory need for increased glucose transport is sharply reduced. The doses required to induce this effect are similar to those that lead to the late kidney damage reported by Stewart and others (1984).

Lung

The lung is a complicated, highly vascularized organ that, because of its special function, operates in a mixed air–fluid environment. It is, in general, a nonturning-over tissue, with some exceptions for specialized cells of the lung. One late effect of irradiation of the lungs is a radiation induced pneumonitis. It has been suggested that this pneumonitis is the result of altered characteristics of the alveolar surface, with accompanying changes in fluid and gas transport across the alveolar subunit. The type 2 alveolar cells of the lung are, indeed, a turning over population that is responsible for the production of the surfactant necessary for maintaining lung function. Reduction in the number of these cells, as well as an effect on, again, the endothelial cells, which to a large degree regulate the transudation of water and macromolecules, is probably responsible for the radiation pneumonitis. The turnover time of the type 2 cell of about four weeks is in accord with the time of onset of the pneumonitis. It has been suggested that the pneumonitis may be attributed to a cell killing of clonogenic precursors for type 2 cells.

After recovery from the radiation pneumonitis, if this is possible at the dose used, a much later and distinctly different syndrome becomes evident. Whereas the pneumonitis has an onset as early as three to six months, with resolution or death in six months or so, in mice, after about a year, a second phase of lung damage sets in. The critical tissue involved in this late lung damage is not known (see Travis and Tucker, 1986). This last phase damage has much in common with other late effects in other organs, that is, fibrosis, inflammation, sclerosis, and loss of elasticity.

Because of this similarity, we are tempted to relate the lesion again to microvascular damage, but there is no clear and convincing evidence for this mechanism.

Central Nervous System

Brain. Surprisingly, there is very limited information about the late response of the brain to irradiation. There is no doubt about the existence of such late effects. Radiation induced necrosis of the brain has been widely reported as a sequel to the treatment of malignant disease in the brain. Little is known about the mechanism of the necrosis, but heated discussions continue about the cause. Some believe that the necrosis is the result of microvascular changes, much as in other organs, despite the fact that the injury is occurring in tissue with unusual specialized vascularization. Others suggest direct effects through demyelination as a possible cause. Only recently has experimental work been undertaken to explore these issues, and the results have only been reported in preliminary form in abstract literature. Much remains to be done before the questions can be answered.

There is one other change in the brain after irradiation that is now a generally accepted observation. Early after irradiation there appears to be a disturbance of the blood–brain barrier that leads to brain edema. Generally, this effect is reversible if the dose is not too high.

Spinal cord. Very different from the brain itself, the spinal cord has been the object of extensive research for some years. Part of the reason for this interest is the well-known fact that the spinal cord is indeed a radiosensitive organ. During radiotherapy the cord must be carefully protected to prevent damage that would lead to functional total or complete transection and paralysis.

A recent and very complete review of the state of the information about spinal cord injury resulting from irradiation is that of van der Kogel (1986).

The functional lesion seen in the spinal cord, either in experimental animals or man, is paralysis of the affected regions of the body after some considerable time, a year in rats and two to three years in man, depending on dose. Some have suggested that the lesion in the cord is the direct result of damage to the vascular endothelium, and this cannot be ruled out. In fact, at the lower end of the dose range at which some paralysis becomes evident, it seems reasonably certain that this damage, which occurs at very late times, is associated with telangiectasia and petechial hemorrhage associated with the area of the lesion. The more rapidly developing paralysis that occurs with higher doses is more complex in its etiology.

Van der Kogel outlines the changes in the spinal cord after irradiation as follows. Within a few weeks to several months after exposure, demyelination occurs associated with the loss of glial cells. Repopulation of oligodendrocytes, stimulated by the glial cell loss, occurs. If glial repopulation is inadequate, there will be white matter necrosis and extensive radiation induced myelitis. While these events are occurring, vascular damage may be a contributing factor. Certainly, neurons are not involved in the process of late injury development in the spinal cord. If glial repopulation is successful, sufficient injury may still be sustained as the result of vascular damage for the onset of the late developing myelitis just described.

The spinal cord may be an example of an organ in which both clonogenic depletion and vascular damage play a role in the development of late damage. It shares this characteristic with the lung.

Eye; Cataractogenesis

Of the anatomical parts of the eye, only the lens is of importance for its radiosensitivity. Cataracts, which are defined as detectable changes in the normally translucent character of the lens, were reported as a result of exposure to ionizing radiation by Abelson and Kruger in 1949. This was the first reported incident of cataractogenesis as the result of occupational exposure to ionizing radiation, even though radiologists for years had known of the special sensitivity of the lens of the eye to opacification as the result of exposure to ionizing radiation.

The lens is an unusual structure in that it possesses no blood supply to provide nutrients and oxygen for the normal oxidative processes of tissue metabolism. It is an encapsulated organ, and it must depend entirely on diffusional movement of oxygen and nutrients for its metabolic survival. The lens is possessed of a slowly turning over clonogenic population of cells. On the anterior surface of the lens there is a layer of epithelial cells, the dividing progenitors of which are located at the equatorial region of this anterior surface. The dividing cells migrate posteriorly and centrally, and during this migration process they undergo a differentiation to form lens fibers. Because of the unusual structure of the lens, there is no adequate mechanism for the removal or destruction of epithelial cells that are damaged by ionizing radiation. The products of division of damaged cells undergo ineffective differentiation as they migrate, and they fail to achieve the level of translucence necessary for the function of the lens as an optic device. These remnants of poorly differentiating lens epithelial cells form the defects in light transmission of the lens seen as cataracts on examination.

Threshold for cataractogenesis. The threshold for cataractogenesis in human beings has been well established by analysis of the occurrence of these lesions in individuals who have undergone radiotherapy. As a result of careful analysis of these patients, some general rules have been established.

If the exposure of the lens occurs as a single brief event, cataracts may be expected to appear if the dose to the lens is in the range 2 to 6 Gy (200 to 600 rad). The severity of the cataract will be dose dependent, and at the bottom of the threshold range, barely detectable changes in the lens will result.

Fractionation of the exposure is effective in protecting the lens. For the lower total doses, fractionation will increase the threshold two- to threefold; but there appears to be an upper dose limit beyond which fractionation loses its protective effect. If the total dose to the lens is in excess of 11 Gy (1100 rad), cataract formation will result, regardless of the pattern of dose delivery.

For man the latent period for cataractogenesis is dose dependent. The latent period may be less than one year at the higher doses and in excess of ten years for lower doses.

One of the most unusual aspects of radiation cataractogenesis is the extreme sensitivity of the process to the LET of the radiation source. From studies in mice, Bateman and Bond (1967) concluded that for high LET neutrons the threshold for cataractogenesis could be as low as 1 cGy (1 rad). Presumably, this large RBE (greater than 20 and possibly as high as 100) will be found for human exposure also, and, indeed, the studies mentioned earlier (Abelson and Kruger, 1949) suggest that the threshold dose for cyclotron produced neutrons is on the order of 0.5 cGy. Careful, long-term studies of occupationally exposed individuals may confirm that the high RBE found by Bateman for mice is applicable to man.

Cataractogenesis is one late tissue change that is clearly not related to degenerative changes in the vascular endothelium, since none is available in this organ. We can almost certainly say that the defect must result from damage to the clonogenic epithelial cells of the lens, resulting in defective differentiation of the daughter cells. Also, cataractogenesis meets all the criteria for a nonstochastic event. There is a definite threshold, about 2 Gy (200 rad), and there is a dose–severity relationship.

FRACTIONATION AND PROTRACTION OF EXPOSURE IN THE MODIFICATION OF LATE RADIATION INJURY

Radiotherapy has always had to be somewhat of an empirical practice of an art dependent more on observations in the clinic than on advanced radiobiological experimentation. The practice of the art simply has not been able to wait for the developments in the laboratory; and application of clinical observations for the improvement in probability of cure and lessened morbidity associated with the treatment has had to move forward without the complete understanding of the underlying biological science. This has nowhere been more true than in the development of treatment schemes for fractionated exposure during the radiotherapy of malignant disease. Long before the development of the concepts of sublethal injury repair, potentially lethal injury repair, and, indeed, even before the development of the in vitro techniques for tissue culture were developed, radiotherapists had come to realize that a multifraction treatment plan, spread over several weeks, had proved to be more effective in the eradication of the tumor than a single dose treatment plan, while at the same time reducing undesirable side effects. Part of the increased effectiveness of fractionated treatment schemes lay in the ability to deliver a larger treatment dose to the tumor while associated morbidity in normal tissues was kept at acceptable levels. This latter observation suggests that the response of tumors to fractionation might be fundamentally different from the response of normal tissues to fractionated radiation.

First Appearance of the Power Law

As early as 1939, Witte was able to visualize that the data of others on the relationship between the dose required to reach the threshold for skin erythema and the dose rate of the delivered dose (or number of fractions) were best described with a power law relationship. There was a linear relationship between the logarithm of the dose required for erythema production and the logarithm of the number of treatments. A separate line described the relationship for each of the dose rates used. These reports by Witte and others were the stimulus for Strandqvist's report in 1944.

Strandqvist Relationship

Strandqvist's improvements on prior models depended on his use of the total time of treatment as the basis for constructing an isoeffect relationship (Figure 11.2). He concluded that, since the power law relationship between dose rate (or intensity) and the overall time of treatment gave an exponent value of > 0.8 to 1.0, it was acceptable to reformat the information on the assumption that I, the dose rate, could be replaced by the quotient of dose, D, and time, T. The rearranged power law according to Strandqvist was then

$$D = \text{constant} \times T^{1-p} \tag{11-1}$$

Strandqvist found $1 - p$ to be approximately 0.2. The results were, however, based on the rather limited data then available. The longest treatment time for the control of skin cancer (squamous carcinoma) was 14

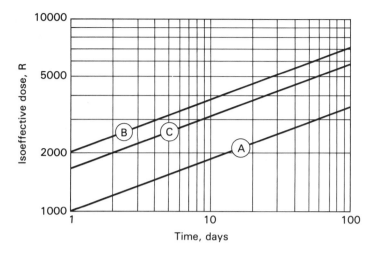

Figure 11.2   The Strandqvist power relationship plotted for: (A) threshold for skin erythema, (B) dry desquamation of skin, and (C) control of skin cancer. The data are from various literature sources, which are reviewed in Fletcher (1980).

days, and only 29 patients represented all the recurrences and complications of the treatment. In spite of many reservations about the quantitative accuracy of the Strandqvist relationship for the establishment of equieffective doses given in different time periods, his results were in use for a number of years in the clinical prescription of dose. This was in spite of a study by Cohen (1949) that convincingly showed that the time exponent for squamous carcinoma and for acute reactions in normal skin were indeed different. He reported that the exponent for normal skin was 0.30 and that for squamous carcinoma was 0.22. In this difference lies, according to Cohen, the ability to successfully treat skin cancer without excessive damage to normal skin.

Nominal Standard Dose

A fundamental assumption of the Strandqvist model is that overall time is the important parameter, and varying the number of fractions or how they are distributed in this overall time has little effect on the biological outcome. Since most, if not all, of his analyses used clinical data that were based on the pattern of five or six daily fractions per week that was, and is, widely used in radiotherapy, there was little sensitivity in the data to the change in the fractionation pattern.

In 1963, Fowler and others published data on the response of pig skin that led them to conclude that overall time, between 5 and 28 days, was relatively unimportant, and that the principal determinant of the isoeffective dose in this time period was the number and size of individual fractions. They concluded that, if time is held constant within the constraint just mentioned, then fewer and larger fractions required a reduction of the total dose for maintenance of isoeffectiveness.

Responding to the collection of new literature suggesting that both overall time and number of fractions were important in the determination of isoeffective dose, as well as the findings of Cohen that the slopes of Strandqvist plots were not necessarily constant, Ellis (1969) proposed a new power law formulation that explicitly recognized the role of number of fractions and the overall treatment time. This formulation proposed that the tolerance dose, D, for a normal tissue was related to the number of fractions, N, and the overall treatment time in days, T, by a power law including a constant, NSD. The power law function, with his choice of exponents, is

$$D = (\text{NSD})T^{0.11}N^{0.24} \qquad (11\text{-}2)$$

The constant NSD was named by Ellis the *nominal standard dose*. It is necessary to determine the constant by empirical methods. For a given organ or tissue, a biologically effective dose is determined (for instance the 50% effective dose for skin erythema) for a given combination of T and N. It is then possible to solve equation 11-2 for the constant. The formulation has clinical attractiveness, since it putatively allows the prescription of dose for different treatment regimes, once NSD has been established. Even though there was almost immediate dispute about the utility and reliability of the formulation itself, as well as the quantitation of the exponents, the Ellis formulation has found widespread application. One great attraction of this construct is that NSD is especially determined for every organ, tissue, and tumor, and, in the best of circumstances, it is specifically determined in the clinic in which the formulation is used, with its machines, methods, and dosimetry.

Summary of the Power Law Relationships

The power law models all have significant shortcomings, and in spite of extensive "fine-tuning" of the exponents and "jiggling" of the mathematical structure of the relationship, there has been no satisfactory accommodation to the reality of biological data. Repair during and after fractionated radiation is the result of both biochemical repair of cellular lesions that alter survival and biological response to damage expressed by repopulation of the tissue as the result of the commencement of cell division among the survivors. The mathematical form of the power law expressions implies more rapid recovery during the earlier part of the fractionation cycle, while it is well known that it takes some time for the proliferative response to set in, reach its maximum, and then subside when cellular population equilibrium is reached. Furthermore, it seems reasonably certain, as Cohen (1949) suggested, that power law exponents are not constants. As later experimental work has shown, and as will be discussed here when the linear–quadratic model is explored, there is enormous variability in the repair kinetics of all tissues, both for the acute and late effects. Given these fundamental biologically based objections to the power law formulations, we must look to alternative models to construct reliable predictions of isoeffect relationships for widely varying tissues and cells. Generally, the newer models based on newer understanding of the survival curve relationships have provided such tools.

Repair and Repopulation after Irradiation

The various studies of postirradiation survival with in vitro clonogenic systems have defined a number of forms of postirradiation repair that are vital in understanding the nature of the response of tissue systems to fractionated radiation. These intracellular repair processes have been described in detail in other chapters of this work, but the importance of each to the biological outcome of fractionated exposure has not been examined. Remember that, in addition to the cellular repair processes, repopulation at the intact tissue or organ level is also of great importance.

The three intracellular damage processes and their subsequent repair are:

1. *Sublethal* damage and its repair
2. *Potentially lethal* damage and its repair
3. *Nonrepairable* damage

As mentioned in the previous paragraph, during protracted radiation and after its completion, the repair processes are complicated and assisted by repopulation of tissues from clonogenic survivors.

In Chapter 7 the various mathematical models for cellular survival were outlined, along with their limitations. Most of the advanced models attempted to deal with two limitations of the multitarget single-hit (MTSH) model. These two limitations of MTSH are the inability of the model to deal with an initial slope of the survival curve that is nonzero and the slowly changing final slope that is seen in some or nearly all clonogenic survival systems. For the purposes of the discussions here, it is convenient to simplify these various models by proposing a nonzero initial slope to the survival curve that is related to the nonrepairable, single-hit component of the cell killing. This cell killing is probably due to the high LET component of the radiation used, and this high LET component is to be found to a greater or lesser extent in all radiation sources, regardless of their overall LET. The reader is reminded that even for high-energy gamma ray interactions, the ends of slowing down electron tracks are relatively high LET.

The original studies of the repair of sublethal damage showed that, if a period of time were allowed to elapse between the first of two irradiations and the delivery of the second, the shoulder of the survival curve tended to reappear, and if adequate time were allowed between the two doses, then the original shoulder was fully reexpressed. A number of workers have visualized the importance of the combination of the initial slope that is nonrepairable and the repair of sublethal damage in determining the response of tissue systems to fractionated radiation. A typical idealized response might be that presented by Withers and Peters (1980).

For the model in Figure 11.3, the effect of fraction size can be visualized. For the very smallest dose per fraction, where cell killing for each fraction is mostly limited to that due to the single-hit component, a limiting slope is that associated with the single-hit component (for example, fraction size W in Figure 11.3). The other extreme limit will be that associated with a single fraction for the entire dose, shown in Figure 11.3 as the limiting slope for large dose per fraction. For intermediate doses per fraction, such as X and Y, the final survival curve for the fractionated regimen is determined by the cell killing for each individual dose, X or Y. There is an effective D_0 that can be measured for each fractionation regimen, represented by the slopes of the straight lines, and for which there would be no visible shoulder to the survival curve.

Cells in organized tissue systems are known to be able to repair potentially lethal damage as well as sublethal damage (see, for example, Hendry, 1985). To the extent that repair of PLD takes place during a fractionated exposure, the simplified diagrams of Figure 11.3 will be modified in somewhat complicated ways. It will be recalled that the effect of repair of PLD on the clonogenic survival curve is to increase the D_0 (that is, decrease the slope of the survival curve). A generalized statement about the effects of PLD repair is that there appears to be a more or less constant increment of dose required for the same level of initial damage. This alteration also appears to be more or less independent of the dose after a certain threshold is passed. It is not possible to be more specific from data available at this time, and, furthermore, no one has successfully modeled the influence of PLD repair on fractionation. For the time being, we must accept that it exists as a perturbing factor that is not exactly quantifiable.

To the extent that clonogenic cell killing is responsible for the late effects of irradiation on normal tissues, it might be expected that simple models derivative of that shown in Figure 11.3 might be valuable in predicting the outcome of fractionation regimes. This is indeed the case. If the underlying mechanism for the late effects attributed to vascular

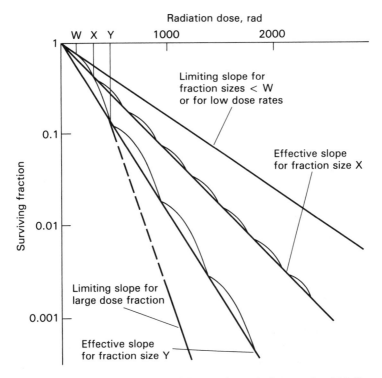

Figure 11.3 Response of an idealized clonogenic survival system in which there is an initial slope less than zero to account for nonrepairable damage and where complete repair of sublethal damage is allowed to occur between fractions. The limiting slope for the smaller fractions will be determined by the slope of the single-hit component and is indicated on the figure for fraction sizes less than W. The survival for fractionated doses with fraction size X, for smaller doses per fraction than for fraction size Y, will always be greater. (From Withers and Peters, in *Textbook of Radiotherapy*, G. H. Fletcher, ed., 1980; by permission of Lea & Febiger, Philadelphia)

endothelial damage is also a clonogenic depletion of endothelial precursors, the models derived from Figure 11.3 will have general applicability.

Douglas and Fowler F_e Formulation

Presently, one of the most popular formulations for the prediction of the modulation of late effects of irradiation by fractionation is that due to Douglas and Fowler (1976). The studies were done on the skin reaction of the foot of the mouse with fractionation schedules up to 64 fractions in a constant overall time for delivery of the total dose. It is important to note that these authors were able to control for and to rule out the effect of repopulation in the short overall time of their exposures. They observed that for a control level of effect or response, when total dose to achieve the desired level of effect was plotted against the number of fractions, the data were very well described as follows:

$$n \times (\alpha D + \beta D^2) = \gamma \tag{11-3}$$

In this equation n is the number of fractions, D is the dose per fraction, and α, β, and γ are constants.

Their interpretation of this formulation was based on three basic assumptions.

1. Repair occurs after single doses of x-rays, although it requires a test dose to detect this repair (the Elkind–Sutton experiment). Equal amounts of repair occur after repeated dose fractions of equal size in any one tissue.

2. The biological outcome depends on the surviving fraction of critical clonogenic cells in that system. The reaction level of the late effect (severity) correlates with this surviving fraction.

3. Every equal fraction will have the same biological effect, independent of its position in the series.

Given the formulation of equation 11-3, for a given tissue and range of dose, a simple function exists that, when multiplied by the dose, yields the logarithm of the surviving fraction.

The equation relating dose per fraction, D, and the number of fractions n for a *constant* surviving fraction S (also, presumably, a constant level of biological effect) is

$$\ln S = n \left(\frac{F_e}{a} D \right) \tag{11-4}$$

Note that in this reformulation of their model Douglas and Fowler have used the constant a, rather than α, as in equation 11-3, to make it clear that it is a different constant. For an appropriate choice of a, F_e is the inverse of the total dose nD for multifraction data. The equation for the single-dose cell survival curve is

$$S = \exp \left(\frac{F_e}{a} D \right) \tag{11-5}$$

where S is the surviving fraction of cells after a dose, D.

Recall that, having appropriately chosen a,

$$F_e = \frac{1}{nD} \tag{11-6}$$

Note that F_e, by its derivation, is the reciprocal of the total dose to produce a constant biological level of effect (isoeffectiveness) with any fractionation scheme in which repopulation is not playing a role. If F_e is plotted as a function of dose per fraction, a straight line is the result, and it is described by

$$F_e = b + cD \tag{11-7}$$

Substituting equation 11-7 in equation 11-4,

$$\ln S = n \left(\frac{b}{a} D + \frac{c}{a} D^2 \right) \tag{11-8}$$

This elegant formulation, which has been found to have wide applicability for late damage in a number of tissues, was derived from first principles by Douglas and Fowler and was not dependent on any assumptions other than the three outlined previously. Can this formulation be reconstructed in a way that relates obviously to our preconceived models for radiation damage? The original authors and others have shown that this is indeed the case.

Using the linear–quadratic formulation, if we can assume constant effect per fraction, the equation for the surviving fraction of critical cells can be written

$$\ln S = -n(\alpha D + \beta D^2) \tag{11-9}$$

And if the effect being measured is directly related to the logarithm of surviving fraction, then minus ln survival can be relabeled E, and we can divide by the total dose nD, obtaining equation 11-10.

$$\frac{1}{nD} = \frac{\alpha}{E} + \frac{\beta}{E} D \tag{11-10}$$

The term on the left is defined as F_e, and equation 11-10 is identical with equation 11-7. Now the constants b and c of equation 11-7 are defined in terms of the well-known constants α and β of the linear–quadratic or molecular model. Since E is an unmeasurable constant, it is not possible to directly determine the two constants α and β, but from an *isoeffect plot*, that is, a plot of the reciprocal of the total dose for isoeffect as a function of dose per fraction, D, we may evaluate the constants α/E and β/E as the extrapolated zero dose per fraction intercept and the slope, respectively, of the isoeffect plot. The ratio of the two constants reduces to the ratio α/β. This ratio has proved to be invaluable in the study of fractionation effects, not only for late effects in normal tissue, but equally for clonogenic survival.

The fact that equation 11-10 provides a good fit to the data for a very wide range of late effects in normal tissue is the strongest evidence available that most late effects have as their underlying mechanism the depletion of some population of critical clonogenic cells. Furthermore, the numerical value of the α/β ratio has proved to have great potential in estimating the repair potential and thus the potential protection provided by fractionation for a wide range of tissues.

The F_e model (Figure 11.4) has grown to be the principal device for the intercomparison of the effects of fractionation on various tissues. It should not be concluded, however, that the effectiveness of this model is necessarily a proof of the biological significance of the linear–quadratic model. Any other cell killing model that presumes an initial slope less than zero appears to be equally useful. Since the MTSH model has an initial slope of zero, its theoretical limitation prevents its application for meaningful results. No version of the power laws discussed in the preceding section will adequately describe the data fitted by the F_e plot.

Significance of the α/β ratio. Since, in its simplest form, in the linear–quadratic model the alpha coefficient describes single-hit, nonrepairable damage in general, and the beta coefficient describes multiple-hit killing, the independent variability in these two terms can have an effect on the fractionation sensitivity of tissues. The simpler mechanistic description of the alpha–beta formulation is no longer considered to be completely accurate, and the newer concepts of lesions and sublesions and their interaction described in Chapter 7 are more acceptable. For the moment though, assuming a constant value for the alpha coefficient, a

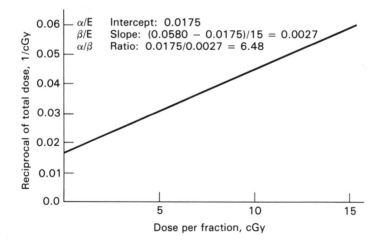

Figure 11.4 A Douglas and Fowler F_e plot for some arbitrary data altered from our own data on gastrointestinal crypt cells. This mythical tissue is taken to have been irradiated at several fraction numbers, and the total dose required to achieve a given biological response is determined. As shown in the theoretical plot, the intercept, which is taken to be α/E, is 0.0175. The slope was then estimated graphically. We can quickly show that the intercept of the line with the zero ordinate is the ratio α/β.

larger and larger value for the beta coefficient will signify a smaller α/β ratio, but a higher proportion of the radiation damage will be repairable. This, in general, is what is found for biological systems. The range of values for α/β is very large, but some significant differences between tumors and normal tissues are evident. Thames and Hendry (see Suggested Additional Reading) have compiled some of these data from various literature sources for tumors irradiated as growing tumors in the animal and for some normal tissue late effects (see Table 3.3, page 76, and Table 3.7, page 94, in Thames and Hendry). The growing tumors have α/β ratios from as low as 5 to as high as 30, but the predominant range is 10 to 20. α/β ratios for late effects in normal tissues range from 1.5 to about 7, but again with a central tendency around 3 to 5. Acute effects in normal tissue generally have α/β ratios around 10. The conclusion to be drawn from these observations, although somewhat gingerly to say the least, is that we would expect to find significantly higher protection for late effects on normal tissues relative to tumors from fractionation. Indeed, these observations have led radiotherapists to their present clinical trials of what has been called *hyperfractionation*, the delivery of several small fractions per day over an extended period to achieve effective total curative doses.

It must be clearly restated that the Douglas and Fowler α/β model, which was derived by them *ab initio* by fitting of data, does not require that the linear–quadratic relationship faithfully represent cell killing in the sensitive tissue being considered. The only requirement is that the model effectively predict the relative sensitivity of various tissues involved in radiotherapeutic practice.

Withers extension of the α/β model.   Withers, Thames, and Peters (1983) extended and slightly modified the Douglas and Fowler model to produce a formulation that can be of daily utility in radiotherapeutic planning. The equations require an independent estimate of α/β, but this is usually obtained by isoeffect–fractionation studies as just described for the methods of Douglas and Fowler.

The Withers, Thames, and Peters isoeffect model makes the same generalized assumptions as for that of Douglas and Fowler. It also starts from the premise that cell killing in the tissue under consideration can be adequately described by the linear–quadratic model. That is, the surviving fraction after N fractions is

$$S = [e^{-(\alpha D + \beta D^2)}]^N \tag{11-11}$$

It will be remembered from Figure 11.4 that there is a dose, W, such that if there is further reduction in the dose per fraction there is no detectable further protection provided by interfraction repair processes. This dose has been called the *flexure dose* by a number of workers, and it is difficult to say where the term was first used. However, it is indeed a useful concept,

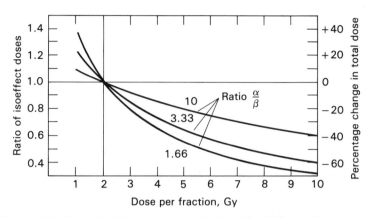

Figure 11.5 Theoretical isoeffect curves developed by Withers, Thames, and Peters (1983) using their modification of the F_e model. The curves are constructed using three different values for α/β and a d_{ref} of 2 Gy. The total equieffective doses are expressed as a ratio and as a percentage of the total dose were it given in d_{ref} fractions of 2 Gy. Repair is assumed to be complete between dose fractions. (Withers, Thames, and Peters, 1983; by permission of Elsevier Science Publishers and the authors)

since it is important, in practice, to know the dose per fraction below which further protection of normal tissue is not provided. The flexure dose has been shown to be a multiple of α/β (Tucker and Thames, 1983). Therefore, characterization of sensitivity of tissue to dose fractionation in terms of α/β may be thought of as dose specification in terms of the useful flexure dose concept.

The most commonly used fraction size in radiotherapy is 2 Gy; therefore, the model expresses change in sensitivity with different fraction size as the ratio of the new total dose D to the reference total dose D_{ref} when a new fraction size, d, is proposed in place of the reference fraction size, d_{ref}. The reader is referred to Withers, Thames, and Peters (1983) for the complete derivation of the equation, but it is the result of simple manipulation of equation 11-11.

$$\frac{D}{D_{ref}} = \frac{(\alpha/\beta) + d_{ref}}{(\alpha/\beta) + d} \tag{11-12}$$

The ratio α/β is that taken as appropriate for the tissue for which an isoeffect dose is required.

Figure 11.5 shows a set of theoretical isoeffect curves constructed by Withers, Thames, and Peters for different values of α/β.

REFERENCES

ABELSON, P. M., and KRUGER, P. G. (1949) Cyclotron induced radiation cataracts. *Science* **110,** 655–657.

BATEMAN, J. L., and BOND, V. P. (1967) Lens opacification in mice exposed to fast neutrons. *Radiation Res.* **7**, 239–249.

COHEN, L. (1949) Clinical radiation dosage, Part II. *British J. Radiol.* **22**, 706–713.

CREASEY, W., WITHERS, H. R., and CASARETT, G. W. (1976) Basic mechanisms of permanent and delayed radiation pathology. *Cancer* **37**, 999–1013.

DOUGLAS, B. G., and FOWLER, J. F. (1976) The effect of multiple small doses of x-rays on skin reactions in the mouse and a basic interpretation. *Radiation Res.* **66**, 401–426.

ELLIS, F. (1969) Dose, time and fractionation; a clinical hypothesis. *Clinical Radiol.* **20**, 1–7.

FLETCHER, G. H. (1980) In *Textbook of Radiotherapy*, 3rd ed. Lea & Febiger, Philadelphia, pp. 202 et seq.

FOWLER, J. F., and others (1963) Experiments with fractionated x-ray treatment of the skin of pigs. I. Fractionation up to 28 days. *British J. Radiol.* **36**, 188–196.

GERACI, J. P., and others (1977) Acute and late damage in the mouse intestine following multiple fractionation of neutrons or x-rays. *Intl. J. Radiation Oncol. Biol. Physics* **2**, 693–696.

HENDRY, J. H. (1985) Survival curves for normal tissue clonogens: a comparison of assessments using in vitro, transplantation or in situ techniques. *Intl. J. Radiation Biol.* **47**, 3–16.

────── (1986) Preface, *British J. Cancer*, Supplement VII, **53**, p. xi.

HOPEWELL, J. W. (1986) Mechanisms of action of radiation on skin and underlying tissues. In *Radiation Damage to Skin, British J. Radiol.*, Suppl. 19. British Institute of Radiology, London, pp. 39–47.

International Commission on Radiological Protection (1977) *Recommendations of the International Commission on Radiological Protection.* Publication 26. Pergamon Press, Elmsford, N.Y.

LEONG, G. W., PESSOTTI, R. L., and KREBS, J. (1961) Liver cell proliferation in x-irradiated rats after single and repetitive partial hepatectomy. *J. Natl. Cancer Inst.* **27**, 131–143.

MICHALOWSKI, A., and others (1983) Some early effects of thoracic irradiation in mice. *Proc. 7th Intl. Congress Radiation Res.*, Abstr. D3–28.

MORAN, A., DAVIS, L., and HAGAN, M. (1986) Effect of radiation on the regulation of sodium dependent glucose transport in LLC-PK1 epithelial cell line: a possible model for gene expression. *Radiation Res.* **105**, 201–210.

PARKINS, C. S., RODRIGUEZ, A., and ALPEN, E. L. (1989) Radiation damage to glucose concentrating capacity and cell survival in kidney tubule cells: effects of fractionation. *Intl. J. Radiation Biol.* **55**, 15–26.

STEWART, F. A., and others (1984) Radiation induced renal damage: the effects of hyperfractionation. *Radiation Res.*, **98**, 407–420.

STRANDQVIST, M. (1944) Studien uber die kumulative Wirkung der Röntgenstrahlen bei Fractionerung. *Acta Radiol.* (suppl.), **55**, 1–300.

TRAVIS, E. L., and TUCKER, S. L. (1986) The relationship between functional assays of radiation response to the lung and target cell depletion. *British J. Cancer* **53,** Suppl. VII, 304–319.

TUCKER, S. L., and THAMES, H. D., JR. (1983) Flexure dose: the low-dose limit of effective fractionation. *Intl. J. Radiation Oncol. Biol. Physics* **9,** 1373–1383.

VAN DER KOGEL, A. J. (1986) Radiation induced damage in the central nervous system: an interpretation of target cell depletion. *British J. Cancer* **53,** Suppl. VII, 207–217.

WITHERS, H. R., and PETERS, L. J. (1980) Biological aspects of radiotherapy. In *Textbook of Radiotherapy*, G. H. Fletcher, ed. Lea & Febiger, Philadelphia.

———, THAMES, H. D., JR., and PETERS, L. J. (1983) A new isoeffect curve for change in dose per fraction. *Radiotherapy Oncol.* **1,** 187–191.

WITTE, E. (1939) Uber die Umrechnung der r-Dosis auf Einheiten biologischer Wirkung bei der protrahiert-fraktionerten Bestrahlung unter besonderer Berucksichtigung der Bestrahlung mit kleinen Raumdosen. *Strahlentherapie* **65,** 630–638.

SUGGESTED ADDITIONAL READING

CASARETT, G. W. (1980) *Radiation Histopathology*, Volumes I and II. CRC Press, Boca Raton, Fla.

RUBIN, R., and CASARETT, G. W. (1968) *Clinical Radiation Pathology*, Volumes I and II. W. B. Saunders, Philadelphia.

THAMES, H. D., and HENDRY, J. H. (1987) *Fractionation in Radiotherapy*. Taylor and Francis, London.

PROBLEMS

1. One of the most widely used treatment schedules for radiotherapy is the administration of daily fractions on weekdays (Monday through Friday) for four consecutive weeks. The most widely used dose per fraction is 2 Gy. Assuming that NSD is 17 and that the exponents given in the text for the Strandqvist and Ellis NSD formulas are appropriate, what change in the daily dose per fraction would be required if the total dose in the standard four-week schedule is 60 Gy and this dose must be given in three weeks instead?

2. Douglas and Fowler found for skin reactions in the mouse foot that the values of α/E and β/E are, respectively, 1.32×10^{-4} and 1.27×10^{-7} when the doses were expressed in rad. Using the formulas in the text, and, as desired, referring to the original reference, deduce the survival curve for the clonogenic cells responsible for maintaining the integrity of the skin of the foot of the mouse.

12

Stochastic Effects:
Radiation Carcinogenesis

INTRODUCTION

Historical

Unfortunately, the propensity for ionizing radiation to produce cancer in exposed human beings was demonstrated within a few short years of Röntgen's discovery of x-rays. Dermatitis of the hands, but with no proven resulting cancer, was reported as early as 1896. The first clearly cancerous changes were reported in 1902 in an x-ray induced ulcer, and clear-cut radiation associated leukemias were reported in 1911 (Stone, 1959). These early cancers were thought to be the result of "excessive exposure" to ionizing radiation, but it was not until many years later that epidemiological studies were able to demonstrate that cancer could be associated with low doses of radiation. It was a landmark study on cancer-related mortality in radiologists in 1946 (Ulrich) that suggested the possibility that there could be a dose–effect relationship for radiation induced cancer that might be a "no-threshold" response. The world's literature on the subject of radiogenic cancer is very large indeed, but a number of very useful general reviews have appeared that summarize the field as of their date of publication. The most recent of these is by Fry and Storer (1987), to which the readers are referred. This reference also provides a good summary of the earlier review literature in the field.

Stochastic versus Nonstochastic Effects

The definition and characterization of *stochastic* and *nonstochastic* effects were given in detail in Chapter 11. A brief discussion of the relationship of these definitions to radiogenic cancer is in order at this point.

In Chapter 11, it was generally possible to describe the dose–response relationships for tissue late effects in the usual pharmacological terms of a sigmoid dose–response curve with some dose below which we would not expect any effect in any exposed individual (the *threshold*). The shapes of the dose–response curve for cancers resulting from ionizing radiation are highly variable, and the shapes depend on many factors other than the radiation dose. In general, however, it has not been possible to demonstrate a threshold dose (nor is it theoretically possible to do so) for radiogenic cancers. Scientific opinion widely holds that there are theoretical reasons why such lack of the threshold should be the true situation, but proof of the null hypothesis is, as ever, not possible. It is taken that changes can occur at any exposure level, and even though for low doses the frequency of the change in an exposed population might be thought to be low, the frequency of occurrence is still not zero. There is still a great deal of discussion as to the actual shape of the dose–response relationship for cancer and for genetic damage, but there is very little scientific opposition to the argument that these are indeed effects without a dose threshold after exposure to ionizing radiation.

Since all the evidence seems to suggest that radiogenic cancer is, indeed, a stochastic effect, it is then possible to express the likelihood of a cancer being produced either for a single individual for a given exposure of that individual or for a population of many individuals, all of whom have received some exposure. For example, if the chance of developing a particular cancer for one subject is thought to be 1×10^{-6} per rad (per cGy) of exposure, then we can just as well project that if 1 million subjects all have an average exposure of 1 rad (1 cGy), the likelihood is that one case of cancer over the normal incidence rate would be expected. Such reasoning gives rise to the concept of *population dose* or *collective dose*, expressed in person-sievert or person-rem. The population dose is the product of the average dose and the number exposed, or the sum of the doses for all the exposed people.

Arguments for and against the stochastic model. The first classification of radiation damage as either stochastic or nonstochastic was suggested by the International Commission on Radiological Protection (1977). Their definitions required that for a stochastic effect the probability of the effect occurring, rather than the severity of the effect, is a function of dose. Nonstochastic effects are those for which severity is a function of the dose and for which there may be a threshold of dose below which there is no effect.

The most impressive counterargument to assigning the preceding definition to radiogenic cancer is that there is an accumulating body of evidence that there is some relationship between severity of the cancer and dose. Fry and others (1976) showed that Harderian gland tumors were seldom invasive and rarely metastasized after low doses of low LET radiation, while after higher doses of low LET radiation, increased invasivity and metastatic potential resulted. Burns and others (1975), Ullrich and Storer (1979), and others have suggested also that there may be cancers for which there is indeed a threshold dose. In spite of these caveats, the stochastic–nonstochastic model remains useful and is in widespread application, especially for risk assessment.

Bases for Our Knowledge of Radiation Carcinogenesis

The three important sources of information for the understanding of carcinogenesis produced by ionizing radiation are as follows:

1. Studies of radiation carcinogenesis in experimental animals.
2. Studies of "transformation" of cells grown in tissue culture, an end point that is thought to be analogous to the processes of the early stages of carcinogenic transformation of cells in living animals.
3. Studies of irradiated populations of human beings exposed either accidentally, for medical purposes, or as the result of atomic bomb detonations in World War II.

Each of these subjects will be discussed in the following sections.

RADIATION CARCINOGENESIS IN EXPERIMENTAL ANIMALS

The induction of cancer in experimental animals had been known well before the unequivocal demonstration of the link between ionizing radiation exposure at low doses, such as those received by radiologists, and cancer in man. In the 1930s, experimenters were demonstrating the increased incidence of various leukemias in mice, and one landmark study of that time, but by no means the first, was that of Furth and Furth (1936). In 1958, Upton and others published extensive information on the dose–response relationships for the induction of myeloid leukemia and lymphoid leukemia in mice. The data for myeloid leukemia are shown in Figure 12.1. The incidence of lymphoid leukemia was much lower for all doses used in this study. The relationship shown in Figure 12.1 is prototypical

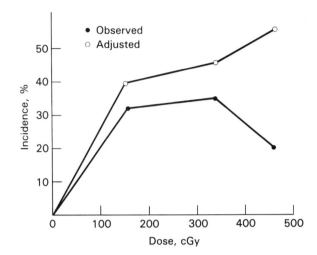

Figure 12.1 Incidence of myeloid leukemia in RF mice as a function of radiation dose (Upton and others, 1958). The lower curve (solid data points) is the observed incidence without correction. When the data are corrected for intercurrent mortality from diseases unrelated to the leukemia, the upper curve (open data points) results. (The data of Upton and others on lymphoid leukemia have not been presented in this figure.)

for many of the dose–response curves of cancer induction for most tumors in experimental animals. Only the dose and incidence scales will be different. These curves illustrate a number of important points that are general for animal tumor systems and probably equally applicable in a qualitative sense for tumors in human beings.

1. There is a nonzero incidence of the tumor in animals that have never been irradiated.
2. The curve of incidence versus dose rises sharply for low doses, and this curve usually has a complicated dose relationship not describable by a first-order equation.
3. There is a maximum value for the incidence obtainable.
4. Often there is a decreasing incidence for doses higher than the maximum, but these decreases often disappear when corrections are made for intercurrent death from other causes.

The correction for intercurrent death is well illustrated in the data of Figure 12.1. Deaths from other causes in the irradiated population remove individual members of the population who might have developed the tumor of interest if they had survived long enough.

The shape and nature of the dose–response curve has probably been among the more important contributions made from studies in experimental animal carcinogenesis. The quantitative description of the risk associated with ionizing radiation exposure for human populations cannot be directly derived from animal studies. That is clear. However, the great range of studies that has been undertaken in animals provides a reasonable understanding of the underlying biology associated with curves of the type seen in Figure 12.1. We have every reason to expect that the qualitative aspects of those curves will apply for human carcinogenesis resulting from radiation exposure.

The data in Figure 12.1 are for significantly higher doses than are of interest for the protection of the public, for example, 25 cGy to a few gray is generally the range of doses used for the animal studies, while the concern for human exposure is in the range of <1 mGy to 1 cGy. The important datum that is required for human radiation protection is the slope of the curve of incidence versus dose at these very low doses. The initial slopes for the curves in Figure 12.1 are essentially unknown because no data were developed for doses below 150 rad (1.5 Gy), and it is unlikely that animal studies will lead to an elucidation of the shape of this low-dose portion of the curve for low LET radiation.

Why is there a maximum for the incidence of tumors in irradiated animals? It has been suggested and confirmed from cell transformation studies that, as the dose of radiation increases, the killing of potentially transformed cells becomes an important consideration. Accordingly, the number of cells remaining alive to ultimately produce a tumor is reduced at very high doses. This explanation is not entirely satisfactory for explaining the continuous plateau in the dose–response curve seen for many tumors. Literally hundreds of studies have been done on animal models for radiation carcinogenesis, and it would be unproductive to review these reports in detail. Upton (1986) has, however, provided us with a very concise summary of the findings of all these experiments, as follows:

1. Neoplasms of almost any type can be induced by irradiation of an animal of suitable susceptibility, given appropriate conditions of exposure.
2. Not every type of neoplasm is increased in frequency by irradiation of animals of any one species or strain.
3. The carcinogenic effects of irradiation are interconnected through a variety of mechanisms, depending on the type of tumor and the conditions of exposure.
4. Some mechanisms of carcinogenesis involve direct effects on the tumor forming cells themselves, but others may involve indirect effects on distant cells or organs.
5. Though the dose–incidence curve has not been defined precisely for any neoplasm over a wide range of doses, dose rates, and radiation qualities, the incidence generally rises more steeply as a function of dose and is less dependent on dose rates with radiations of high linear energy transfer

(LET)—such as alpha particles or fast neutrons—than with radiations of low LET—such as x-rays and gamma rays.

6. The development of neoplasia appears to be a multicausal and multistage process, in which the effects of radiation may be modified by other physical or chemical agents.

7. At low to intermediate dose levels, the carcinogenic effects of radiation often remain unexpressed unless promoted by other agents.

8. At high dose levels the expression of carcinogenic effect often tends to be suppressed by sterilization of the potentially transformed cells or by other forms of radiation injury, resulting in saturation of the dose–incidence curve.

9. The distribution in time of radiation induced tumors characteristically varies with the type of tumor, the genetic background and age of the exposed animal, the conditions of irradiation and other variables.

10. Because of the diversity of ways in which irradiation can influence the probability of neoplasia, the dose–incidence relationship may vary accordingly.

The foregoing conclusions are quoted directly from Upton (1986), and, as for nearly all matters related to radiation carcinogenesis, there will be disputes among specialists as to the precise tone of each quote. The "bending over" of the dose response curve is easily explained by the cell killing hypothesis; but, as we will see shortly, when the cell transformation model is used, even after correction for cell killing, the curve of transformants per surviving cell still shows a saturation-type curve in some instances. Statement 7 could imply the existence of the threshold effect for radiogenic carcinogenesis. This has been noted through private communication from R. J. M. Fry. In item 10 it might be more informative to suggest that the biology of a specific tumor, including its relationship with host factors, is often more important than the radiation that has been given.

Clonal Theory of Carcinogenesis

Essentially all the observations made in the preceding sections about experimental radiation carcinogenesis are in accord with the widely held concepts of modern theories of carcinogenesis in general. The theory holds that the disease, cancer, which is characterized by three properties, (1) transformation of a cell that brings about a state of unresponsiveness to growth control mechanisms of the intact organism, (2) the ability of these transformed cells to invade surrounding tissues, and (3) the ability of these cells to migrate to other locations in the body and to establish a new growing tumor (metastasis), is the result of a heritable change in the genome of a somatic cell. There is strong experimental support for the clonal model of carcinogenesis. Chromosomal alterations, the role of

DNA repair, misrepair, and replication, and the heritability in somatic cells of the transformed properties all follow patterns that are very similar, if not identical, to the patterns seen in radiation induced cancer. The stochastic nature of the incidence of radiation induced cancers is also supportive of such a model. There is strong evidence from analysis of isozyme patterns within tumors, the consistency and long-term persistence of karyotypic changes in the cancer cell, and special immunoglobulin production patterns that the cancer has arisen from a single cell.

If a heritable, nonreversible change in a single cell may give rise, ultimately, to a neoplasm, then it is required to alter only the DNA of a single cell in such a way that the properties necessary for neoplastic growth are the result, and only a single radiation interaction should suffice, under the proper conditions, for this transformation to occur and, ultimately, to lead to cancer.

Such a simple model, which requires an appropriate change in the DNA of a somatic cell, is not adequate, however, to explain the many interactions that are known to modify the likelihood of a given transformation event developing into a full-blown malignant neoplasm. Both experimental studies on animals and observations on human patients have shown the importance of a number of factors, other than the clonal transformation event, that modulate the final result of the transformation of a single cell. Growth and function modulators in the intact organism have been shown to be particularly important. Tumors of the mammary glands, the ovary, the adrenals, the pituitary and the Harderian gland of the mouse, as well as several other tumor systems, have all been shown to have a significant interaction with the level of hormones to which the normal tissues are usually responsive. The presence or absence of oncogenic viruses has been shown to be of great importance, particularly for murine leukemias, as shown in the work of Kaplan and reviewed by him in 1964 (Kaplan, 1964). Large within-species (that is, strain-related) differences in susceptibility to carcinogenesis induced by radiation, as well as large differences among species, indicate an important role for the genetic makeup of the irradiated subject. Immunocompetence is surely an important modifier in the ultimate determination of whether the transformed cell will survive to be expressed as a neoplasm. Many other factors, such as the state of cellular proliferation and the age dependence of sensitivity to cancer induction, are also intimately involved in the process. Figure 12.2 is intended to indicate some of these various interrelationships.

The general concept of clonal origin for tumors is no longer widely disputed, subject to the interactions of all the modifying influences suggested in Figure 12.2, but new and interesting ideas have recently been presented by Alexander (1985), suggesting that even initiation may require the transformation of several cells to provide an appropriate en-

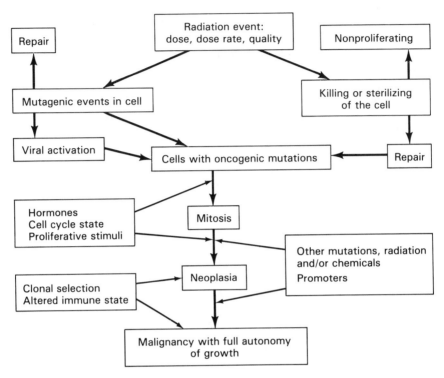

Figure 12.2 The various events from the initial transformation of a single cell into a clonogenic, neoplastic phenotype. Many factors can influence the course of subsequent events until there is full expression of the phenotypic behavior in the fully autonomous malignant cell capable of invasion and metastasis.

vironment for the development of the proliferation characteristics of tumors. These ideas of Alexander are supported by several scattered but well-established observations that some tumors might be multiclonal in origin.

Latency of Tumor Development

All the human data that have been evaluated, as well as the animal experimental data, have shown that there is always a delay between the irradiation of the subject and the ultimate appearance of the neoplasm. For rodent populations, the usual subjects for the study of experimental radiation carcinogenesis, the latency between irradiation and appearance of the tumor can be from a few months to most of the life-span of the animal (2 to 3 years). For human populations, the latency can be as short as 2 to 5 years, as it is for leukemogenesis, or as long as 30 years, as it is for some solid human tumors. The reasons for this latent period are not at all clear, but the existence of a long latency indicates that a number of changes are necessary for ultimate tumor development, either in the transformed cell or in the host relationship to these transformed cells.

Initiation Promotion Hypothesis

A proposed mechanism first suggested by Berenblum (1941) has become part of the classic literature of carcinogenesis. He proposed that carcinogenesis was the result of a two-step process. The first step, which he called *initiation*, was an irreversible and necessary first step. This stage of the process is now associated with the somatic genome alteration process (clonal origin), which may include somatic mutations or activation of an oncogenic virus. These processes or any necessary combination of them may be interrelated. In more modern terminology, this step could be interpreted as derepression or activation of one or several oncogenes.

The second step in oncogenesis, labeled *promotion* by Berenblum, is the result of exposure of the transformed cell to agents that ultimately lead to expression as a visible neoplasm. Promotion is a process that must necessarily happen after initiation and could be related to further as yet ill-defined changes in the already transformed cell, as well as organismic changes such as altered immunosuppressive capability or altered cell replication characteristics. It is this promotion step that is thought by many investigators to be responsible for the long time period from the transformation event until the appearance of a visible tumor.

It is important to understand one of the necessary axioms of the initiator–promoter concept. That is that the initiator need act only once, and its continued presence or activity is not required, while the presence of the promoter, in principle, would be required for some longer time. However, there are known cases where the initiator is also the promoter, and ionizing radiation seems to act in both roles.

Dose Rate Effects

A general observation for radiation carcinogenesis is that decreasing the dose rate of the irradiation process reduces the effectiveness of the radiation for cancer induction. The models just described suggest that the observed reduction of effectiveness with lowered dose rate is intimately connected with the ability of the irradiated cell to appropriately repair damage to the DNA molecule. Since the greater effectiveness of high LET radiation is also related to DNA repair or lack of it, we would expect that high LET radiations would also be more effective than low LET radiations for carcinogenesis. This is indeed the case, presumably because the lesions associated with the high LET radiation are more difficult to repair. Whatever the underlying mechanism, it remains the case that lowering the dose rate will generally decrease the effectiveness of low LET radiation for tumor production. For this stochastic process, however, there is a dose level below which the dose rate should be irrelevant. If the total dose delivered is such that the cell "sees" only one ionizing event, then the time during which multiple events might be delivered is irrelevant.

If the time over which the irradiation exposure is protracted is more than a few hours, other complications are introduced into the biological outcome. Repair at the level of the DNA molecule is generally thought to be effective on a time scale measured as a few hours. If protraction is longer relative to DNA repair time, not only will repair occur, but re-population of tissues to replace cells killed by irradiation will offer new target cells at risk for transformation. If the time of irradiation is long relative to the life-span of the animal, the changing age-dependent sensitivity to radiogenic cancer formation becomes an important factor and can cloud the issue of the effects of protraction.

In Figure 12.2, an attempt has been made to outline most of the important interactive processes in radiation carcinogenesis.

TRANSFORMED CELL IN VITRO

If, as has been suggested in the foregoing sections of this chapter, cancers are of clonal origin, and all the cells of a tumor are the progeny of a single cell that has undergone a transformation, then an experimental procedure that would allow the quick, reliable determination of this alteration in phenotype would be of great value. In terms of the mechanisms of carcinogenesis, the transformation of a single cell into one that is potentially malignant represents the initiation event and may be regarded as the first step in the process described in Figure 12.2. It is essential to realize, however, that such an assay of initiation or transformation is indeed only an identification of the first step. The in vitro determination of the transformation step does not recognize the importance of the subsequent steps that depend on the responses of the intact organism. These other determinants, such as hormonal or immunologic factors, as well as the physiological processes of vascularization, are of great importance but cannot be observed in a simple transformation system.

The in vitro transformation system meets many criteria, however, that suggest that it is, indeed, a model system reflective of the cellular changes that occur in the two-stage initiation–promotion description of transformation. There is a high degree of correlation for the transformation potential of a wide range of chemicals and physical agents and their potential for the production of cancer in the intact animal. Both initiation and promotion can be demonstrated in the system by appropriate experiments. At least for mouse cells transformed in vitro, there is a demonstrable presence of activated oncogenes that can, with modern molecular biological methods, be transfected into recipient cells, which in turn are transformed.

In Vitro Cell Transformation Systems

It would be ideal, for transformation studies, to have cell lines that were pertinent for human and animal cancers. Since well over 90% of human cancers are of nonmesenchymal origin, it would be preferable to use cell lines of epithelial origin. However, the two principal systems in use are of fibroblastic origin. These two systems are significantly different in their characteristics. One of these, cells from Syrian hamster embryo primary fibroblast culture, represents a cell line that is not immortal in in vitro culture. These cells are obtained by subculturing cells that grow out of a primary culture of minced hamster embryo cells. The hamster system has the advantage that the cells are euploid in their karyotype and have not undergone the limited transformation that is required for a cell line to acquire the property of unlimited growth (immortal culture). The hamster cells, though, when assayed as transformed are not tumorigenic when transplanted into syngeneic recipients. The tumorigenic trait is acquired only after multiple tissue culture passage for transformed lines arising from the hamster embryo culture. The radiation transformation of Syrian hamster embryo cells was first studied by Borek and Sachs (1967).

For hamster embryo cell lines, the transformation process is measured by using a variant of the colony assay. Cells are plated at low density on a plastic tissue culture plate. These cells are treated either by exposure to the potential carcinogen just before plating, or they are exposed on the culture dish after plating. The cells are allowed to grow to macroscopic colonies, usually about 10 days for this tissue system. At the end of the growth period the cells are fixed on the plate and stained. If no transformation events have taken place, the cells will have grown in an orderly fashion and individual colonies will be flat, with the cells growing in a single layer. Clones that are transformed will grow in a "piled-up" fashion and with very disorderly arrangement. These latter are scored as transformed clonogens. Such transformation foci are easily distinguished in the fixed and stained plate as deeply staining, dense accumulations of cells. It is essential in the transformation assay to allow for the death of cells (or, more precisely stated, loss of clonogenic potential of the cells) as the result of exposure to the carcinogen. Any cell that is incapable of further division is incapable of expressing its transformed potential. The results of the transformation assay are usually corrected for loss of clonogenic potential and expressed as the transformation frequency per surviving cell.

The second cell system widely used for transformation assay is one based on immortal cell lines that have arisen from mouse embryonic tissue culture. Mouse fibroblasts that have been grown for many generations in in vitro culture are the cell lines of choice for this purpose. Cell lines are chosen that are highly sensitive to contact inhibition. *Contact inhibition* is the property of a cell line such that, when a growing clone grows to dimensions where cells are in contact on all sides with contiguous cells, growth stops. These cell lines are immortal and aneuploid, so they have undergone some of the steps necessary for ultimate tumorigenic trans- formation. They do still possess some of the traits of nontransformed cells, such as contact inhibition and the failure to produce tumors in syngeneic animal hosts.

In contrast to the colony assay used for measurement of transfor- mation in the hamster embryo cell system, transformation in the mouse fibroblast cell line assay uses an end point called the *focus assay*. The preliminary treatment is the same as that for the hamster system; how- ever, the cells are allowed to grow to total confluence on the plate. This usually takes four or five weeks of culture. The cells are fixed and stained, and foci of transformation, seen as heavily staining overgrown areas on a background layer of lightly stained cells, are counted as transformed cells. The transformation frequency is corrected for the fraction of cells dying as a result of the treatment with the carcinogen. To express the data in this fashion, it is necessary to independently determine the sur- viving fraction of cells as a function of the dose of the chemical or physical agent being used as a transformer. The surviving fraction for the dose of carcinogenic agent used is read from the survival curve, and the number of plated cells in the transformation assay is corrected for the expected death of cells. Spontaneous transformation, that is, without treatment, is a rare event in these cell lines, and no correction need be made.

Whereas transformed cells isolated from the hamster embryo culture do not have the ability to establish tumors in appropriate host animals without further treatment and passage, the transformed mouse cells fre- quently are tumorigenic in appropriate hosts. A wide range of phenotypic states can be found in the transformed mouse cells, from nontumorigenic early initiation states to full-blown tumorigenicity on transplant.

The important difference between the mouse and the hamster sys- tems is that for the former some degree of transformation has already taken place to offer a cell line immortality in vitro. Therefore, with the mouse line, transformation steps caused by carcinogens are presumably late stages in the multistage process of transformation. The hamster embryo system represents transformation at its earliest stages. Both sys- tems provide important insights into the stages of carcinogenesis by ra- diation, but at different points in the process.

The mouse cell lines most often used for transformation studies are the C3H/10T-$\frac{1}{2}$ line and the Balb/c/3T3 line.

Dose–Response Relationship for Transformation

There are significant differences in the dose–response relationship for the hamster embryo cell line and the mouse cell lines.

Mouse cell lines. Figure 12.3 is a typical representation of the dose–response relationship for a mouse cell line, in this case, the C3H/10T-½ line. Data are provided in this figure both for x-rays and for high LET neutrons. For the moment the x-ray curve is the object of attention. There is still significant dispute about the exact nature and shape of this curve at low doses. The present consensus is that at very low doses, below 30 or so rad (cGy), the slope is linear or nearly so. As the dose increases, particularly in the range of 150 to 400 rad (1.5 to 4.0 Gy), there may be a quadratic component. At high doses all workers agree that the frequency

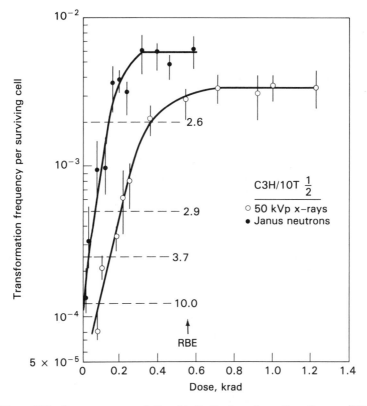

Figure 12.3 Dose–response relationship for the transformation of mouse C3H/10T-½ cells in the in vitro transformation assay. Open symbols are for transformation by 50 kVp x-rays, and the solid symbols for fission neutron irradiation. The RBE at several levels of transformation are indicated. (Reproduced from Han and Elkind, 1979, by permission of the publisher and the authors)

per surviving cell reaches a plateau. If the frequency of transformation were plotted without correction for survivorship, there would be a descending arm on the curve, reminiscent of the findings for in vivo animal tumor systems.

Hamster embryo fibroblast system. The dose–response relationship for this system appears to be a great deal more complicated than that for the mouse lines. Borek and Hall (1973) found that for low doses, below 100 rad (1 Gy), there appeared to be a linear relationship between transformation frequency per surviving cell and dose. For larger doses they found that there was a decreasing likelihood of transformation per unit dose, and therefore the curve tended to have a plateau region. Larger doses (above 200 rad; 2 Gy) again showed somewhat of an increase in transformation frequency per unit dose. These findings are presently without theoretical explanation.

Effects of Dose Fractionation and Dose Rate

A wide range of experiments has been done in both the mouse and hamster systems evaluating the potential for repair of the transformation process. With one significant exception, the findings have been uniform, that either protraction of the dose or fractionation of the dose leads to a lessened frequency of transformation. Most workers have concluded that there is significant repair of whatever lesion is induced in the DNA, if time for such repair is provided. The significant exception to the rule that fractionation reduces yield of transformants is that of Borek (1977), in which split dose administration is reported to have increased transformation rates with two split doses of radiation if the total dose was less than 100 rad (1 Gy). If the total dose exceeded 100 rad, the split dose combination was less effective than a single dose of the same total size. A difficulty in interpretation of these divided dose experiments with the Syrian hamster embryo system with its complicated dose–response relationship is that repair following the first dose may, in effect, reduce the dose equivalence of the first dose so that the second dose is at such a position on the dose–response curve that the effectiveness of the second dose is enhanced. When appropriate consideration is made for this fact, all experiments confirm that there is significant repair and/or recovery at all dose levels.

High LET Radiation Effects

Studies of the relative biological effectiveness of high LET radiations have been done in many laboratories with neutrons, alpha particles, and other charged particles, ranging from helium ions to silicon ions and higher. The findings are, again, consistent, and the data in Figure 12.3 are typical. These data demonstrate that the high LET radiations are more effective than x-rays, as might be expected, and that the shape of the dose–response curve for the neutrons, as compared to x-rays, is such that the RBE increases as the dose is decreased. Generally, at low dose levels, with all the high LET sources used, RBEs in the range of 10 to 20 are observed. It has also been shown that the peak effectiveness is seen for high LET beams of approximately 150 to 200 keV/μm.

Consistent with findings in other biological systems, repair has been shown to be sharply decreased or nonexistent when high LET beams are used as the source of radiation. Scattered reports have appeared in which evidence has been given that fractionation of high LET radiation will increase rather than decrease the yield of tumors. No satisfactory explanation has yet been given for these observations, although they have been verified often enough that there is no doubt of their validity.

Promoters and Cell Transformation

Ideally, the cell transformation system should be equally suitable both for experimental examination of the initiation event and for the examination of the posttransformation promotion effects that are inherent in the two-step model of carcinogenesis. Indeed, in spite of the difficulty of maintaining the necessary prolonged contact of promoter and the transformed cells, promotion has been successfully carried out in at least the mouse system. TPA (12-O-tetradecanoyl-phorbol-13 acetate) is an ester of phorbol that has been shown to be an effective promoter ingredient of croton oil, the classic tumor promoter. Kennedy and others (1978) showed that TPA could be demonstrated to be a promoter for 10T-$\frac{1}{2}$ cells that had been transformed with radiation. Repeated exposure of the irradiated cells was required for the six weeks of expression used in their experiments. TPA was added weekly or twice weekly in the exchange of medium for the growing cells. TPA itself could not produce transformation in the absence of preirradiation. The results are shown in Figure 12.4. Since the dose–response relationship for the cells untreated with TPA showed a very low transformation frequency at low doses while the promoter treated cells had a linear relationship of transformation frequency with dose, the relative effect of the promoter was much greater at low dose.

These studies show the importance of the multistep process of transformation leading to tumorigenic potential for the ultimate transformed cell, but unfortunately very little insight is provided as to the mechanisms by which the promoter effects the increase in transformation frequency.

The transformation system for cells grown in vitro has been a useful tool for the examination of at least part of the process of carcinogenesis by radiation. It must, however, be borne in mind that the system lacks

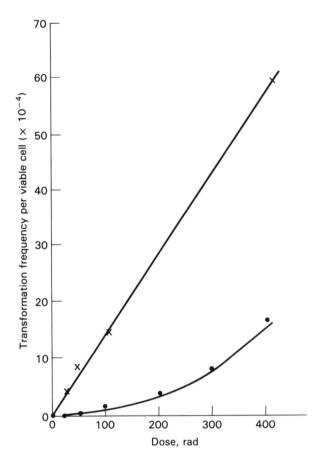

Figure 12.4 Transformation frequency for mouse 10T-½ cells after x-ray alone (solid symbols, •) or by x-rays followed by treatment with the phorbol ester promoter TPA (crosses, ×). The cells were exposed to the TPA for six weeks after irradiation. (Reproduced from Little, 1981, by permission of the author and Academic Press)

the complicated responses and the modulators that exist in the intact animal system. The hormonal status of the host, its age, sex, and genetic makeup, have all proved to be extremely important parts of the radiation carcinogenesis process. In vitro systems cannot effectively examine these interlocking processes that are host dependent.

RADIATION CARCINOGENESIS IN HUMAN POPULATIONS

Radiogenic cancer in human populations is now such a well-documented event that there remains only the quantitation of the dose–response relationship between radiation exposure and the incidence of the many types of cancer that are the result of such exposure. This goal of quantitation of risk, however, still remains unfulfilled. Extensive epidemiological analysis of populations that have been exposed to ionizing radiation for one reason or another have been carried out. The best detailed reports and analyses of these studies are to be found in the scientific reports of the United Nations Scientific Committee on the Effects of Atomic Radiation (UNSCEAR) and the Committee on the Biological Effects of Ionizing Radiation of the National Academy of Sciences (BEIR). The latest relevant UNSCEAR scientific report was published in 1977, and that of the BEIR Committee was published in 1980. An important new analysis that has not yet been incorporated into the overall analyses conducted by UNSCEAR and BEIR is that of Preston and Pierce (1988). These authors have reanalyzed the data on excess cancer among the populations at Hiroshima and Nagasaki with newly available data for the radiation doses delivered at both of these cities. The new dose estimates significantly lower the doses and also reduce the contribution from neutrons (high LET radiation) to the total dose. The result is a significant increase in the risk coefficients for both the development of excess leukemia and excess of other solid tumors. For the present, these new data are not used in the following sections on risk assessment, since the BEIR and UNSCEAR formulations have not yet responded to them, but there will be a closing remark suggesting what impact the newly derived risk coefficients might have.

A number of groups of exposed individuals have been available for epidemiological evaluation of radiogenic cancers, among which the following are the most significant.

OCCUPATIONAL EXPOSURE

1. Radiologists exposed in the course of their practice, particularly before modern procedures and equipment were in place.
2. Workers in uranium and other mines exposed to radon and radon daughters in the workplace.
3. Radium dial painters who ingested radium in the course of painting luminous dials. These were mostly women who were young at the time of their exposure.

MEDICAL EXPOSURE

1. Patients treated for ankylosing spondylitis (a chronic inflammatory disease involving the spine and adjacent tissues) with either x-rays or radium.
2. Women treated for benign (nonmalignant) disease of either the pelvic region or the breast.

3. Women patients in tuberculosis sanitoria who were subjected to multiple fluoroscopic procedures to the chest in the course of treatment of their disease.

4. Infants and children who were treated for two benign conditions, enlarged thymus glands or tinea capitis (ring worm of the scalp).

5. A group of children who were exposed in utero for diagnostic purposes during the pregnancy of the mother.

NUCLEAR WEAPON DETONATION RELATED EXPOSURES

1. The survivors of the nuclear weapon explosions in Japan in the cities of Hiroshima and Nagasaki.

2. The inhabitants of several islands in the Marshall Island chain who were exposed to downwind fallout from nuclear testing in the Pacific.

3. Workers and military personnel possibly exposed on site at the nuclear weapons testing ranges in Nevada and the Pacific.

4. Residents of the western states who were exposed or potentially exposed to downwind fallout from the atmospheric nuclear weapon testing at the Nevada test range.

This is by no means a complete list of all the human populations available for the study of the induction of cancer as the result of exposure to ionizing radiation, but it is provided as a measure of the extent of the epidemiological data base that is available for the quantitative analysis of the relationship of the incidence of additional cancers in the population and the dose of ionizing radiation. Some of these population subgroups have been studied more exhaustively than others, for example, the survivors of the nuclear weapon detonations in Japan. The Atomic Bomb Casualty Commission and its successor agency, the Radiation Effects Research Foundation, has followed the surviving population for over 40 years. The other groups that have been of great importance in the establishment of radiogenic cancer risk criteria are the ankylosing spondylitis patients and the tuberculosis patients subjected to repeated chest fluoroscopy.

Each group studied has severe limitations on the utility of the derived risk estimates. For all the groups, it has been necessary to develop dose estimates after the fact. For all the groups for which measurable increased cancer incidence is found, the dose range over which the increases are observed has a lower limit, which is always very high by comparison with regulatory exposure limits. The dose rate at which the exposure occurred was highly variable, from very high (nearly instantaneous delivery of the total dose) in the case of the Japanese victims to working lifetime exposure in the case of the uranium miners. The volume distribution of the dose within the irradiated subject is also highly variable, ranging from nearly whole body exposure in the Japanese to small volume partial body exposure in some of the diagnostic medical procedures.

APPROACHES TO RISK ESTIMATION

There are two important considerations in analyzing the potential for radiation exposure to produce an excess number of cancers over that which would be expected in the control population. These are the shape of the dose–response relationship from which we can derive *risk coefficients* and the nature of the assumption made as to the underlying mechanisms of radiogenic cancer. For the latter issue, it is necessary to decide whether the effect of the radiation is to simply add a fixed number of cases per unit dose, independent of the natural incidence of the disease, usually called the *absolute risk*, or is it necessary to adjust the number of cases produced by a given dose of radiation, depending on the natural incidence of that disease at the time of irradiation. The latter is referred to as the *relative risk model*. The absolute risk model is simply additive to the natural incidence. The relative risk model is multiplicative.

Shape of the Dose–Response Relationship

In the ideal case, suppose that a population of individuals, all of the same age and sex, were irradiated with a series of increasing doses, and the excess incidence of cancer was determined in that group. The dose–response curve could be plotted and a functional fit to the data could be made that would be predictive for a similar group in future exposures. Such a procedure is not possible, and existing epidemiological evidence must be analyzed. From the biophysical models of radiation interaction that were described in detail in Chapter 8, at least two dose–effect relationships have come into widespread use in the interpretation of increased incidence of radiogenic cancer. The two relationships are discussed next.

Linear dose–response relationship

$$I_D = I_n + a_1 D \qquad (12\text{-}1)$$

where I_D is the observed incidence of cancer for dose D, I_n is the normal incidence of the disease in the absence of radiation exposure, a_1 is the *linear risk coefficient*, and D is the dose.

Linear–quadratic dose–response relationship

$$I_D = I_n + a_1 D + a_2 D^2 \qquad (12\text{-}2)$$

where all the variables are as in equation 12-1, with the addition of α_2, which is the *quadratic risk coefficient*.

Equation 12-1 is simply a degenerate form of equation 12-2, in which the quadratic risk coefficient is taken to be zero. Another alternative is that the linear risk coefficient is zero. In that case equation 12-2 degenerates to a simple quadratic expression. This alternative has not found wide acceptance, but it has been considered by some as a possible fit to some radiogenic cancer data, especially for high doses of low LET radiation.

As in the case of cell transformation studies, which were described earlier in this chapter, if doses are high enough to cause cell death, those cells that have lost their clonogenic potential cannot give rise to a growing neoplasm, and corrections must be made for their loss. It has become the practice in estimating radiogenic cancer risks to assume that loss of clonogenic potential is described by the molecular model of cell killing. The procedure is then to correct the radiogenically induced excess incidence by multiplication by a term equal to the surviving fraction of clonogenic cells. This then provides us with a general expression for estimation of the increased incidence of radiogenic cancer:

$$I_D = I_n + (\alpha_1 D + \alpha_2 D^2)(e^{-\beta_1 D - \beta_2 D^2}) \tag{12-3}$$

where all other parameters are as in equations 12-1 and 12-2, and β_1 and β_2 are the linear and quadratic coefficients for loss of clonogenic potential. To avoid misunderstanding, it is important to note that β_1 and β_2 are the α and β coefficients of the molecular model.

The expression in equation 12-2, in which the cell killing is presumed to be negligible, can be assumed to be related to the frequency of transformation through a multiple-step process to cells that are then capable of expressing the neoplastic phenotype. The two risk coefficients, α_1 and α_2, have the properties associated with the linear and quadratic coefficients in the molecular model of radiation action (sometimes referred to as the α/β model or the L/Q model). The linear coefficient will predominate in the action of high LET radiations, while the quadratic coefficient will predominate for low LET radiations except at very low dose and/or dose rate. When $D = \alpha_1/\alpha_2$, the contributions of the linear and quadratic terms to the total transformation will be equal.

It has been pointed out very clearly in the report of the BEIR Committee (1980) that the bias introduced into estimates of the risk coefficient for cancer production depends strongly on the assumed shape of the dose–response curve, as well as on the range of doses over which data are available. Since most available data are in the high-dose range, if data are fitted to a linear model over the range of dose information available, and if the underlying "true" relationship is best described by the linear–

quadratic model, then the predicted cancer incidence at low doses will be higher than actual. The bulk of experimental data from animal carcinogenesis, as well as from cell transformation experiments, would lead us to expect that the linear–quadratic model is the best descriptor of radiogenic human carcinogenesis. There are, however, no data on human beings that will unequivocally support either the linear or linear–quadratic model. Indeed, the quadratic model cannot be excluded on analysis of available data on human populations.

The 1980 BEIR report subjects the data on incidence of leukemia in survivors of the Hiroshima–Nagasaki bomb detonations to a comparative analysis of the goodness of fit of these data to the three model alternatives. The chi-square test for goodness of fit does not provide a basis for rejecting any of the three models. Assuredly, all the Japanese data are presently being subjected to intensive reanalysis as mentioned earlier (Preston and Pierce, 1988) as the result of new information about the doses received, but only the quantitative value for the risk coefficients are likely to change significantly. Figure 12.5 is a graphic presentation of the three models for a mythical set of data, which are presented to indicate the impact of the model applied on estimates of excess radiation induced cancers at low dose. The values for a_1 and a_2 for the best fit to the Japanese leukemia incidence data before the recent reanalysis were used to calculate the curves shown. A good bit of editorial license is involved, since no account was taken of the additional linear component that would be contributed by the neutron radiation at the two locations. For this reason, the excess incidence as shown should not be given serious consideration.

Latent Period

It has been observed that the time from the irradiation exposure until the appearance of a detectable cancer in any given organ may be very long, and this latent period will vary greatly among the organs in which cancer may occur. This fact must be carefully noted when epidemiological studies are undertaken to demonstrate the dose–response relationships for radiogenic cancers. Among human cancers the shortest latency is for leukemias, which can be observed in excess incidence as early as two or three years after the radiation exposure. On the other hand, latency for cancer of the female breast and of the lung appears to be as much as 30 years. It is obvious that the age at irradiation is important, even if there were no age-dependent variation of the sensitivity to cancer induction. The individual irradiated late in life will probably not experience an excess tumor induction simply because he or she will not live beyond the latency period.

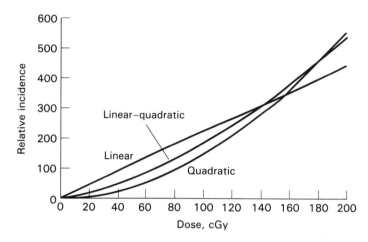

Figure 12.5 Comparison of hypothetical fits to data for excess cancer incidence as a function of dose. It is assumed that the data contributing to the fit exist principally in the dose range from 100 to 160 cGy. Because the data are edited to make the point of the graph, the incidence values should not be given serious consideration. The units are, however, excess cases per million per year.

Absolute Risk versus Relative Risk

Two general epidemiological models are in use for the extrapolation of radiogenic cancer risks from the existing data (data that are acknowledged to generally be at high doses) in a manner that would provide reliable estimates of increased incidence of the disease at the much lower doses that are of interest in radiation protection. The *absolute risk model* is based on the assumption that the risk associated with a specific dose level is a determinate number that is wholly independent of the risk for the spontaneous occurrence of the same disease in the absence of radiation exposure. For this model the number of excess cancers induced by radiation simply adds to the number predicted from natural incidence rates for the population. In other words, the risk is *additive* to the natural risk.

The *relative risk* model is based on an assumption that the likelihood of a radiogenic cancer is directly related to the natural incidence of the disease. Since the natural incidence of most cancers increases throughout life, the relative risk model suggests that the number of radiogenic cancers from a given dose will increase with age, at least as long as there remains adequate residual life-span for the predicted latency for tumors in the organ in question. In other words, the risk is *multiplicative* of the natural risk at the time of irradiation.

The terms relative and absolute risk are in wide use in the epidemiological community, but are not always well understood outside of that professional group. For those wishing a more detailed development of the

two approaches, the text by Armitage and Berry (1987) is highly recommended for a more detailed development than the following.

In a mathematical formulation the two models can be expressed as
follows:

$$\text{Absolute risk:} \quad I_D = I_n + f(D) \quad\quad\quad (12\text{-}4)$$

where $f(D)$ describes the excess disease as a function of dose. At $D = 0$,
$f(D) = 0$.

$$\text{Relative risk:} \quad I_D = I_n \times f'(D) \quad\quad\quad (12\text{-}5)$$

where $f'(D)$ is a multiplier on natural incidence. At $D = 0$, $f'(D) = 1$.

In these expressions, $f(D)$ and $f'(D)$ are functions of dose in a suitably
chosen formulation that describes the excess incidence of cancer as it is
related to dose. For example, the linear–quadratic expression given in
equation 12-3 could be the source of data for the development of these
functions. It is necessary to make certain conversions in the excess frequency information for use in the risk expressions given here. As an
example, suppose that for dose, D, the linear–quadratic expression provides an estimate of excess cancers for the population at risk and that
the estimate for dose, D, is that 100 additional cancers would be expected
in a population of 10^6 individuals. The risk per individual is then 10^{-4}
for the development of a radiogenic cancer. The value of $f(D)$ for the
absolute risk model would then be (assuming that I_D and I_n are stated
in terms of cases per million people exposed to dose, D) $10^{-4} \times 10^6$, or
100 cases, and I_D would be equal to $I_n + 100$.

The estimate of $f'(D)$ for the relative risk model is a little more
problematical. The formalism is that, since $I_D = I_n \times f'(D)$, then $f'(D)$
$= I_D/I_n$. Using the values just applied to the absolute risk model, I_D is
equal to $10^6 + 100$ and $I_n = 10^6$; therefore, evaluating, $f'(D) = 1.0001$.
The difficulty with this value derived for the relative risk is that it is
unique to the age and/or sex distribution of the population at risk. For
human epidemiological studies, the distribution of ages and sex is usually
strongly skewed by the existing circumstances associated with the radiation exposure. For example, the patients treated for ankylosing spondylitis are generally over 30 and frequently male. It remains a thorny
problem of how to derive appropriate relative risk coefficients for populations of a specific age distribution. The usual compromise is to use the
value as just derived for the model calculation as a universal risk coefficient for dose D.

The epidemiological tables of risk versus dose for a given population,
for instance the Japanese survivors, cannot give us much insight into
which of these models are likely to be most representative of the real
circumstance. What is recorded is excess incidence of a given tumor type,
possibly on a population with a typical life table distribution of ages. The

TABLE 12.1   Lifetime Risk of Cancer Mortality Induced by Low
             LET Radiation Excess Deaths per Million per Rad[a]

	Absolute	Relative
Single exposure to 10 rad	77	226
Lifetime exposure, 1 rad/yr	67	182

[a] The authors warn that the results are averaged over dose, and the
data should not be used to predict incidence rates for exposure to 1 rad
(1 cGy).

data are known as a function of dose, not as a function of underlying
natural incidence rates. We must therefore turn to animal experiments
for some insight into which model is most appropriate. Storer, Mitchell,
and Fry (1988) examined the goodness of fit of data on nine animal tumor
systems. Of these sets of data, five data sets required the relative risk
model and did not fit the absolute risk model. For two data sets, either
the relative risk or the absolute risk model was a satisfactory fit. For one
other set, neither model was a satisfactory fit. In no case was it possible
to reject the relative risk model in preference to the absolute risk model.

How sensitive are the projections of excess cancer risk as a function
of the relative and absolute risk models? BEIR (1980) has given projections
for several radiation exposure scenarios. One of these is exposure to a
single dose of 10 rad (10 cGy). The second is for a lifetime exposure to
one rad (1 cGy) per year. Using the L–Q dose–response model, the excess
cancers given in Table 12.1 were predicted.

ORGAN-SPECIFIC RADIOGENIC CANCER IN HUMAN BEINGS

The estimation of risk coefficients for cancers of specific organs is presently
in a state of intense reexamination. As mentioned earlier, it has been
discovered that the estimates of the doses received by the survivors at
Hiroshima and Nagasaki were in error as the result of incorrect assumptions about the radiations from the weapons that were detonated
at these sites. The reestimations of the dose are now complete, with, in
general, reductions in the neutron doses with their assumed high RBEs
for cancer induction and some modifications downward in the gamma ray
dose. As of this writing, the reestimation of the risk coefficients is only
just appearing. These early analyses suggest that the absolute risk coefficients will be not quite double the values that have been used by BEIR
and UNSCEAR. Further data will be released as the analyses continue.
A recent report from the annual meeting of the National Council for
Radiation Protection and Measurements (1988) is completely given over

to an analysis of the impact of the new dosimetry on risk estimates. In this report it is noted that, in addition to the increase in the risk coefficients, there is a better fit of the excess cancer data to a linear dose model than with the earlier dosimetry, and the data in the two cities are consistent.

In spite of the uncertainties, it is important to examine the relative sensitivities of individual organs to radiogenic cancer induction, and for this reason the presently published values for risk coefficients will be used. The principal current sources for the risk coefficients for radiogenic cancer in an organ by organ fashion are the United Nations Scientific Committee on the Effects of Atomic Radiation (report of 1977), and the BEIR Committee (report of 1980). An overview of these risk estimates was presented by A. C. Upton in a 1982 review in *Scientific American* (see Suggested Additional Reading).

The central theme we find when examining risk coefficients for specific organs is that the sensitivity for radiogenic cancer induction varies widely. As mentioned earlier, the latency for the onset of these diseases also varies widely, from less than 5 years for leukemia to 30 years or so for the female breast. As time has passed and as new data have accumulated, our view of relative sensitivities has changed greatly. When the data on leukemia incidence in the Japanese survivors and in other groups were evaluated, its short latency led observers to predict that the bone marrow would be the most sensitive organ for radiogenic cancer. That view has changed drastically as time has permitted the expression of radiogenic cancers in other organs. The two organs now believed to be the most sensitive for radiogenic cancer are the female breast and the lung. The risk coefficients for the various organs are given in Table 12.2,

TABLE 12.2 Organ-specific Cancer Rates in Human Beings

Organ	Risk (Fatal Cancers)[a]	Risk (All Cancers)[a]
Skin	2	None
Lymphatics	15	30
Thyroid	10	120
Esophagus	15	30
Female breast	100	200
Lung	130	140
Stomach, liver, and colon	50	60
Pancreas and other abdominal organs	15	30
Bone	15	30
Bone marrow (leukemia)	40	50

[a] *Lifetime risk* as number of cases observed per 10,000 person-sievert.

which provides some very revealing insights into both the incidence of radiogenic cancer and the effectiveness of modern treatment. For example, female breast cancer is by far the front-runner in induced cancers, but nearly half of them are not fatal. Lung cancer, on the other hand, shows the intractable nature of control for this disease, with the incidence and fatality rate being nearly equal. Cancers of the thyroid rank third in sensitivity to induction of radiogenic cancers, but clearly they are only rarely fatal.

A new concept in dosimetry is reintroduced in Table 12.2. The person-sievert, usually called the collective dose, represents the product of individual doses and the number of persons exposed. In the simplest case, for example, 10,000 people each exposed to 1 Sv of radiation would result in a collective dose of 10,000 person-sievert (10,000 person-sievert $= 1 \times 10^6$ person-rem). Since, in practice, the doses vary significantly among exposed persons, it is convenient to take the sum of the products of dose times number exposed at that dose for all levels of exposure in the group being studied. The justification for such pooling of the dose lies in the stochastic nature of the radiogenic cancer induction process and the usual need to develop risk estimates from groups of individuals all of whom have not had the same dose.

REFERENCES

ALEXANDER, P. (1985) Commentary: Do cancers arise from a single transformed cell or is monoclonality a late event in carcinogenesis? *British J. Cancer* **51**, 453–457.

ARMITAGE, P., and BERRY, G. (1987) *Statistical Methods in Medical Research.* Blackwell Scientific Publications, Oxford. Chapter 16, Statistical Methods in Epidemiology.

BEIR COMMITTEE (1980) *The Effects on Populations of Exposure to Low Doses of Ionizing Radiation.* National Academy of Sciences, Washington, D.C.

BERENBLUM, I. (1941) The mechanism of carcinogenesis, a study of the significance of cocarcinogenic action and related phenomena. *Cancer Res.* **1**, 807–814.

BOREK, C. (1977) Neoplastic transformation following split doses of x-rays. *British J. Radiol.* **50**, 845–846.

———, and HALL, E. J. (1973) Effect of split doses of x-rays on neoplastic transformation of single cells. *Nature* **252**, 450–453.

———, and SACHS, L. (1967) Cell susceptibility to transformation by x-irradiation and fixation of the transformed state. *Proc. Natl. Acad. Sci. U.S.A.* **57**, 1522–1577.

BURNS, F. J., and others (1975) The effect of 24-hour fractionation interval on the induction of skin tumors by electron radiation. *Radiation Res.* **62**, 478–487.

FRY, R. J. M., and STORER, J. B. (1987) External radiation carcinogenesis. *Advances Radiation Biol.* **13**, 31–90.

——, and others (1976) In *Biological and Environmental Effects of Low-Level Radiation*, Volume I, pp. 213–227. International Atomic Energy Agency, Vienna.

FURTH, J., and FURTH, O. B. (1936) Neoplastic disease produced in mice by general irradiation with x-rays. 1. Incidence and type of neoplasms. *Am. J. Cancer* **28**, 54–65.

HAN, A., and ELKIND, M. M. (1979) Transformation of mouse C3H/10T-½ cells by single and fractionated doses of x-rays and fission spectrum neutrons. *Cancer Res.* **39**, 123–130.

INTERNATIONAL COMMISSION ON RADIOLOGICAL PROTECTION (1977) *Recommendations of the International Commission on Radiological Protection.* Report No. 26. Pergamon Press, Elmsford, N.Y.

KAPLAN, H. S. (1964) The role of radiation in experimental leukemogenesis. *Natl. Cancer Inst. Monogr.* **14**, 207–220.

KENNEDY, A. R., and others (1978) Enhancement of x-ray transformation by 12-O-tetradecanoyl-phorbol-13-acetate in a cloned line of C3H mouse embryo cells. *Cancer Res.* **38**, 439–443.

LITTLE, J. B. (1981) Influence of noncarcinogenic secondary factors on radiation carcinogenesis. *Radiation Res.* **87**, 240–250.

NATIONAL COUNCIL FOR RADIATION PROTECTION AND MEASUREMENTS (1988) Proceedings of the Twenty-third Annual Meeting. *New Dosimetry at Hiroshima and Nagasaki and Its Implications for Risk Estimates.* NCRP, Bethesda, Md.

PRESTON, D. L., and PIERCE, D. A. (1988) The effects of changes in the dosimetry on cancer mortality risk in atomic bomb survivors. *Radiation Res.* **114**, 437–466.

STONE, R. S. (1959) Maximal permissible standards. In *Protection in Diagnostic Radiology.* Rutgers University Press, New Brunswick, N.J.

STORER, J. B., MITCHELL, T. J., and FRY, R. J. M. (1988) Extrapolation of the relative risk of radiogenic neoplasms across mouse strains and to man. *Radiation Res.* **114**, 331–353.

ULLRICH, R. L., and STORER, J. B. (1979) Influence of γ radiation on the development of neoplastic disease in mice. II. Solid tumors. *Radiation. Res.* **80**, 317–324.

ULRICH, H. (1946) The incidence of leukemia in radiologists. *N. Eng. J. Med.* **234**, 45–46.

UNITED NATIONS SCIENTIFIC COMMITTEE ON THE EFFECTS OF ATOMIC RADIATION (1977) *Report to the General Assembly.* United Nations, New York.

UPTON, A. C. (1986) Introduction. In *Radiation Carcinogenesis*, A. C. Upton and others, eds. Elsevier Science Publishing Co., New York.

——, and others (1958) A comparison of the induction of myeloid and lymphoid leukemia in x-irradiated RF mice. *Cancer Res.* **18**, 842–848.

SUGGESTED ADDITIONAL READING

FRY, R. J. M., and STORER, J. B. (1987) External radiation carcinogenesis. *Advances Radiation Biol.* **13,** 31–90.

UPTON, ARTHUR C. (1982) The biological effects of low-level ionizing radiation. *Scientific American* **46,** No. 2, 41–49.

————, and others, eds. (1986) *Radiation Carcinogenesis.* Elsevier Science Publishing Co., New York. Part of the series "Current Oncology."

WALBURG, H. E. (1974) Experimental radiation carcinogenesis. In *Advances in Radiation Biology*, Volume 4. Academic Press, New York, pp. 210–245.

13

Stochastic Effects: Genetic Effects of Ionizing Radiation

INTRODUCTION

The emphasis in the earlier chapters of this text has been on the overriding importance of the action of ionizing radiation, predominantly on DNA, leading to damage to the nucleotide bases, breaking of the sugar phosphate backbone, and possibly through covalent cross-linking damage to proteins associated with DNA or other proteins in the locale. The loss of clonogenic potential with sustained vegetative activity, that is, metabolism, has been the central interest of the text to this point. It has been stated that the loss of the ability of an otherwise actively dividing somatic cell to continue its division and production of progeny is almost certainly due to defects in the DNA of that cell that interfere with the replicative process. Since all the genetic information carried by the cell is in the same DNA molecules in which this damage is declared to exist, it is only reasonable to suggest that the same mechanisms that lead to loss of clonogenic potential in somatic cells will also lead to possible loss of, or alterations in, the genomic information carried in the cell. This loss of genomic information is expressed as a change in the gene complement of the cell; and if the genomic change exists in the gametes of the radiated subject, these changes may be passed on to future generations at the time of reproductive activity.

Mutations in germ cells are generally referred to as genetic mutations, while mutations in somatic cells are referred to as somatic mutations. The latter class of changes is not transmissible to offspring, but they can reveal their presence as altered function or phenotype in the cellular progeny of the altered cell. An important class of such somatic mutations is the production of cancer, presumably through alterations in the genome of somatic cells.

The reader is reminded that two classes of damage have been shown to result from ionizing radiation exposure, stochastic and nonstochastic. Both terms have been defined in detail previously. It will be seen that the changes in the DNA of the gametes that are classed as genetic mutations are, indeed, of the stochastic class. They occur on a probabalistic basis with no threshold for the occurrence, and the severity of the phenotypic change resulting from them is unrelated to dose.

Two general types of genetically related events can occur as the result of ionizing radiation damage to DNA. The classification is at best artificial, since there is a continuum in the severity of damage that can be sustained in the genome as a result of radiation exposure. The two classes of change in the genome that have been identified are structural changes visible in the chromosomes at metaphase and gene mutations, which are not generally associated with visible structural change in the metaphase chromosomes.

A general and presumably self-evident principle must be stated. For a cell, either gamete or somatic dividing cell, to express the changes that have occurred in its genome, it must be capable of division. In other words, it must continue to have clonogenic potential.

STRUCTURAL CHANGES IN CHROMOSOMES

Severe damage to the DNA molecules of the cell may be visible by microscopic examination of the cell at the time of cell division. During mitosis the cells can be treated in such ways that the chromosomes can be revealed in their metaphase format, and they can be examined individually for structural defects. Some of the earliest work on chromosome changes resulting from radiation damage was carried out on the chromosomes of the salivary gland of the fruit fly; but since these studies had to be carried out on the offspring of irradiated parents, they are of limited utility, since all nonviable chromosome forms produced in the irradiated sperm will have been eliminated in the course of production of the larva that is examined. The first quantitative studies of chromosomal abberations in the first mitotic cycle following irradiation were done by Sax, starting as early as 1938. These data, collected on *Tradescantia* microspores, are reported in some detail in Sax (1950). Lea, and Lea and Catchside (Lea, 1946) were also prolific contributors to the literature on chromosome alterations in *Tradescantia*.

A major forward step was the development in the 1960s of techniques to examine the postirradiation chromosomes of mammalian and, indeed, human lymphocytes (Bender and Gooch, 1962). It must be borne in mind, however, that no matter which class of cells might be examined the metaphase chromosome represents only a brief portion of the life cycle of the chromosome, and during the rest of the cell cycle the chromosome cannot be examined except by special new techniques that have been developed for the examination of interphase chromosomes (Waldren and Johnson, 1974). During this "invisible" period, much of importance is happening: the molecule is being replicated to form the chromatid pairs seen at metaphase, and much structural change activity involving the packing of the DNA molecule and the regulation of its relationship to control proteins is proceeding. In spite of the limitations, the examination of the metaphase chromosome for structural change is fruitful and has given valuable insight into radiation induced structural alteration in chromosomes.

Chromosome Breakage

Chromosome breakage is one of the more important outcomes of ionizing radiation exposure of cells, and, indeed, ionizing radiation is most effective in producing this class of structural change. Another chromosome defect, other than breakage, that is important in genetic diseases, but that is of little consequence as a result of radiation damage, is maldistribution of chromosomes, such that daughter cells have an incorrect complement of intact chromosomes. Since this latter change is generally not an important outcome of radiation exposure, no more will be said about it.

Chromosome breakage can occur before replication of the DNA of the cell, in which event, at replication, the structural defect will also be replicated (or it will fail in attempted replication), and the defect will be seen in the metaphase chromosome in both chromatids. If breakage occurs after DNA replication the damage will usually be seen as an asymmetric change in one of the two chromatids. As a population of dividing cells is exposed to ionizing radiation, the cells are in all possible stages of progression through the cell cycle. Depending on the location of the affected cell in the cycle, three different types of chromosome abberations can be found. Each of these will be observed in the metaphase chromosome array as that portion progresses forward through the cell cycle time to mitosis. Time from the irradiation event to the collection of mitoses will determine the position of the cell cycle that is being observed. For example, if for a given cell line the length of the S period is, say, 6 hours, and the length of G_2-M is 1 hour, then collection of mitoses at 7 hours after irradiation will provide access to changes that occurred as the result of the irradiation of cells that were on the verge of entering the S period at the time of irradiation.

The three principal types of changes that will be observed through observation of metaphase chromosomes are *subchromatid*, *chromatid*, and *chromosome* changes. Cells that were already in prophase at the time of irradiation will have aberrations of the subchromatid type. This class of aberrations is hard to distinguish, and generally no attempt is made to score changes within a chromatid region. Cells that are in the post-DNA replicative stage (G-2) at the time of irradiation will have mainly chromatid type aberrations on their appearance at metaphase. Aberrations in the chromosomes of cells that were in G-1 at time of irradiation (pre-DNA synthesis) will be of the chromosome type, since the changes will generally be replicated if possible as the cell proceeds through DNA synthesis.

There is a strong driving force of chemical bond energy for the fragments of DNA to rejoin. Many of the breaks that are induced, and probably the majority, will be restored to their normal condition (restitution), and at the time of examination no defect will be seen. This presumption from chemical evidence has been confirmed through the technique for examination of interphase chromosomes called *premature chromosome condensation* and referred to previously (Waldren and Johnson, 1974). In some cases the rejoining will be either to a site that is incorrect or the incorrect end of the broken fragment will join at the point of the initial lesion, causing a transposition of nucleotide order in the final product. Since ionizing radiation is producing many breaks in the genome, it is possible that multiple breaks may occur in the same chromosome or chromatid, and it is also possible that breaks in nearby chromosomes or chromatids may give rise to interactions between these entities. Some of the types of interactions that might occur are diagrammed in the following. The symbolism used is due to Gaulden as found in Dalrymple and others (1973). An alphabetic marker indicates position on an unreplicated chromatid. The heavy dot indicates the location of the centromere, the point of attachment of the spindle fibers that are so important in the distribution of chromosomes to the two daughter cells at the time of cell division. Several types of structural change in chromosomes are known to be the result of irradiation of the cell. These structural alterations are described as follows.

Single-hit breakage. The diagram assumes a strand break as the result of radiation damage at the point marked by the arrow.

A B C · D E F G H I   yields   A B C · D E F   and   G H I
        ↑

The great majority of these lesions will be restored to the original state with no alteration. Those that are not restored can undergo a number of incorrect rejoinings. For example, consider the following.

Inversion of the fragment or rejoining at the wrong end.   The italic fragment, *GHI*, is the fragment formed in the previous example. It can rejoin to its original site in a number of ways.

$$A \ B \ C \cdot D \ E \ F \ I \ H \ G \qquad I \ H \ G \ A \ B \ C \cdot D \ E \ F$$

$$G \ H \ I \ A \ B \ C \cdot D \ E \ F$$

Or the large fragment can join end to end to form a circular chromosome, leaving a small fragment without a centromere. The latter is generally referred to as an *acentric* fragment. The importance of the acentric fragment is that, since it does not have a centromere, even though it undergoes replication normally, the two resulting chromosomes will not be capable of movement to the two daughter nuclei through the action of the mitotic spindle machinery. The fragment and its replica will be sorted randomly into the daughter nuclei. The circular product can replicate normally, and, at mitosis, the spindle machinery will sort the two chromosomes normally to the two daughter cells. The genetic contents of the circular chromosome and the fragment can both presumably be expressed, but, in the case of the fragment, its absence from one daughter cell will cause the loss of its genetic information.

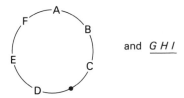

and *G H I*

Two-hit breakage.   In an unreplicated chromatid, two or more breaks may occur in the same chromosome; again the arrow indicates the location of the two strand scissions in DNA.

$$\underline{A \ B \ C \cdot D \ E \ F \ G \ H \ I} \quad \text{yields} \quad \underline{A \ B \ C \cdot D}, \ \underline{E \ F} \quad \text{and} \quad \underline{G \ H \ I}$$

If these fragments are not correctly restored, all possible recombinations can result; for example,

$$\underline{A \ B \ C \cdot D \ G \ H \ I}$$

is a recombination of fragments that has omitted a small fragment from its normal position between D and G. This is normally referred to as an *interstitial deletion*, and $\underline{E \ F}$, an acentric fragment, is left without a rejoining site, since the normal ends of the molecule do not readily accept an addition. The remnant acentric fragment is called a minute fragment or *minute*. The three fragments may rejoin in any of the possible alternative arrangements, of which there are a number, including rejoinings of all three in a disordered array. An example at random is

$$E \ F \ A \ B \ C \cdot D \ I \ H \ G$$

Reordering of the fragments without loss is usually referred to as *inversion*. Depending on the length of the larger fragments, they may also form ring structures, as shown previously.

Multiple hits in replicated chromosomes.   If ionizing radiation damage to the chromosomes occurs after replication is complete, a wide variety of chromosome abberations may occur. Exchange of parts of the DNA between the two chromatids is a possibility, and this can occur in very complex ways. Exchange of parts between two chromosomes is also a possibility. Only simple examples are shown in Figures 13.1 and 13.2.

Figure 13.1 shows one of the possible results of two chromatid breaks in a single chromosome. This lesion occurred after replication. A break at essentially the same location in the two chromatids is called an *isocentric break*. The possible results of such a lesion are shown. The two small ends without a centromere may join as shown in Figure 13.1C, and the products are a chromosome that has lost a very large part of its genomic information but still possesses a centromere; it leaves an acentric remnant that will be incapable of appropriate distribution at mitosis.

An exchange between two chromosomes can give rise to a *dicentric* product, which is frequently seen as the result of irradiation. Movement of parts of chromosomes to another chromosome is called *translocation*. Frequently, a translocation of a portion of a chromosome will leave no detectable structural change, and all genetic information is intact. This change is referred to as a *balanced translocation*.

In the examples of Figure 13.2 of multiple hits in replicated chromosomes the two chromatids are shown in juxtaposition. In the true state

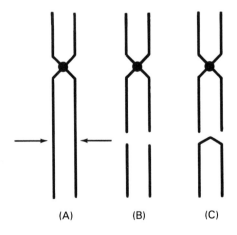

(A)   (B)   (C)

Figure 13.1 An isochromatid break. Both chromatids are broken at the points indicated by the arrow in part A. If restitution does not occur, a joining as shown in part C is possible. The result is a large acentric fragment.

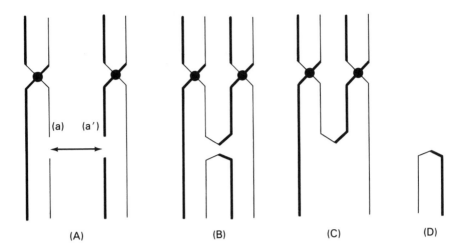

Figure 13.2   A break in one arm of each of two chromosomes is shown in this representation. In panel A the breaks are formed in one chromatid of each chromosome. In panel B the rejoining of the two breaks is such that the two chromosomes are now connected. Panels C and D show the two products, a dicentric chromosome and an acentric remnant. (After Revell, 1974)

of the chromosome, the centromere (·) should be considered as a common junction point for the two chromatids.

The product in Figure 13.2C contains two centromeres, hence the name *dicentric*. One of the special characteristics of mitosis for cells that contain dicentric chromosomes is the formation of dicentric bridges that are fatal to the cell or its postmitotic daughters. As the chromosomes proceed through the mitotic process, spindle fibers attach to the centromere region of the chromosome, one to each chromatid, and, presumably, with the spindle fibers so arranged that the two chromatids will sort to the appropriate daughter nuclei. In normal, undisturbed mitosis, it occasionally happens that the two spindle fibers attached to the centromere may be such that if no correction happened the two chromatids would both go to one daughter nucleus after disjunction. These attachment errors are usually corrected during metaphase and before chromatid movement. In the case of the dicentric chromosome, however, it is possible for spindle fibers to attach to both centromeres of the same chromosome such that that chromosome is moved simultaneously to opposite poles of the mitosis. Such attachments are not recognized as errors by the cell, and the dicentric chromosome remains stranded in the mid-region of the dividing cell. During cytokinesis, the cell wall forms around the chromosome bridge. Such an event, as mentioned previously, would be expected to be fatal, and dicentrics would be expected to be eliminated quickly from the cell population. This is the case.

Another important chromosome aberration that is seen with many agents that cause chemical damage to DNA is the exchange of DNA fragments from one chromatid to another between the two chromatids of a single chromosome. If the net exchange is such that each chromatid still has the same apparent amount of DNA, that is, the exchange is symmetrical, the process is called a *sister chromatid exchange*. Although this process has great importance for chemical mutagens, it appears to play a very minor role in chromosome aberrations resulting from ionizing radiation, and it will not be discussed further here.

Breakage Hypothesis and Exchange Hypothesis

The models presented previously provide very simplified explanations of the complicated processes that lead to chromosome aberrations produced by ionizing radiation, but they are indeed simple and probably entirely oversimplified.

The *breakage hypothesis* was proposed by Sax in 1940 (see Sax, 1950) when knowledge of the molecular biology of the genetic material of the cell was in its infancy. This model was capable of further development with the application of Lea's target theory to the problem. The target theory model would suggest that aberrations that require one break (hit) should follow a dose–response relationship that is first order in dose, and it follows that those aberrations requiring two or three breaks (hits) should follow a dose–response relationship that is, respectively, second and third order in dose. Development of more modern data has shown that the target model does not entirely explain the data for chromosome aberrations. Revell (1955) found that for a lesion called "achromatic gaps," which were defined by him as nonstaining regions of the chromatid, the dose–response curve was first order in dose as predicted by the breakage hypothesis. He found, on the other hand, that the frequency of true chromatid deletions, which should follow a first-order relationship, were best described by the dose to the 1.71 power. This and other data led him to propose what he called the *exchange hypothesis*.

The exchange hypothesis suggests that all chromosome aberrations arise from an exchange process that has much in common with, and may be similar to, the meiotic crossing over that occurs normally during the division stages of the gametes. Radiation initiates the complex process by damaging the chromatids, and if damaged regions come into proximity, the exchange process proceeds. The exchange hypothesis versus the breakage–reunion hypothesis argument is still in full swing after many years. The readers are referred to Revell (1974) for a complete description of the competing models and to any textbook of genetics for a description of the normal crossing over process.

In any case, it is now known that the processes of the production of chromosome aberrations is very complicated, and both the breakage–reunion model and the exchange model are inadequate in light of the findings of modern molecular biology. It is beyond the scope of this text to examine the molecular biology of chromosome aberrations in detail, but the simple breakage model must, to a great extent, be put aside.

GENE MUTATIONS

The structural changes in chromosomes that were just discussed are relatively gross disturbances in the genetic material of the cell, and they often lead to such metabolic or structural disturbances in the cell that no, or only a few, divisions will occur before cell death. Many more subtle changes can and do occur that are not visible in the morphology of the chromosome, but that make their presence known through alteration of some genetically controlled function of the cell. These changes are generally called *gene mutations* and are taken to be rather smaller changes than those associated with chromosome aberrations. The smallest or simplest of these changes is the alteration of a single base in the nucleotide sequence of DNA. This change may arise as the result of chemical change in the base resulting from radiation damage or as the result of the incorrect insertion of a base in repair or replication. These small changes are called *point mutations*, and they give rise to a change in one codon that will misread for a single amino acid in the gene product protein eventually synthesized. A second significant type of mutation can arise as the result of a minor deletion, addition, or substitution. This type of mutation is called a *frame shift*. It can be visualized as follows, using a nonsense codon, CAT.

Suppose the following sequence exists:

CAT CAT CAT CAT *CA*T CAT CAT CAT . . .
×

and a CA is deleted at the ×. The sequence now reads

CAT CAT CAT CAT TCA TCA TCA TCA . . .

and all codons from the point of deletion onward are now incorrect.

The mutations that are produced by radiation are in no way different from those produced by other agents or from those mutations that are otherwise occurring through natural misadventure. The essential difference is the frequency. In the absence of radiation, the mutation rate for a single gene may be on the order of one mutation or less per gene site per million cells that are formed. Radiation can easily raise this rate by a thousand fold or more. The other important characteristic for all mutations, whether they are the result of radiation action or other mutagenic agents, is that they will overwhelmingly be recessive in character and thus will be observed phenotypically only when the cell line or animal has the homozygous presence of the altered gene. Finding radiation induced mutations then presents certain problems. Either we must find ways to examine an extremely large number of progeny for phenotypic change or methods must be developed for the detection of the recessive changes. Both methods have been used.

Microorganisms and cultured cells are relatively easy to use as systems for the measurement of mutation frequency after ionizing radiation. Biochemical mutations are those most frequently used, since it is possible to detect the inability of a mutant to grow unless a certain metabolite is added as a preformed nutrient in the medium. The normal, unmutated cell is capable of synthesizing the specific mutant. Growing the cell on an incomplete medium allows the detection of the mutant.

DeSerres, Malling, and Webber (1967) carried out a series of studies on the fungal organism *Neurospora crassa*. They studied the mutation of two adjacent genes, both of which control the synthesis of adenine. A mutation in either or both of these genes causes the organism to lose its ability to synthesize adenine, and the mutant will only grow on a medium supplemented with preformed adenine. The mutation frequency data, as a function of dose, did not fit a linear relationship with dose, but rather were a best fit to a quadratic form. The actual final formulation was as follows:

$$\text{Yield*} = 0.3 + 5.39D + 0.16D^2 \qquad (13\text{-}1)$$
$$(\text{* mutants per } 10^6 \text{ surviving cells})$$

where D is the dose in kilorad (1 kilorad = 10 Gy).

The interpretation by deSerres and his group of these data was that the total yield of adenine requiring mutants was made up of a portion produced by point mutations that required one hit, and another portion of the mutations was produced by chromosome deletions that required two hits. In light of modern data, this is certainly an oversimplified explanation, but it represents elegant work in the field and provides interesting insights to compare the formulation of equation 13-1 to the linear–quadratic molecular model.

Müller's Sex-linked Recessive Test

In 1927, Müller developed a truly elegant test system to reveal the presence of lethal mutations anywhere on the X chromosome of the fruit fly, *Drosophila melanogaster*. Since these mutations will almost always be recessive in nature, they will not be disclosed by the more simple experiments associated with changes in phenotypic expression. The model system, which earned the Nobel prize for Müller for being the first to show that radiation could produce mutation events, is as follows:

1. Construct a female fly with an X/X chromosome that carries three gene markers on one of the X chromosomes: C, of only procedural interest,* a lethal gene ℓ, and a morphological marker, B. The latter symbol, B, which indicates bar-eyed, is useful as a sorting tool to identify flies that carry the ℓ gene, since the same chromosome carries both genes. The notation for this sex chromosome combination in the female fly is CℓB/ +, indicating that one of the X chromosomes, +, is sufficient, if present, to prevent the lethal but recessive action of ℓ.

2. Cross the CℓB/ + female with wild-type males that have been irradiated at doses chosen by the investigator. The competent, wild-type male is coded + /Y to indicate that it also has an X chromosome (+), which will prevent the action of ℓ.

If the irradiated, wild-type male has not suffered a radiation induced mutation in the X chromosome, the bar-eyed daughters of this cross will all carry the sex chromosome, CℓB/ +. The + / + females will not be bar-eyed and are excluded from further study.

3. Now mate *individually* the bar-eyed daughters collected from this first mating with the irradiated male with normal, unirradiated males, and examine for surviving *male* offspring.

The possible outcomes are as follows: *The original irradiated male parent of the first mating has no mutation in X:*

CℓB/ + female × + /Y male yields: males: CℓB/Y, + /Y

females: CℓB/ +, + / +

The CℓB/Y males will die. There is no X+ present. The + /Y males will survive and will not be bar-eyed.

It is clear that in the absence of a mutation in the irradiated male parent, some male offspring will survive from these crosses, but they will not be bar-eyed. As will be seen, the survival of any males will not be important.

---

* The C marker is a gene that suppresses crossing over, a process that occurs normally at meiosis and one that would negate the experiment if it were to occur with any significant frequency.

The original male parent of the first mating has a lethal mutation somewhere on X. The F1 bar-eyed daughters of the cross will be $C\ell B/\ell_m$ where ℓ_m is a lethal mutation on the other X chromosome:

$C\ell B/\ell_m$ female $\times$ $+/Y$ male yields males: $C\ell B/Y$, $C\ell_m/Y$

No males survive from this cross. The lack of surviving males from the cross is the sign that a mutation occurred in the irradiated male grandparent, and the number of such crosses lacking male survivors is the number of the original irradiated males with mutations. There is no need to enumerate survivors of either sex.

This elegant experiment allowed Müller to construct dose–effect curves for this mutation type in Drosophila. The work on Drosophila was carried forward by Stern and others. The data of Spencer and Stern (1948) unequivocally demonstrated the linear dependence of the mutation rate on dose for the particular model of recessive mutations in Drosophila. It must be borne in mind that the Müller recessive lethal model does not permit us to examine the sensitivity of the female for radiation mutagenesis.

The conclusions that can be drawn from the cumulative studies on Drosophila over the years can be summarized as follows:

1. From 25 rad (0.25 Gy) to 4000 rad (40 Gy), the rate of production of sex-linked recessive lethals is constant per unit dose. That is, the plot of excess mutation frequency as a function of dose is linear.

2. There appears to be no threshold below which an effect would not be predicted. But even the lowest dose used, 25 rad (25 cGy), is high by radiation protection standards. We cannot unequivocally say that there is no threshold, but it seems highly probable for Drosophila.

3. There is no effect of dose rate over a wide range of dose rate. There is also no effect of fractionation of the dose if account is taken of the elapsed time and of the stage of the male gamete that is irradiated.

4. The sensitivity of the male gamete varies greatly with the stage of development at which it is irradiated. The following table lists the relative sensitivities of the various stages of the male gamete.

Stage	Relative Sensitivity
Spermatogonia and early spermatocytes	1
Meiotic division stages	8
Spermatids	2
Spermatozoa	3–6

5. The relative sensitivity of the female gamete has been measured by another method (the specific locus method to be described shortly). The values are as follows

    Oocytes and oogonia   1
    Late oocytes   2

6. The induction of mutations by ionizing radiation in Drosophila will occur at the rate of 1.5×10^{-8} to 8×10^{-8} per locus per rad (per cGy).

These reports on mutation rates in Drosophila were very influential in the establishment of guidelines for radiation exposure of human beings. It was taken as established precept for man that there would be no threshold and that no matter how the dose was accumulated there would be no reduction in effectiveness of the dose. Concerns about the reliability of Drosophila data for projections to human beings led to the establishment in the early 1960s of a large project at the Oak Ridge National Laboratory to establish the sensitivity of a mammalian species for radiation mutagenesis. The mouse was chosen as a test animal. The papers by L. B. Russell and W. L. Russell are too numerous and extensive to cite here, but the reader is directed to an extensive review of the work by Searle (1974).

Specific Locus Test in Mice

The Russells were able to make use of an experimental mouse colony in which nine mouse specific locus end points were available for them to study. The general problem stated earlier still holds. The great majority of mutations produced by any agent will be recessive, and a special methodology is necessary to accomplish the measurement of mutation frequency. The *specific locus method* is suitable for this purpose and is particularly valuable for studies in animal species where numbers of subjects, breeding frequency, and length of parturition rule out studies of the Müller type.

Specific phenotypic expressions are chosen that can be easily detected in the offspring. Examples might be "curly wing" in Drosophila or "kinky tail" in mice. Mice or flies are bred to be homozygous recessive for a trait. The mice or flies are mated with normal subjects in which the marker trait is known not to exist in heterozygous form. Figure 13.3 indicates how the specific locus method operates. Emphasis is on the use of this method in mice, but it is obviously available for any species in which the appropriate phenotypic markers are available.

The frequency of appearance of the recessive phenotype is a measure of the mutation frequency for the locus for which the test is designed. It is also possible to measure the relative sensitivity of the various stages

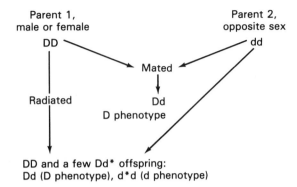

Figure 13.3 General structure of the specific locus model. This approach was used in all the studies of mutations in mice. Either the male or female can be the irradiated partner.

of gamete development by choosing an appropriate time for mating after the radiation has been delivered. The longer we wait before the mating, the earlier will be the stage of maturation of the gamete that is being measured.

The observations and conclusions from this study with the specific locus test in mice are very different from those just described for Drosophila. For seven of the loci studied in the mouse, the average mutation rate per locus, per rad (per cGy) was found to be 2.2×10^{-7}. Compare the figure just given for Drosophila of 1.5 to 8×10^{-8}.

Of even greater significance was the discovery of a dose rate effect for induction of mutations in mice. Studies with male mice were undertaken at dose rates of 90 rad/min (90 cGy/min) and at 0.009 rad/min (0.09 mGy/min). There were very large reductions in the mutation frequency at the lower dose rate, indicating a large capacity for repair of induced mutations in the mouse. Interestingly, though, when the dose rate was lowered even further to 0.001 rad/min (0.01 mGy/min), almost one-tenth of the 0.009-rad/min rate, no further reduction in the mutation frequency was seen for the same total dose. The authors postulate a repair process that is dose rate, not dose, dependent and that it is saturatable at the 0.009-rad/min (0.09-mGy/min) dose rate.

The Russells have also been able to demonstrate a dose rate effect in the female mouse, with a factor of 8 lower mutation rate per unit dose for the 0.009-rad/min (0.09-mGy/min) dose rate compared with the highest dose rate. If the matings are undertaken immediately after irradiation of the male, the spermatozoon is the gamete stage that is being evaluated. In this last case, no dose rate effect is demonstrable.

Radiation that is received by human beings beyond that attributable to background is, to a great extent, delivered either in brief episodes, widely separated, as medical radiographic exposures or continuously as exposure to natural background. There is no possible way to interpret the dose rate effects just described in terms of the exposures usually received by human beings.

An extensive summary of the results of the study of specific locus mutation frequencies in the mouse has been provided by Searle (1974). In briefer form, this summary is abstracted here.

FOR THE MALE GAMETE

1. The mutation frequency for the various loci studied varied significantly (differential repair?).
2. The average mutation rate in mice is 10 to 15 times higher than for Drosophila.
3. There is no evidence that spermatogonia containing these induced mutations are less viable than the normal. No change in the mutation frequency was observed for matings taking place as long as two years after irradiation. Therefore, these mutations will persist throughout reproductive life.
4. The mutation frequency varied widely depending on the stage of development of the gamete at the time of irradiation.
5. There is no relationship between the spontaneous rate for a locus and its susceptibility to radiation mutagenesis.
6. Dose rate is an important independent variable.

FOR THE FEMALE GAMETE

1. Mature and nearly mature follicles have the highest mutation frequency for the stages of the female gamete.
2. Early or immature oocytes appear to be very resistant to radiation induced mutations.
3. Dose dependence is also seen in the oocyte, and it extends to much lower dose rates than for the male before saturation of repair sets in.
4. On the whole the female gamete is significantly less sensitive than the male gamete.

GENE MUTATIONS MEASURED IN CULTURED MAMMALIAN CELLS

The work of deSerres, Malling, and Webber (1967) on the measurement of mutation rates in a fungal organism, *Neurospora crassa*, was mentioned at the start of the section on gene mutations. Over the last two decades the same types of studies have been undertaken on cultured mammalian cells. Clearly, the mutation rates deduced are not for mammalian gametes, but rather are for somatic cells grown in culture and undergoing continuous population growth. Three classes of studies have been undertaken. One of these is the study of auxotrophic mutants (as in the deSerres case, an auxotrophic mutant has a particular growth requirement). Another group of studies has used temperature-sensitive mutants of mammalian cell lines that will grow optimally at specific temperatures other than the normal incubator temperature. Finally, a group of mutants has been developed through irradiation that is resistant to certain drugs, such as nucleotide base analogs or specific metabolic inhibitors.

Susuki and Okada (1976) measured the mutation frequency for an auxotrophic strain of the mouse L5178Y lymphoma line that required alanine for normal growth. The mutation end point was the disappearance of the dependence on alanine for growth (conversion to prototrophy). The measured mutation frequency with x-rays was 3×10^{-7} per rad (per cGy).

Among many other investigators using acquired drug resistance as the selection tool for identification of mutations, Knapp and Simons (1975) used acquired resistance to the drug 6-thioguanine in the same mouse L5178Y line. They obtained a mutation frequency for x-rays of 1.5 to 3 $\times 10^{-7}$ per rad (per cGy).

These mutation frequencies for cultured somatic mammalian cells are to be compared with the mouse-specific locus mutation frequencies of, for example, mouse spermatogonia of 2 to 3×10^{-7}, essentially an identical rate.

REFERENCES

BENDER, M. A., and GOOCH, P. C. (1962) Types and rates of x-ray induced chromosome aberrations in human blood irradiated in vitro. *Proc. Natl. Acad. Sci. U.S.* **48**, 522–532.

DALRYMPLE, G. V., and others (1973) *Medical Radiation Biology* (Chapter 5). W. B. Saunders Co., Philadelphia.

DESERRES, F. J., MALLING, H. V., and WEBBER, B. B. (1967) In *Brookhaven Symposium on Biology* **20**, 56.

KNAPP, A. G. A. C., and SIMONS, J. W. I. M. (1975) A mutational assay system for L5178Y mouse lymphoma cells, using hypoxanthine-guanine-phosphoribosyl-transferase (HGPRT) deficiency as a marker. *Mutation Res.* **30**, 97–110.

LEA, D. E. (1946) *Actions of Radiation on Living Cells* (Chapters V and VI). Cambridge University Press, New York.

MÜLLER, H. J. (1927) Artificial transmutation of the gene. *Science* **66**, 84–87.

REVELL, S. H. (1955) In *Proceedings of the Radiobiology Symposium, Liège, 1954*, Z. M. Bacq and P. Alexander, eds., pp. 243–253.

————— (1974) The breakage-and-reunion theory and the exchange theory for chromosomal aberrations induced by ionizing radiations: a short history. *Adv. Radiation Biol.* **4,** 367–416.

SAX, K. (1950) The effects of x-rays on chromosome structure. *J. Cell. Comp. Physiol.* **35,** Suppl. 1, 71–82.

SEARLE, A. G. (1974) Mutation induction in mice. *Adv. in Radiation Biol.* **4,** 131–207.

SPENCER, W. P., and STERN, C. (1948) Experiments to test the validity of the linear R-dose/mutation frequency relation in *Drosophila* at low dose. *Genetics* **33,** 43–71.

SUSUKI, N., and OKADA, S. (1976) Isolation of nutrient deficient mutants and quantitative mutation assay by reversion of alanine-requiring L5178Y cells. *Mutation Res.* **34,** 489–506.

WALDREN, C. A., and JOHNSON, R. T. (1974) Analysis of interphase chromosome damage by means of premature chromosome condensation after x- and ultraviolet irradiation. *Proc. Natl. Acad. Sci. U.S.* **71,** 1137–1141.

SUGGESTED ADDITIONAL READING

REVELL, S. H. (1974) The breakage and reunion theory and the exchange theory for chromosome aberrations induced by ionizing radiations: a short history. In *Advances in Radiation Biology*, Volume 4. J. T. Lett, H. Adler, and M. Zelle, eds. Academic Press, New York.

SEARLE, A. G. (1974) Mutation induction in mice. In *Advances in Radiation Biology*, Volume 4. J. T. Lett, H. Adler, and M. Zelle, eds. Academic Press, New York.

14

High LET
Radiation Effects
and
Relative Biological
Effectiveness

INTRODUCTION

Earlier in this text, and particularly in Chapters 4 and 5, the concept of the rate of energy transfer per unit of track length of the high-kinetic-energy electron was discussed at great length. It is now time to examine in greater detail the differences in the physical, chemical, and biological events that are associated with radiations that do indeed have a high rate of such linear transfer of energy compared to x- and gamma rays. There are several important sources of high LET radiations in the environment. One of the most important of these from the health protection point of view is the neutron radiations that are produced in the course of power generation from nuclear fission. If nuclear fusion ever becomes a significant industry in the world, it too will be an important source of high LET radiations. Man in the space environment will also be faced with significant sources of high LET radiation, particularly from the high-energy charged particle environment (Tobias and Todd, 1974). Significant high LET radiations are also generated from radiation generating equipment in modern physics research (high-energy accelerators) and from advanced equipment used in radiotherapy of cancer and other diseases.

STOPPING POWER AND LINEAR ENERGY TRANSFER

Linear Energy Transfer

Linear energy transfer was explicitly defined in Chapter 1, using the definition of the International Commission of Radiation Units and Measurements (ICRU), as the average energy dE_L *locally* imparted to the medium by a *charged particle* of specific energy traversing a distance dx.

$$\text{LET} = \frac{dE_L}{dx} \qquad (14\text{-}1)$$

In what way is LET, as defined, different from the usually used term in physics, stopping power? *Stopping power* is defined as the energy loss per unit thickness (read length) and sometimes is redefined as a density corrected stopping power. The latter is obtained by dividing the stopping power dE/dx by the density, ρ.

The differences are specific and important. It should be noted that the stopping power used by physicists, dE/dx, is not confined to a description of energy loss from charged particles, as is the ICRU definition of LET, and stopping power can, indeed, describe the rate of energy loss from a photon beam. Another important constraint on the ICRU definition of LET is that it describes the *rate of transfer of energy to the medium locally*. These constraints give to the concept of LET special utility for the understanding of the biological action of radiation as it relates to the pattern of energy deposition. As has been suggested so many times in this text, for biological damage to be sustained in a cell, energy must be deposited by some means in that cell. In that context, the definition of local in the LET formulation is essential, in that it limits the consideration of energy deposition events to a radius from the line of the particle track that is appropriate for the biologically effective deposition of energy.

In practice, the use of the LET involves the development and acceptance of a number of important assumptions. When we examine the detailed structure of individual charged particle tracks, every track is different from its predecessors. Also, it is a very rare case where we can examine the LET of a population of charged particles in which each particle has the same initial energy. Furthermore, the concept implicit in the LET is a continuous deposition of energy, the "continuous slowing down approximation" of the Bethe–Bloch equation to which reference has been made earlier. Finally, the LET, as it is used in radiation biophysics, is the average rate of local deposition of energy from a real spectrum of charged particles as they traverse a medium. Since part of the different biological effectiveness of high LET particles will be seen to be due to the closeness of ionization events along the particle track, we must not lose sight of the fact that the continuous slowing down approximation is aptly named as an approximation.

It also must be said that there are general references throughout the practice of radiobiology that indicate that gamma or x-rays have an associated LET, even though the definition limits the LET to charged particles. The same discrepancy is also seen in reference to neutrons passing through tissue. This widespread and useful usage, which is observed in this text also, is simply a recognition that, in the case of gamma or x-rays, the transfer of energy occurs from electron tracks, and in the case of neutrons, from recoil protons or other charged particles produced as the result of interactions in the tissue. Zirkle (1940) and Zirkle and Tobias (1953), when they first discussed the importance of linear energy transfer and then introduced the concept of LET, were certainly aware of these limitations on the definition, but they found it a useful construct to include all energy loss processes in a single descriptive parameter.

Before discussing the effects of this parameter, LET, on the energy transfer mechanisms and biological effect of radiations, it would be useful to first define the LET values associated with radiations that are commonly available, and, second, to define what is intended by those who loosely describe "effects of low LET radiation" and the "effects of high LET radiation."

In the radiobiological literature it can be taken that the term "low LET radiation," without another descriptor, is meant to include the usually available gamma ray sources such as ^{60}Co and ^{137}Cs, as well as usual laboratory x-ray machines that have peak energies in the range of 50 keV or more. The electrons generated in tissue by interaction with the x-rays from typical 250-kVp x-ray generators have mean LET values of about 2 to 3 keV/μm, while the gamma ray sources have mean LET values in the range of 0.2 to 0.5 keV/μm. Experiments done with these radiation sources will generally be classified as low LET experiments. A wide range of natural radiation sources will provide radiations with LET values far in excess of those just discussed. The group of sources that are of particular importance because of their high LETs are the radionuclides, which decay with the emission of high-energy alpha particles. For example, a ^{222}radon alpha particle will have an LET of about 100 keV/μm. Neutrons will also generally be a high LET radiation because of the short range and the significant mass of the recoil particles produced in their interaction with tissue. Finally, a wide range of radiations can be produced on modern accelerators, including neutrons of high energy, as well as accelerated heavy charged particles. The range of LET values associated with the latter can exceed 10,000 keV/μm.

To recall the interactions of high-kinetic-energy electrons as they slow down, δ rays are one of the possible products of these slowing down reactions. The δ rays are electrons recoiling from interactions with the fast electron that have imparted to the δ ray sufficient energy to have an ionization track of its own. A δ ray of 0.1 to 1.5 keV, the energy that might be expected from track interactions of electrons generated from ^{60}Co gamma rays, has associated LET values of 33 and 9 keV/μm, respectively. These LET values are associated with a physical range in tissue for the δ rays of 0.003 and 1.0 μm, respectively. Obviously, then, even for low LET radiations, a high LET component will frequently, if not always, be associated with them.

BRAGG PEAK OF IONIZATION

In Chapter 5 the formulation of the coulombic interaction of a charged particle with another charged particle in the medium was developed. This formulation, which was particularly directed toward an understanding of the interactions of secondary electrons with the medium, used a mostly classical approach. It was demonstrated that the rate of energy deposition for high-energy-charged particles was proportional to the inverse of the velocity of the particle squared or, stated another way, inversely proportional to the kinetic energy of that particle. For stopping electrons, this aspect of the rate of energy deposition as a function of trajectory is not particularly important, because the electron, with its small mass, follows a tortuous path to its final stopping point, and the change of LET with the slowing down electron will be masked by the average LET of other slowing down electrons in the volume. For alpha particles from radioactive decay, for recoil protons from neutron scattering interactions, and for charged particles produced in accelerators, the nature of this pattern of energy deposition is indeed important, since the particle trajectory is only little influenced by the interactions that it undergoes, and the rate of energy loss will have a geometric significance. To reiterate that statement a little more simplistically, since the electron has a tortuous path, and all the secondary electrons do not stop at the same location in the absorbing medium, the high rate of energy transfer at the terminus of the electron track will be distributed more or less randomly in the absorber. For heavy particles, with specific range and minimum scatter, the fact that the majority of the energy is deposited at the end of the track has great physical and biological significance.

Continuously Slowing Down Approximation

The complete description of the rate of energy loss from a charged particle passing through an absorbing medium is given by the Bethe–Bloch formulation (Evans, 1955) for any particle having a rest mass

$M \gg m_0$. This complex formula will not be developed here, and the readers are referred to Evans for the details. A more widely used form of the Bethe–Bloch expression is that provided by Curtis (1974):

$$\frac{-dE}{dx} = \frac{0.307 Z^{*2} Z}{A\beta^2} \left[\ln \frac{2\, m_0 c^2 \beta^2}{(1 - \beta^2) I} - \beta^2 - \frac{C}{Z} - \frac{\delta}{2} \right] \qquad (14\text{-}2)$$

In equation 14-2, Z^* (read, z-star) is the effective charge on the projectile particle, making adjustment for the charge of the particle as it picks up electrons at low velocity, and for the case of an atomic nucleus not fully stripped of its electrons, Z^* is corrected for the charge shielding influence of these electrons. Z and A are the atomic number and mass of the absorbing medium, β is the ratio of the particle velocity to the speed of light, m_0 is the electron rest mass, I is the mean ionization potential of the absorbing medium, C is the sum of electron shell corrections, and the last term, $\delta/2$, is a condensed medium density correction. C/Z and the density correction are very small under most conditions and can be ignored. The important implication of the Bethe–Bloch expression is that, even in this form with its necessary relativistic corrections, the rate of energy transfer is still mainly determined by the inverse of the velocity of the particle squared, which in this formulation is represented as β^2. If, as the charged particle slows down, it captures electrons, changing Z^*, then the rate of energy loss might be better described as depending on Z^{*2}/β^2.

Bragg Peak

The rate of energy loss of a heavy charged particle as a function of its residual energy is shown in Figure 14.1. The peak energy loss occurs just as the particle approaches the end of its range (the residual energy is small). The LET distribution versus residual energy of Figure 14.1 is characteristic of a heavy charged particle and has come to be known as the *Bragg peak*. Formally, the Bragg distribution is more generally shown as a plot of ionization against penetration into the absorber. Such a plot would be the mirror image of the one shown here. The Bragg ionization curve is named in honor of W. H. Bragg (1912), who first systematically examined it.

For very energetic particles (hundreds of MeV), the track in tissue can be significantly long; but for the more usual case of an alpha particle from radioactive decay or a recoil nucleus from a neutron interaction, these dimensions will be very small. For example, the alpha particle from ^{210}Po decay has an initial energy of 5.3 MeV, and it will have a track length in tissue of only 35 μm, only of the dimension of a few cell diameters. Under these conditions of rapidly changing LET with distance of penetration, it is necessary to refine the definition of LET. If the sensitive

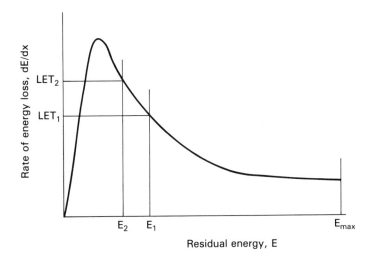

Figure 14.1 Rate of energy loss for a heavy charged particle as a function of its residual energy. The arbitrary scales show that the energy loss follows approximately as the inverse of the effective velocity squared. E_1 and E_2 are residual energies associated with LET_1 and LET_2, respectively. Note that the rate of energy loss is nearly constant at the initial energy of the particle (E_{max}).

target of the cell has dimensions that are appreciably larger or smaller than, for example, the distance from E_1 to E_2 in Figure 14.1, then LET variation within the region, or even over a somewhat larger region, will be significant. Two conceptual parameters have been developed for this application.

Examining Figure 14.1, it is clear not only that the LET depends on the depth of penetration in the absorbing medium, but that, if the sensitive target dimensions are such that the entrance and exit depths correspond to E_1 and E_2 with their corresponding LET values, the change in LET is not so great, and the LET can be assigned a value between narrow limits. In the limit, however, only when ΔE and ΔLET are very small can the LET be stated with precision. From these considerations, two alternate forms of LET estimate have come into general usage. These are the *track segment average LET*, $(LET_{av})_S$, and the *average LET*, averaged either over (1) track dimension, the *track average LET*, $(LET_{av})_T$, or (2) averaged over energy deposited, the *energy average LET*, $(LET_{av})_E$.

Formulations for LET

The track segment LET can be understood clearly by reference to Figure 14.1. For the designated E_1 and E_2 in that figure, the following relationship defines the *track segment LET*:

$$(\text{LET}_{\text{av}})_S = \frac{1}{E_2 - E_1} \int_{E_2}^{E_1} [\text{LET}(E)] \, dE \tag{14-3}$$

To be explicit, the track segment LET is the average of the LET integral over the energy range E_2 to E_1, and if E_2 and E_1 are in regions of the curve where LET is changing slowly, that is, near E_{max}, then

$$(\text{LET}_{\text{av}})_S \approx \frac{1}{2}[(\text{LET})_1 + (\text{LET})_2] \tag{14-4}$$

When LET studies of biological effectiveness can be performed with monoenergetic, charged heavy particles of sufficient energy such that the ΔE is small passing through a thin biological target, equation 14-3 (or 14-4 for a thin target specimen) is an unambiguous estimate of the density of ionization along the track core. Some studies to be described shortly will meet these criteria. For the more usual case, an LET estimate must be used such as the *track average* or *energy average*. These parameters are strongly dependent on the averaging method used. In principle, equation 14-1 can define either energy average or track average for a moving particle. In practice, the differences in the definitions of energy average and track average can be best visualized as differing by the localization of the energy absorbed.

The energy average LET can be computed, say, for a neutron of some initial energy by summing the initial energies of all charged particles set in motion by the neutron over a track length, L. The energy average LET, $(\text{LET}_{\text{av}})_E$, will then be the quotient of these two quantities. On the other hand, the track average LET would be obtained as the sum of all energy deposited over the total track length of the initial particle and would not account for energy that was not locally deposited in the dimension L. The $(\text{LET}_{\text{av}})_T$ will generally be a smaller value than $(\text{LET}_{\text{av}})_E$. A comparison for some radiations is shown in Table 14.1.

SIGNIFICANCE OF LET TO BIOLOGICAL DAMAGE

The general effectiveness of a given radiation in producing lesions or irreparable damage in bioactive molecules depends to a great extent on the spatial distribution of the ionization and excitation events that occur in the medium. A useful oversimplification that allows us to visualize the effect of LET on biological effectiveness is to imagine the two possible extremes. For low LET radiations, gamma rays for example, the average spacing between energy transfer events along the track of the charged particle (the scattered electron) will be on the order of hundreds of nanometers. Under these circumstances, as was discussed in the earlier

TABLE 14.1 Comparison of Track Average LET, $(LET_{av})_T$, and Energy Average LET, $(LET_{av})_E$, for Some Typical Radiation Types

Radiation	Track Average (keV/μm)	Energy Average (keV/μm)
[60]Co gamma rays	0.27	19.6
250-kVp x-rays	2.6	25.8
3-MeV neutrons	31	44
Radon alpha rays	118	83
14-MeV neutrons	11.8	125
Recoil protons	8.5	25
Heavy recoils	142	362

Data are taken from Randolph (1964).

sections on radiation chemistry, there is little or no chance for the volume elements containing the products of the ionization event from interacting with each other. Therefore, there can be little or no cooperative interaction between these events leading to unrepairable or misrepairable damage to the DNA molecule. This generalization will be true for both the direct and the indirect action of radiations. For high LET radiations, the formation of regions of ionization will be close together and will, in the limit, form a continuous chain, or column, of ionization damage, again either by action in water or in the biomolecules themselves.

To examine a simple calculation that will make this point, assume a gamma ray, the secondary electrons of which provide a track average LET of 0.25 keV/μm. Assume also that the average energy transfer per spur is 60 eV. What will be the spacing of spurs along the track?

$$\frac{250 \text{ eV}}{60 \text{ eV}} \approx \text{4 events per μm, or an average of 250-nm spacing of these events}$$

Now make the same assumption for a radon alpha ray with a track average LET of 118 keV/μm. For approximate purposes, we can assume an average energy transfer of 60 eV per event.

$$\frac{118,000 \text{ eV}}{60 \text{ eV}} \approx \text{2000 events per μm, or an average of 0.5-nm spacing for these events}$$

Clearly, in the latter case the products of the ionization reaction in a single spur will interact with the products in the neighboring spur, and if the concept of cooperative damage between or among two or more lesions in the strands of DNA is correct, the latter form of radiation will be more effective, as it indeed is.

Direct versus Indirect Action

High LET radiation will not have a preferential action on biomolecules over water. All the usual water radiochemistry will happen whenever ionization occurs in that medium. All the products will form and diffuse away from the site of formation in the same fashion as after photon irradiation. The products of radiolysis of water will be capable of all the chemical damage that is known to result from low LET radiations in aqueous systems. There will, however, be significantly more interaction of the water radicals with each other because of the event distribution described in the previous paragraph. Of greater importance with high LET radiations is the high likelihood that an ionizing event will occur directly in the important target bioactive molecule, and that more than one lesion (or sublesions, to use the more modern terminology of the cell inactivation models) will be formed in close enough proximity for cooperative interaction of the lesions. This is simply because of the very high density of the ionizing events and the subsequent high likelihood that the ionization event will fall in the volume of the important biomolecule if the track crosses the cell or nucleus at all. If the important molecule is indeed DNA, its large size provides even more assurance that interactions will occur within the ionization track.

A number of radiation effectiveness modifiers are known to act through modification of the indirect effect. One of the most important of these is oxygen, and we will discuss the role and relevance of oxygen for high LET radiations shortly. Other scavengers also compete for the water radiolysis products. Each of these radiation response modifiers is seen to be nearly inactive for very high LET radiation and to have limited activity for moderately high LET radiation. This evidence strongly suggests that for high LET radiations the direct effect of the radiation predominates, but we must not forget that the indirect action is most certainly still taking place in the aqueous milieu and, to the extent that it does occur, the radiation modifiers will intervene in the reactions. The indirect effect simply has little relevance in the presence of the more effective direct action. The reader is reminded that the oxygen effect may be demonstrated in the irradiation of bioactive molecules in the dry state. In the dry state the direct effect must assuredly predominate. Why, then, is it not possible to demonstrate the oxygen effect on the products of the direct action of high LET radiations? Oxygen will indeed react, if it is present; but since the damage induced by a high LET track will generally be extensive, the damage fixation by oxygen is not a necessary component of the process.

RELATIVE BIOLOGICAL EFFECTIVENESS

There must be a comparison indicator in order to consider the effects of various radiations and to evaluate the role of LET. The term that has generally been adopted for this purpose is *relative biological effectiveness* (RBE). It is defined explicitly, using the ICRU definition, in Chapter 1 and elsewhere in several locations in the text.

In brief, the RBE is the ratio of absorbed doses to produce the same biological end point, using a reference radiation and a comparison radiation. A lower absorbed dose for the comparison radiation than for the reference radiation for the same end point will give an RBE greater than 1 (see Chapter 1 for the explicit definition). The expression for RBE is

$$\text{RBE} = \frac{\text{dose for given end point (reference radiation)}}{\text{dose for given end point (test radiation)}} \qquad (14\text{-}5)$$

There is a very serious limitation to the RBE as a concept. It assumes that the dose–response curves for the biological effect are described by identical functions, with parameters such that a simple multiplicative dose modification would describe the difference in response for the two radiation types. Stated in another fashion, the RBE, as defined, should be independent of the response level at which it is estimated. Unfortunately, this is rarely the case, as will be seen. An example is shown in Figure 14.2. The authors whose work is shown in simplified form in Figure 14.2 examined a number of cell lines with the same two radiation sources and found a reasonable degree of similarity among the responses of all these cell lines. The key problem is to decide on the appropriate biological end point to use for determination of the RBE. In their particular case, they settled on the ratio of the D_0 values and found that the RBEs calculated in this way were in the range of 2.0 to 2.2. The extrapolation number ratios, on the other hand, varied among the cell lines studied from slightly more than 2 to greater than 6. These differences clearly delineate the difficulty in using the RBE as a parameter that could be a single-valued indicator of relative effectiveness of a given radiation. Because low LET curves generally seem to fit reasonably well to multitarget single-hit types of curves or to linear–quadratic curves with nonzero values of both α and β, while the high LET curves seem better fits to single-hit types of curves or to linear–quadratic curves with very small or zero values for β, there will rarely, if ever, be an RBE that can be used as a dose modifying adjustment factor at all levels of response. The one special case for which a unique RBE can be derived, good at all response levels, is when both curves are linear in the plot of logarithm of response versus dose. In this special case the RBE is the appropriate ratio of the D_0's or of the inactivation coefficients.

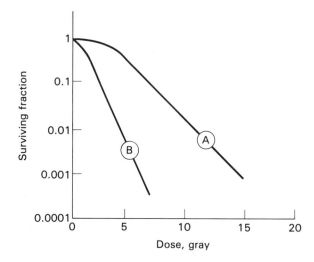

Figure 14.2 Survival curves for hamster fibroblasts irradiated with 280-kVp x-rays (curve A) and fast neutrons (curve B). The estimated LET is about 30 keV/μm for the neutrons. The D_0 (x-rays) is 1.45 Gy (145 rad). The D_0 (neutrons) is 0.65 Gy (65 rad). In various experiments the extrapolation number for neutrons was 1.1 to 2. For x-rays it varied from 4 to nearly 10. (After Schneider and Whitmore, 1963)

For radiation protection purposes, RBE values are chosen in conservative fashion to assure worst-case protection. These values so chosen are then used in estimating *quality factor* for radiation protection purposes (see definition in Chapter 1).

Given the limits described for RBE, it is still the only convenient descriptor that has been available. Using the RBE for this purpose, the RBE/LET relationship can be described in a moderately successful fashion.

DEPENDENCE OF RBE ON LET

Biophysical Descriptions of RBE/LET Relationships

Returning to considerations from target theory, if a single active event will produce the end point being considered, then additional events occurring in the irradiated active volume would be unproductive of further biological inactivation in that volume. With high LET radiations, the likelihood of multiple events occurring, being equivalent to "wasted hits," would increase with increasing LET. For a number of systems, including bacteria, some viruses, bioactive molecules in solution, and a few other systems, the expected decrease of effectiveness with LET is observed.

Figure 14.3 is an example of such a relationship. There is a more or less smooth, monotonic decrease in the effectiveness of the radiation as its LET increases, in accord with target theory predictions. There are, however, a wide range of other examples, using end points such as clonogenic survival of mammalian cells, chromosome aberrations, mitotic delay, late effects in normal mammalian tissues, and cancer production, for which Figure 14.3 would be not at all a correct description. Generally, the cellular inactivation or other end points in these latter systems is not describable by single-hit target theory but by shouldered curves. There is at least one system, survival of mouse spermatogonial stem cells, for which the survival curve is apparently single hit in nature, yet the LET/RBE dependence for this system is not as described in Figure 14.3 (Alpen and Powers-Risius, 1981).

An LET/RBE relationship that is much more generally applicable for the wide range of end points in cellular systems that are sigmoid in their dose–response relationship is that described by Todd (1965) and others.

The curve is characterized by an ascending portion, with the RBE increasing monotonically until it reaches a peak value. For many systems studied, the LET associated with maximum RBE is around 200 keV/μm. At LET values higher than this the RBE is seen to decrease in a fashion similar to that seen for the organisms described previously and typified in Figure 14.3. Figure 14.4 is a graphic representation of the data of a number of investigators as collated by Blakely and others (1984).

Several models have been presented in the past to explain the findings graphically shown in Figure 14.4. None of these has been completely satisfactory. The earlier models generally called on mechanisms that re-

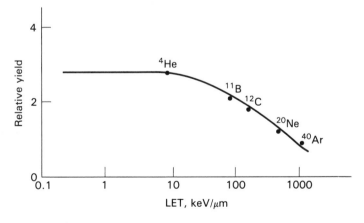

Figure 14.3 Inactivation of trypsin by radiations of differing LET. The author, Brustad (1967), used beams of charged particles produced by an accelerator. (From data of Hendrickson, 1966)

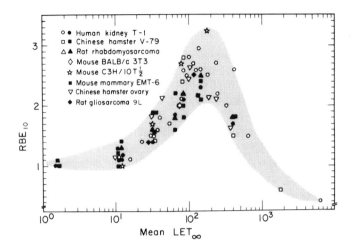

Figure 14.4   Range of RBE values found for eight different cell lines examined by several investigators (Blakely and others, 1984, by permission of the authors and Academic Press). Blakely intended that the shaded zone would encompass the data on all cell lines included, and it represents the range of RBE values to be found associated with a given LET. The ordinate is RBE measured for a survival fraction of 0.1. The same end point was used for all the cell lines.

sponded in opposite directions to increasing LET. The model of Zirkle and Tobias (1953), for example, proposed two reactive species that were formed such that one increased and one decreased with LET. These models have generally fallen into disfavor, and a model based on the cooperative effects on DNA presently is more widely accepted.

This model proposes, as does the molecular model described in Chapter 7, that damage to DNA occurs most effectively when radiation is delivered in such a fashion that just enough ionization is deposited in or near the DNA strands that both strands are broken. Ionization densities below this critical value will cause easily repairable damage, while ionization densities above the critical value will simply cause additional ineffective damage, and therefore the killing effectiveness of the radiation will be decreased. The peak effectiveness (RBE) will deposit just enough energy to cause the critical damage without overkill or wasted ionization energy.

CELL CYCLE DEPENDENCE OF RADIOSENSITIVITY

Most in vitro studies of clonogenic survival of cell lines are done on exponentially growing populations. In that case, any variation in sensitivity of the cells at various parts of the cell cycle, G-1, S, G-2, M, are effectively masked. The first evidence that there was a differential sensitivity for cells in various parts of the cycle was the data from Elkind and Sutton's studies of the repair of sublethal injury (Elkind and Sutton, 1960). The first experiments specifically designed to test the sensitivity of the cell as a function of postmitotic cell age with synchronized cells were those of Terasima and Tolmach (1961). Since then a large number of reports have appeared describing the cell age specific radiosensitivity of various cell lines. The Terasima and Tolmach study with HeLa s3-91V cells showed a very significant increase in the sensitivity of the cells, which were in mitosis at the time of irradiation. The next most sensitive period was the period of DNA synthesis (S phase). Unfortunately, the cell age specific sensitivity varies widely with cell line. Only the high sensitivity in mitosis is common to all lines.

Within the last few years, very careful studies of the cell age specific radiosensitivity to high LET radiations have been done (Blakely, Chang, and Lommel, 1985). The common finding is that most of the cell age specific changes in radiosensitivity disappear when the cells are irradiated with high LET radiations.

OXYGEN EFFECT AND HIGH LET

One of the most well recognized modifying agents for radiation response was discussed in Chapter 6. The ability of molecular oxygen to potentiate the effectiveness of radiation was described there as the *oxygen effect*. One parameter that has been developed to quantitate the oxygen effect for enhancement of the effectiveness of radiation in causing loss of clonogenic effectiveness was described in Chapter 6. It is labeled the *oxygen enhancement ratio* and is defined as the ratio of the dose required for a given level of cell killing (or other biological end point, for that matter) in a fully oxygenated medium to the dose required to achieve the same level of killing in a fully anoxic medium, that is, radiation in pure nitrogen. The reader is referred to Chapter 6 for more complete background on the OER, but, as a reminder, the mathematical description is

$$(OER)_{0.1} = \frac{\text{dose in } N_2 \text{ for s.f. of } 0.1}{\text{dose in } O_2 \text{ (or air) for s.f. of } 0.1} \tag{14-6}$$

Note: The level of response must be specified. In the example it is taken as 0.1.

The formulation in equation 14-6 is for a surviving fraction (s.f.) of 0.1, but any appropriate level can be chosen. For example, in radiotherapy, where each individual radiation fraction given the patient is in the range of 1 to 2 Gy (100 to 200 rad), the OER is usually taken for a survival fraction more appropriate for that dose, for example, 0.5.

The models generally accepted for the potentiating action of molecular oxygen generally all call on a mechanism that suggests that oxygen intervenes in a competitive fashion with the reaction of radicals in important molecules such as DNA to cause their fixation before they are restored by other scavengers in the cell. The product, a peroxidated radical, is more stable and therefore more long-lived than the original DNA radical, and it is more likely to proceed to irreversible damage or misrepair of the DNA molecule.

Since the oxygen effect in aqueous media calls on the indirect action pathway for a good deal of its effectiveness, we must then presume that its potentiating action will be suppressed to the extent that damage occurs via direct action of radiation on DNA. This is particularly true if that direct action causes damage so extensive that restitution through intracellular scavengers is not effective. Barendsen (1968) and Blakely and colleagues (1984) have shown that the OER decreases monotonically as the LET is increased. These data were collected with elegant experiments using monoenergetic charged particles for which the LET was well defined, and indeed was a track segment LET as defined earlier, $(LET_{av})_S$. The Barendsen data are drawn in Figure 14.5. It can be seen that the OER decreases monotonically to a value of 1.0 at an LET of approximately 200 keV/μm. The extensive survival curve data presented by these workers, along with the OER–LET relationship shown here, demonstrate several interesting points. The shoulder portion of the survival curve decreases as the LET increases, and this decrease in the shoulder of the survival curve is accompanied by a decreasing OER. If the shoulder of survival

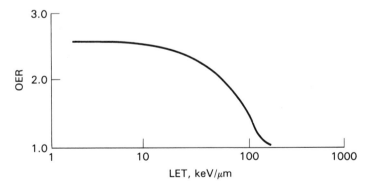

Figure 14.5 For a mammalian cell line of human origin, Barendsen measured the oxygen enhancement ratio for a range of LET values. All the data were collected with monoenergetic charged particles at all LET values above that for 250-kVp x-rays, for which a track average LET of 1.3 keV/μm was assumed. (Data from Barendsen, 1968)

curves is related to the amount of repairable damage that is presumably induced by the indirect effect, the decrease in the extrapolation number with increased LET is not surprising, since for high LET radiations we attribute most of the lethal action to the direct effect. That the picture is not entirely that simple is demonstrated by the fact that the OER has not fallen to its minimum of 1.0 for survival curves where the extrapolation number is 1.0. For example, for 4-MeV alpha rays with a track average LET of 110 keV/μm, the extrapolation number for the survival curve measured either in oxygen or nitrogen is 1.0, and there is clearly no shoulder for either the nitrogen or the oxygen curve. However, the OER for this radiation quality is 1.3. In other words, there is still a significant potentiation of radiation effect by the presence of oxygen in spite of the lack of shoulder on the survival curve. How can we interpret these observations? Probably the best explanation is that there is indeed still some contribution to cell killing from the indirect effect, and this cell killing might be attributable to δ rays that are formed during the slowing down interactions of the alpha particle. These δ rays will, at least for part of their track, have LET values that are a good deal lower than that for the initial alpha particle.

HIGH LET, DOSE RATE, AND FRACTIONATION

The survival curves for high LET radiations have shoulders that get smaller and smaller as the LET increases, that is, the extrapolation number approaches 1.0. To the extent that the extrapolation number reflects repair capacity, we would expect that, as the LET increases, repair capacity would decrease accordingly, and that is what is observed. Reference was made in Chapter 8, when describing the Elkind–Sutton type of divided dose experiments, that with high LET radiations there was only limited reappearance of the shoulder during the delay before the second dose is administered. The damage, either if it is damage that would be classed as sublethal injury or if it is potentially lethal damage, appears to be poorly repaired, if at all, after high LET radiations.

The generalization can be made that both fractionation of the dose as well as low dose rate will not reduce the effectiveness of high LET radiations, but that complete suppression of repair does not occur until the LET approaches 100 keV/μm or more.

It should be noted that a number of reports in the literature indicate that fractionation of the high LET radiation exposure actually enhances the effectiveness of the high LET exposure. One of the first of such reports was that of Ainsworth and others (1976) regarding the potentiation of the effectiveness of neutron irradiation by fractionation. Another report by Ngo (see Blakely and others, 1984) suggests that potentiation of survival of cells grown in vitro could be demonstrated with divided exposures to high LET argon ions.

REFERENCES

AINSWORTH, E. J., and others (1976) Life shortening, neoplasia and systemic injuries in mice after single or fractionated doses of neutron or gamma radiation. In *Biological and Environmental Effects of Low Dose Irradiation*, Volume I. International Atomic Energy Agency, Vienna, pp. 77–92.

ALPEN, E. L., and POWERS-RISIUS, P. (1981) The relative biological effect of high Z, high LET charged particles for spermatogonial killing. *Radiation Res.* **88**, 132–143.

BARENDSEN, G. W. (1968) Responses of cultured cells, tumors and normal tissues to radiations of different linear energy transfer. *Curr. Topics Radiation Res.* **4**, 293–356.

BLAKELY, E. A., CHANG, P. Y., and LOMMEL, L. (1985) Cell cycle dependent recovery from heavy ion damage in G1 phase cells. *Radiation Res.* **104**, Suppl. 8, S145–S156.

———, and others (1984) Heavy ion radiobiology: cellular studies. *Adv. Radiation Biol.* **11**, 295–389.

BRAGG, W. H. (1912) In *Studies in Radioactivity*, Chapter 22, Section 5, and Chapter 25, Section 2. Macmillan and Co. Ltd, London.

BRUSTAD, T. (1967) Effects of accelerated heavy ions on enzymes in the dry state and in aqueous solution. In *Proc. Intl. Symp. Biol. Interpretation of Dose from Accelerator Produced Radiations*. Berkeley, Calif., pp. 30–44.

CURTIS, S. B. (1974) Radiation physics and current hazards. In *Space Radiation and Related Topics*. Academic Press, New York, Chapter 2, pp. 21–99.

ELKIND, M. M., and SUTTON, H. (1960) Radiation response of mammalian cells grown in culture. I. Repair of x-ray damage in surviving Chinese hamster cells. *Radiation Res.* **13**, 556–593.

EVANS, R. D. (1955) *The Atomic Nucleus*. McGraw-Hill Co., New York, pp. 592–599.

HENDRICKSON, T. (1966) Effects of irradiation temperature on the production of free radicals in solid biological compounds exposed to various energy ionizing radiations. *Radiation Res.* **27**, 694–709.

RANDOLPH, M. (1964) Genetic damage as a function of LET. *Ann. N.Y. Acad. Sci.* **114**, 85–95.

SCHNEIDER, A., and WHITMORE, G. (1963) Comparative effects of neutrons and x-rays on mammalian cells. *Radiation Res.* **18**, 286–306.

TERASIMA, T., and TOLMACH, L. J. (1961) Changes in x-ray sensitivity of HeLa cells during the division cycle. *Nature* **190**, 1210–1211.

TOBIAS, C. A., and TODD, P. (1974) *Space Radiation and Related Topics*. Academic Press, New York.

TODD, P. W. (1965) Heavy ion irradiation of human and Chinese hamster cells in vitro. *Radiation Res.* **61,** 288–297.

ZIRKLE, R. E. (1940) The radiobiological importance of the energy distribution along ionization tracks. *J. Cell. Comp. Physiol.* **16,** 221.

————, and TOBIAS, C. A. (1953) Effects of ploidy and LET on radiobiological survival curves. *Biochem. Biophys.* **47,** 282.

SUGGESTED ADDITIONAL READING

TOBIAS, C. A., and TODD, P. (1974) *Space Radiation and Related Topics.* Academic Press, New York.

15

Metabolism
of Radionuclides
and Biological Effects
of Deposited Radionuclides

INTRODUCTION

The text has, to this point, dealt entirely with the delivery of ionizing radiation from external sources. There is, however, another enormously important contributor to the dose received by biological systems. This source is all man-made as well as naturally occurring radionuclides. Obviously, these radionuclides can also contribute to the dose received from externally located sources. Naturally occurring radioactivity in the earth's crust will always be a significant contributor to externally delivered dose, and it cannot be more than marginally controlled.

Those radionuclides that are able to enter the living cell by either metabolic or other processes give rise to localized dose that may be very high and that may be qualitatively different in its effect from the dose delivered externally, since often these deposited radioisotopes are sources of high LET radiations.

Only in rather unusual circumstances will living systems, particularly human beings, be exposed to high LET radiation from external sources. Such circumstances might include occupational exposure around large experimental accelerators or the occupational exposure of astronauts to high-energy, heavy charged particles in extraterrestrial space. Such exposures to high LET radiations will be of little significance to the large proportion of the population. On the other hand, when radionuclides are incorporated into living systems, the high LET radiations produced by their disintegration, which would otherwise be unimportant, become very significant. Alpha particles, for example, which are incapable of delivering significant dose when external to the living system, can deliver very large local doses to cells in which they may be deposited after absorption into the body.

The focus of this chapter will be the exploration of the health risks arising from the incorporation of radionuclides into the cells and organs of human beings.

PATHWAYS OF ENTRY OF RADIONUCLIDES

The metabolic fate of any radionuclide is not affected by the fact that it is capable of radioactive decay and the release of energy. Its pathway in any metabolically active system, cell, organ, or man, will be determined by the reactivity of the parent element, as fixed by its position in the periodic chart of the elements, and by the chemical reactivity of the chemical and physical form in which it exists. By the latter is meant its solubility, its valence state, its salt form, and the chemical milieu in which it exists, for example, the pH of the medium. There is a minor exception to the generalization that radioisotopes are treated in identically the same manner as the stable, metabolically involved nuclides. For the lighter elements and, in particular, tritium, a mass effect can be observed. For tritium the mass difference is a factor of 3, and it should not be surprising that there will be certain kinetic discriminations resulting from this mass difference. For all other nuclides, this mass effect is unimportant.

Complex chemistry and complex metabolic chains from the state of the radionuclide in the inorganic world to man are beyond the scope of this text, since it is the intention to focus principally on effects in human beings. However, to remind the reader of the complexity of the pathways leading to deposition in an organ of the human body, the example of the biological pathways for iodine and its radioactive isotopes will be briefly outlined. Note that in all the ensuing discussions the nuclide of iodine that is involved in the pathway is irrelevant, as is its radioactivity.

Elemental iodine existing in the environment may enter into the human body by the simplest available route, that is, by *inhalation* of the nuclide. By transfer across the alveolar surface of the lung, the nuclide is taken up in the circulation and distributed according to the usual metabolic pathways for blood-borne iodine or iodide. This inhalation route of entry is, in all except very limited and unusual conditions, the route of least importance.

The alternate pathway is by *ingestion*. The aerosol-borne or free vapor state iodine will be taken up by green vegetation, either by deposition on the surface of the foliage or by metabolic transport into the cellular contents. The green vegetation may be eaten by man, leading to absorption through normal gastrointestinal processes. Again, because human beings eat only small quantities of green field crops, this will be a limited source of the nuclide. Much more important is the ingestion of milk, either bovine or from other sources. The cow eats large quantities of field crops, and the absorbed iodine from these foodstuffs of the cow appear in milk. Milk that is then ingested by people becomes an important source of the iodine nuclides. Would the flesh of the cow be an important source of iodine nuclides? For this special set of nuclides the answer is no, since all the important radionuclides of iodine are short-lived, and they would have disappeared by radioactive decay before the meat enters the consumption chain.

The reader is directed to the very extensive literature that exists on the food chain relationships for many of the important radionuclides (Whicker and Kirchner, 1987). One important generalization can be made, however. The example just given is for an element that has a well-known active metabolic role in the mammal. Iodine is used in the synthesis of the iodinated hormones of the thyroid gland, so any radioactive form of the element will be treated indiscriminately by the body. If the element involved is not usually part of normal metabolic processes, its metabolic fate will be determined by the metabolic activity and pathways of its nearest neighbor in its column in the periodic chart of the elements that does have a usual metabolic role. For example, rubidium and cesium, both of which have no normal metabolic role, will tend to behave metabolically like potassium, their nearest neighbor with an important metabolic function. In these cases the normal nuclide will compete with the foreign nuclide, with the normal nuclide, in this case potassium, having a discriminatory edge in the competition. In the case of most foreign nuclides, discrimination factors have been developed for estimating the relative importance of the competing nuclides in displacing the natural metabolically active element.

A third pathway also exists, but it is not normally of importance. This pathway is *injection*. Direct entry of the nuclide into the bloodstream will occur when radionuclides are used in the practice of nuclear medicine for diagnostic or therapeutic purposes. Another possible injection modality is by the accidental penetration of the skin with materials that have surface contamination with radionuclides. This latter possibility is usually only associated with occupational uses of radioactive materials.

After once entering the body by any of the pathways mentioned, there is a complicated network of interactions among these pathways. A simplified diagram of these interactions is shown in Figure 15.1. We see that there are four principal body compartments in which the interchange of radioactive materials entering the body may occur. These are (1) the lungs and upper respiratory tract, (2) the mouth and gastrointestinal tract, (3) the circulatory system for blood, including the heart and kidney,

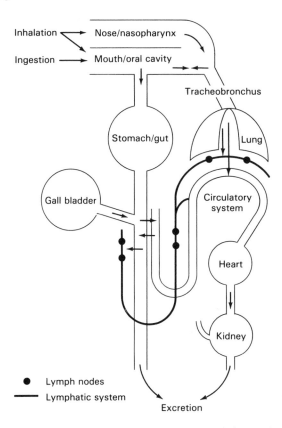

Figure 15.1 Pathways for the absorption, transport, and excretion of radio-nuclides in the mammal. Inhalation may be by either nose/nasopharynx or by way of the mouth. Ingestion is via the mouth, but the two-way arrows indicate that inhaled material may return from the tracheobronchial tree to the oral cavity and be swallowed to become part of the ingestion pathway. The text describes the processes in detail.

and (4) the lymphatic circulatory system, including the associated lymph nodes. The injection pathway mentioned briefly has little practical significance as a route of entry for radionuclides into the body except in special cases. Injection would simply amount to direct placement of radionuclides into the blood circulation. The other entry pathways, ingestion and inhalation, account for essentially all the nondeliberate entry of radionuclides into the body.

Ingestion

Material may be taken into the oral cavity in either soluble or insoluble form and, once having entered, it will not be expelled via the mouth. In addition to that which enters the oral cavity directly, there is an additional contribution from the normal "cleaning" activity of the upper portions of the tracheobronchial tree. Particulate material that is inhaled via nose or mouth will enter the respiratory organ and, depending on the particle size, will deposit either on the walls of the upper tracheobronchus or penetrate the deeper lung. Particulate matter deposited in the upper tracheobronchial region will be cleared upward by the ciliary action occurring on the luminal surfaces of the upper regions of the organ. The material will be expelled into the mouth incorporated in mucus, and it will generally be swallowed. This material makes a significant contribution to the total ingestion burden, particularly when exposure to a heavily dusty or a wet aerosol environment exists.

Once ingested, radioactive material will follow the normal pathway of other substances transiting the gastrointestinal tract. The most important consideration in determining the amount of a radionuclide absorbed after ingestion is its solubility in the chemical environment of the tract, strongly acid in the stomach to alkaline in the upper small intestine. That which is not transported across the gastrointestinal epithelial surfaces will ultimately be excreted, as shown in the diagram, in the feces. Absorption from the gastrointestinal tract can be directly into the bloodstream or, less likely, into the lymphatic circulation. An additional complication to the gastrointestinal pathway is the possible secretion into the gastrointestinal tract of bile containing radionuclides that have been collected in the liver from one or both circulatory pathways. This route is indicated in Figure 15.1 by the arrow into the lumen of the small intestine from the gall bladder.

Inhalation

For purposes of consideration of the transport and localization of radionuclides in the human body, we can divide the respiratory system into three regions that have distinctly different characteristics for retention of aerosols that are introduced, whether radioactive or not. These regions are the nasal passages, (NP), the tracheobronchial tree, (TB) and the deep lung parenchyma (P). The penetration of aerosols into these regions is determined entirely by the aerodynamic properties of the aerosol particles inhaled. The usual term to describe the aerodynamic properties of aerosol particles is the *activity median aerodynamic diameter* (AMAD). The exact definition of AMAD is *the diameter of a unit density sphere with the same terminal settling velocity in air as that of the aerosol particle the activity of which is the median for the entire aerosol.* For purposes of these discussions, if the reader simply assumes that we are dealing with unit density spheres of the median size shown, there will be no great distortion in understanding of models. In Figure 15.2 (International Commission on Radiological Protection, 1978), the relationship between the

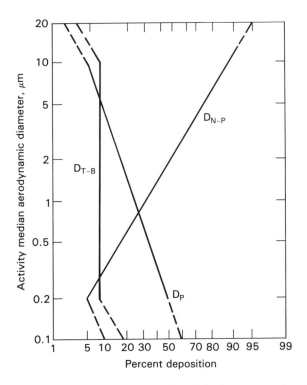

Figure 15.2 Relationship between aerosol particle size (expressed as AMAD) and the penetration of the particles into the lung. The labels on the lines are: D_{TB}, the fraction deposited in the tracheobronchial tree; D_{NP}, the fraction deposited in the nasal passages; and D_P, the fraction deposited in the deep parenchymal regions of the lung. (Reproduced with permission of the International Commission on Radiological Protection from ICRP publication 30, Part 1, adopted in July 1978)

AMAD and the penetration of particles into the respiratory tree is shown in some detail.

Since a reasonably large fraction of material deposited in the nasopharynx and in the upper tracheobronchial tree is removed from the lung and often swallowed, the most important curve in Figure 15.2 is that for D_P, the fraction retained in the deep parenchyma. It is interesting that the fraction retained in the tracheobronchial tree is essentially independent of AMAD over a very wide range. This is because of the competing behavior of particles in the other two compartments. Large particles, AMAD over 1 μm, are principally retained in the nasopharynx, and particles of less than 1 μm penetrate into the deep lung.

After penetration of the smaller airborne particles into the lung, followed by sedimentation onto the air-side surfaces of the lung, the ultimate fate of the radioactive material incorporated in the particle will, as for ingestion, depend on the physical chemical state of the material. Substances that are soluble at the essentially neutral pH of the lung surface will quickly dissolve into the aqueous media of the lung surface and will be transported across the alveolar wall into the vascular system. The plentiful surfactants available on the lung surface facilitate this process effectively. If the radioactive material is in an insoluble or poorly soluble form and in a form which cannot be altered by the chemical environment of the lung surface, the particle will be phagocytized by the lung macrophages, which have this function. These phagocytized particles will ultimately reside in nearby lymph nodes. The particle will remain at this location for some time. If there is some small solubility of the radionuclide deposited in the lymph node, transport of this soluble fraction into the lymph as it passes through the node will occur. Ultimately, the collected lymph empties into the vascular circulation and the dissolved material will distribute as if it were directly transported into the bloodstream.

Once into the vascular system, radionuclides that entered by inhalation and were ultimately transported into the vascular system will be metabolized in the same fashion as nuclides absorbed from the gastrointestinal tract.

METABOLISM OF RADIONUCLIDES

In the introduction to this chapter a brief synopsis of the fate of radionuclides within the body was offered. The point was made that radionuclides that are simply isotopes of naturally occurring constituents of the body will be treated in the same fashion as the naturally occurring element. The example given was iodine, an element occurring naturally in the mammalian body, which enters into the metabolic processes of the thyroid gland and which is distributed in varying amounts throughout the body tissues. Other examples of elements for which there are important radioactive counterparts are carbon (^{14}C), potassium (^{40}K), iron (^{55}Fe and ^{59}Fe and other iron nuclides), sodium (^{24}Na), calcium (^{45}Ca), and hydrogen (^{3}H). For those elements that normally occur in the body, an introduced radioisotope will ultimately come to equilibrium specific activity throughout all tissues of the body. The speed of attainment of this equilibrium specific activity will depend on the turnover and biological availability of the element in question, and depending on competing excretory processes, equilibrium may, in practice, never be achieved. For example, tritium (^{3}H) will come to very rapid equilibrium with all the hydrogen in body water in a few minutes, but it will interchange only slowly with organically bound hydrogen. Iodine, the example discussed earlier, will come rapidly to specific activity equilibrium with the iodine of the blood plasma, since this iodine is loosely bound to plasma albumin; but iodine cannot exchange with the bound iodine molecules of the thyroid hormones. Since other competing processes, excretion and radioactive decay, are continuing while the radionuclide is coming to specific activity equilibrium with its stable elemental counterpart, the mathematical description of the process is complicated and often difficult to quantitate.

Radionuclides that are isotopes of elements normally foreign to the metabolic composition of the body are usually treated metabolically like their "near neighbor" in the periodic chart of the elements. Since chemical activity is to a great extent determined by valence electrons, we would expect that the nearest neighbor in the vertical column of the chart would be the determinant of the metabolic fate of the foreign isotope. Examples were given earlier. Rubidium and cesium isotopes are metabolically distributed as if they were potassium. Radium and strontium isotopes are treated as if they were calcium. Astatine isotopes are distributed as if they were iodine.

The important distinction for the metabolic distribution of foreign isotopes is that they will generally be at a metabolic disadvantage when competing with their normally metabolized counterpart. A term called the *discrimination ratio* has been developed to characterize the competition between a given radionuclide and its normally occurring metabolic counterpart. If 100 atoms of a radionuclide are competing with 100 atoms of its normal counterpart for metabolic incorporation or for metabolic transport, and it is found that 5 atoms of the radionuclide are acted upon for every 50 atoms of the normal counterpart, then the discrimination ratio is 5/50 = 0.1. This discrimination ratio is all-important in developing and applying mathematical models to the transport and deposition of radionuclides.

DETERMINATION OF DOSE WITH INTERNALLY DEPOSITED RADIONUCLIDES

The determination of dose to a tissue from internally deposited radionuclides is an especially difficult and vexing problem. Within the body in which the radionuclide has been deposited, there is nonuniform deposition of the isotope in various organs, depending on their metabolic activity, biological accessibility, and the competing excretion of the isotope in urine and feces. While these events are in progress, the isotope is undergoing its normal physical decay, described by its decay constant or half-time. In addition, the radionuclide administered or accidentally incorporated may be undergoing decay that gives rise to radioactive daughter products that also contribute to dose.

The complications of dose determination for internally deposited radionuclides include, among others, some of the following concepts.

1. There is a variable and organ-dependent rate for the radioisotope to come to specific activity equilibrium, if ever, in the pool that is important for determination of dose in that organ.

2. Each organ *may* have a very different pattern of distribution of the isotope within it. An example is the thyroid gland and the isotopes of iodine. Since iodine is an important constituent of the thyroid hormones such as thyroxine, which contains four atoms of iodine per molecule, the concentration of iodine in this organ will be intense, and the dose to the organ received from iodine radionuclides will be affected not only by the half-life of the radionuclide, but also by the turnover time for the hormone and the urinary excretion of the isotope. In addition, within the gland the iodine will be specifically localized in those portions of the gland in which the synthesis and storage of the thyroid hormones are occurring.

3. As the radioisotope is incorporated into various components of body chemistry, the rate of excretion via urine and feces will be controlled by the metabolic fate and excretory management of these newly synthesized compounds.

4. Within the organ of interest, at the microscopic level and dimensionally, at a level where the physical range in tissue of the radioactive emissions of the isotope are significant, the distribution of the isotope is not homogeneous. An example just mentioned is the localization of iodine in the thyroid. Another important example is the deposition of strontium, which follows pathways for calcium metabolism. The strontium is not deposited generally throughout the substance of calcium containing bone in a homogeneous fashion; rather, it is deposited on the more rapidly turning over surfaces of remodeling or growing bone. Beta particles emitted by strontium incorporated in the general corpus of bone would be sharply attenuated by surrounding bone, but in reality the strontium incorporated in bone surfaces is subjected to very little attenuation by surrounding bone and can efficiently irradiate nearby bone marrow.

Absorbed Dose

The simplest approach to dose determination is to use the physical model that assumes a uniform distribution of the isotope and furthermore assumes that it is known how many disintegrations will take place in the volume of distribution of the isotope that will contribute to dose in that organ in the time period of interest. The formulation and notation is that used in ICRU Report 32 (International Commission on Radiation Units and Measurements, 1979).

The *source region* is defined as that volume of tissue from which the energy released by radioactive disintegrations arises. The *target region* is defined as the volume in which the dose is to be determined. The source and target volumes may be the same or different. The mean absorbed dose to the target region as the result of radioactive decays arising from the source region is given as follows:

$$\overline{D} = \tilde{A}nE\phi m^{-1} \qquad (15\text{-}1)$$

In this expression, $\overline{D}$ is the mean absorbed dose to the target region from nuclear transformations in the source region. $\tilde{A}$, the time integral of the activity A, is the number of nuclear transformations (radioactive decay events) in the source region in the time period of interest. The mean number of ionizing events per nuclear transformation is n, and the mean energy per particle is E. The term ϕ is the fraction of the particle energy emitted by the source volume that is delivered in the target volume. This term is called the *absorbed fraction*. The reciprocal of the mass of the target region is m^{-1}. It is clear that the product of n and E is mean energy emitted per nuclear transformation, and it is conveniently formulated as $\Delta = nE$. Another term is defined as the *specific absorbed fraction*, Φ, the quotient of $\phi/m = \Phi$. A new formulation for the mean absorbed dose is then

$$\overline{D} = \tilde{A}\Delta\Phi \qquad (15\text{-}2)$$

Note that time is implicit in equations 15-1 and 15-2 and that equation 15-1 describes dose to a finite volume or mass, while equation 15-2 has no such limit.

It is suggested to the reader that it would be helpful to review in Chapter 3 the definitions of activity and specific activity.

Absorbed dose under equilibrium conditions. In Chapter 5, where general principles of dosimetry were discussed, including energy absorbed and energy emitted, it was proposed that there can be an *equilibrium* condition that is satisfied when energy emitted is equal to energy absorbed in any given volume element. The same principle may be applied for radionuclides under a special condition. That condition is one in which the radionuclide is uniformly distributed in an infinite homogeneous absorbing material. The absorbed dose under these equilibrium conditions can be defined as

$$\overline{D}_{eq} = \tilde{C} \Sigma_i \Delta_i \qquad (15\text{-}3)$$

where $\tilde{C}$ is the time integral of C, the activity, A, per unit mass, integrated over time for the period of interest. The term Δ_i is the mean energy per nuclear transformation for the ith particle type, and there may be more than one particle of the ith type; therefore, $\Delta_i = kn_iE_i$, where k is simply an appropriate constant for the choice of units. The terms are summed, as indicated by Σ_i, over all particles from 1 to i.

Absorbed fraction, ϕ, and specific absorbed fraction, Φ. The *absorbed fraction*, ϕ, is defined as the quotient of the absorbed energy delivered per unit mass in target region v from the source region r. For the ith type of radiation, the notation is $\phi_i = \phi_i(v \leftarrow r)$. Since ϕ is defined by a target volume, it is useful to use the concept of *specific absorbed fraction*, the absorbed fraction per unit mass of target. This concept is useful to describe dose to a massless point in space or a surface, for example. Specific absorbed fraction is formulated as

$$\Phi_i(v \leftarrow r) = \frac{\phi_i(v \leftarrow r)}{m_v} \tag{15-4}$$

Reciprocity theorem. A general theorem holds for the conditions where the radionuclide is uniformly distributed in regions of absorbing material that are homogeneous and infinite, or if, for the same homogeneous distribution, the radiation is absorbed without scatter. The theorem states that for any pair of regions, r_1 and r_2, the specific absorbed fraction is *independent* of which region is designated source and which is designated target.

Absorbed dose equation. The general equation for dose from a distributed radionuclide can be written using the concepts just developed as follows:

$$\overline{D}(r_1 \leftarrow r_2) = \tilde{A}_2 \, \Sigma_i \, \Delta_i \Phi_i(r_1 \leftarrow r_2) \tag{15-5}$$

One very important special case for equation 15-5 is when the source volume and target volume are identical. That is, the radionuclide is deposited in an organ, and it is the goal to estimate the dose to that organ. A simplified form of equation 15-5, which uses the time integral of specific activity $\tilde{C}$, where $\tilde{C} = \tilde{A}_v/m_v$ and $\phi_i = m_v\Phi_i$. The summation operator Σ_i has its usual meaning.

$$\overline{D}_{(v,v)} = \tilde{C}_v \, \Sigma_i \, \Delta_i \Phi_i(v, v) \tag{15-6}$$

Biological Realities of Dose Estimation

It was stated previously that, for equations 15-1 through 15-6, time was implicit in the expressions. The time of exposure to the radionuclide is contained in the two integral expressions $\tilde{A}$ and $\tilde{C}$, which are, respectively, the time integral of activity and specific activity. The time histories of activity and specific activity are, however, usually very complicated, and they certainly do not follow the usual rules of radioactive decay, since biological incorporation and elimination are proceeding while radioactive decay is taking place.

There are two generally important situations for which the determination of dose from radionuclides deposited in the body is important. A very large number of radionuclides is used in medical diagnosis and treatment under the rubric of nuclear medicine. In these applications the radionuclides are usually injected in a single brief dose. Since the isotope is usually injected, questions of transport and deposition within the body are less critical. Determination of dose to the organs of the patient is important for these applications. The second important application for dose estimation from deposited radionuclides has to do with the determination of dose to organs or tissues from radionuclides incorporated into the body subsequent to the release of these radionuclides into the environment. In some cases the isotope is already generally present in the natural environment. For these cases, determination of the time integral of activity or specific activity is extremely complicated. Exposure may be continuous or intermittent and may be from several routes of entry, for example, food, water, or inhaled aerosols.

One of the simplest models for estimation of the time integral of activity assumes instantaneous administration and first-order exponential removal and radioactive decay. This model has given rise to the concept of *biological* or *effective* half-life.

Biological half-life. As just outlined, when a radioactive nuclide is present in the body, the concentration or activity of the isotope per unit volume will decrease with time through two mechanisms. The isotope is excreted in urine or feces (and in rare cases, such as for carbon and hydrogen, through expired air as carbon dioxide and water), and it undergoes its normal physical decay with the characteristic decay constant of the radionuclide.

Let λ_p be the physical decay constant, which is the fraction of the radionuclide disappearing by physical decay per unit time. Let λ_b be the constant that describes the fraction of the isotope removed by excretion or other biological means per unit time. Then the effective constant for the fraction removed per unit time (λ_{eff}) will be the sum of the two constants, λ_p and λ_b.

$$\lambda_{\text{eff}} = \lambda_p + \lambda_b \tag{15-7}$$

From equation 3-3, the decay constants may be rewritten in terms of half-times as follows:

$$\lambda_{\text{eff}} = \frac{\ln 2}{T_{\text{eff}}} = \frac{0.693}{T_{\text{eff}}} \qquad (15\text{-}8)$$

The equivalent expressions for λ_b and λ_p are identical. T_{eff}, T_p, and T_b are the effective, physical, and biological half-times, respectively. Substituting in equation 15-7,

$$\frac{1}{T_{\text{eff}}} = \frac{1}{T_p} + \frac{1}{T_b} \quad \text{giving} \quad T_{\text{eff}} = \frac{T_p T_b}{T_p + T_b} \qquad (15\text{-}9)$$

An example for the calculation of the effective half-time is the following.

The radionuclide ^{51}Cr is used for the determination of the survival time of red blood cells since it irreversibly attaches to red cells in vitro, and these labeled cells are efficient markers, when reinjected, for the survival of a population of red cells from which they are drawn. The biological half-time, T_b, can be taken to be 30 days. The physical decay half-time, T_p, for this isotope is 27.7 days. The effective half-time may be calculated using equation 15-9 in either form as follows:

$$\frac{1}{T_{\text{eff}}} = \frac{1}{27.7d} + \frac{1}{30d} \quad \text{giving} \quad T_{\text{eff}} = \frac{27.7 \times 30}{27.7 + 30} \quad \text{days}$$

$$T_{\text{eff}} = 14.40 \text{ days}$$

The great oversimplification of the effective half-time is the assumption that the decay constant for the effective half-time can be described as the simple combination of two exponentials, one describing physical decay and one describing biological removal. In only a very few cases can that be shown to be a true representation of fact. The ^{51}Cr example just given is one of these. Since physical decay is known to be a constant fraction of the activity remaining per unit time, the assumption that is made in this model is that the fraction of the isotope excreted per unit time is also a constant, a condition rarely met.

Figure 15.3 shows data that are more characteristic of the complicated time–activity functions that might be found after the administration of a single dose of a radionuclide (Galt and Tothill, 1973).

With the complexities present in the time–activity relationships apparent in the data shown in Figure 15.3, it is necessary to adopt a more general formulation for $A(t)$ and $C(t)$. Short of developing a complex and comprehensive model for the biological movement of a given chemical form of a radionuclide, it has become common practice to express the activity and specific activity relationships to time as a product of exponentials, combining the radioactive decay constant with a series of exponential expressions that arbitrarily describe the time relationships. This formulation is

$$A(t) = \exp\left(-\lambda t\right) \Sigma_j \, q_j \exp\left(-\lambda_j t\right) \qquad (15\text{-}10)$$

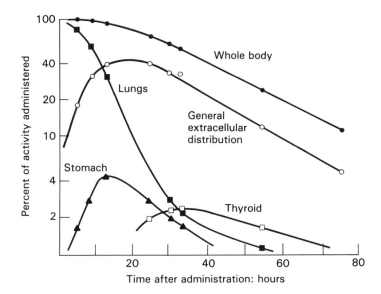

Figure 15.3 Complexities of the time course of activity in individual organs after the single administration of a radionuclide. The example is the data following administration of ^{131}I labeled macroaggregated albumin. (Data from Galt and Tothill, 1973; by permission of the British Institute of Radiology)

q_j is the starting value of activity of the jth component of $A(0)$. The physical decay constant is λ and λ_j is the biological disappearance constant of the jth component. This simple model assumes that the dose contributed during the rising biological uptake portion of the curve is negligible. Integration of equation 15-10 gives the value of $\tilde{A}(t)$. Where is $A(0)$ in equation 15-10? The beginning activity, $A(0)$, is partitioned among the various q_j's, and after the summation of all the biological decay terms in the λ_j's, the physical decay is calculated with the term $\exp(-\lambda t)$.

$$\tilde{A}(t) = \Sigma_j \frac{q_j}{\lambda + \lambda_j} [1 - \exp(-(\lambda + \lambda_j)t)] \qquad (15\text{-}11)$$

If the activity is allowed to go to zero in time t, or nearly so, equation 15-11 simplifies to

$$\tilde{A}(t) = \Sigma_j \frac{q_j}{\lambda + \lambda_j} \qquad (15\text{-}12)$$

It is left to the reader to rewrite these equations in terms of half-life by the usual substitution: $\lambda = \ln 2/T$.

Actual absorbed fractions. The radiations emitted in radioactive decay will be of two types, those absorbed locally, such as alpha and beta particles, and those with long range, such as gamma rays, bremsstrahlung, and related penetrating radiations. For the former, beta and alpha radiations, the absorbed fraction can be taken as 1.0 for any rational organ size, since these radiations have a range of a fraction of a millimeter to tens of millimeters. For the penetrating radiations, the estimation of absorbed fraction is more difficult. Generally accepted principles of gamma ray attenuation are used for arriving at estimates of the absorbed fraction; but since the calculations are tedious, they have been done once and for all and presented in tabular form in appropriate resources. The reader is particularly referred to tables in ICRU Report 32 (International Commission on Radiation Units and Measurements, 1979) for examples of such resources.

Compartmental analysis. The most elegant and appropriate approach to complex dosimetric problems, particularly those associated with the transport and localization of radionuclides absorbed from the environment, is to develop complete compartmental analyses of the activity as a function of time for all relevant body compartments. It is beyond the scope of this text to explore these models, but the reader is referred to the lung model, the gastrointestinal model, and the bone model in ICRP Publication 30, Part 1 (International Commission on Radiation Protection, 1978).

RELATIVE BIOLOGICAL EFFECTIVENESS AND INTERNALLY DEPOSITED RADIONUCLIDES

To this point there has been no discussion of the relative biological effectiveness of the various radiations emitted during radioactive decay. From the foregoing discussions, it should be clear that the local deposition of energy from nonpenetrating radiations will be of utmost importance in the determination of dose in an organ, particularly if these radiations make up a large part of the emitted energy of the radionuclide. Since this is frequently the case, there must be a modification of the dose formulations to allow for the variability of biological effectiveness for different end points and different radiations. For this purpose, the concept of *dose equivalent* has been formulated (ICRU Report 25, International Commission on Radiation Units and Measurements, 1976). The dose equivalent, H, is defined by

$$H = DQN \tag{15-13}$$

where D is the absorbed dose, Q is the quality factor, and N is the product of any further factors recommended by scientific commissions or bodies as necessary modifying factors to be used in addition to describe the biological effectiveness of the dose. The quality factor, Q, is intended to be an adjustment for the influence of radiation quality on the specified biological outcome. There is extensive controversy in the scientific community about the exact value of Q to be used for any given value of LET. For alpha particles of a few MeV, Q has until recently been assigned a value of 10, but it has been proposed that biological evidence supports a value for these radiations approaching 20.

RADIONUCLIDES OF BIOLOGICAL IMPORTANCE

Tritium (^{3}H)

Tritium is the mass 3 isotope of hydrogen. The principal source of tritium in the environment is from nuclear weapon testing in the atmosphere. It is also a by-product of the operation of nuclear power reactors.

Exposure of individuals to elemental tritium is very unlikely, given the reactivity of this element. Exposure to water containing tritium as a partial replacement for the other isotopes of hydrogen is a more likely occurrence. When such exposure occurs, either by ingestion or inhalation, rapid equilibration with body water takes place. Generally, it has been shown that tritiated body water is a good label for total body water, and elimination of the radionuclide follows the kinetics of body water turnover. The biological half-time of 10 days is essentially the same for tritiated water as for all body water. The pool of dilution for tritiated water has been shown to be essentially the soft tissue mass of the body, taken as 63 kg for "standard man."

Because of the large volume of dilution of tritiated water in the body and the very low energy of the beta particles from tritium, the biological hazards from this radionuclide in water form are low. On the other hand, tritium taken into the body incorporated into organic compounds may give significant doses to sensitive cellular components. Feinendegen and Cronkite (1977) have suggested that the exposure limits for tritium incorporated in DNA precursors should be as low as one-tenth to one-fiftieth of that allowed for tritium in water.

Noble Gases: Krypton and Radon

Of the noble gases, only krypton and radon are of environmental significance. Xenon is used in nuclear medicine as a diagnostic agent. All the noble gases are characterized by a lack of chemical reactivity, and this limits the hazards associated with them. They all have the property of preferential solubility in body fat.

Krypton has a relatively long half-life, 3.9×10^3 days, and it is produced in large amounts by the nuclear power industry. Most of the krypton is released at the time of reprocessing the nuclear fuel when the cladding of the fuel rods is ruptured. Since the isotope is not incorporated into biological systems, it is associated with relatively limited risk. The dose to the bone marrow is less than one-thousandth of that received by the lungs when the gas is inhaled.

Radon is widely distributed in the environment and will be considered as a special case in Chapter 16. Radon itself is not different from krypton or the other noble gases, but because of its ubiquitous presence in the environment and because of the special conditions relative to its radioactive daughters, it is of unusual significance.

Alkali Metals: Sodium, Potassium, Rubidium, and Cesium

Lithium has no known isotopes of half-life longer than a second, therefore it can be neglected.

Isotopes of sodium are used in nuclear medicine diagnostic procedures, but these are of no environmental importance.

Potassium, on the other hand, has a naturally occurring isotope, ^{40}K, that is present in all living systems and that contributes a significant fraction of the annual background dose to human beings.

Cesium and, to a very much lesser extent, rubidium, are produced by nuclear fission. Cesium has widespread distribution in the environment as the result of atmospheric testing of nuclear weapons. The common chemical forms of cesium are rapidly and completely absorbed across the gastrointestinal tract, and the element tends to behave as if it were potassium. It is localized intracellularly, and since muscle represents the largest proportion of body intracellular space, muscle also represents the largest proportion of the cesium content of the body. The biological half-time varies rather widely among individuals between 50 and 150 days. Its excretion is characterized by a two-component model (Cryer and Baverstock, 1972). Since cesium distributes effectively in all soft tissue of the body, it is also present in meat animals, and the radionuclide may enter the human body via the food chain in a complicated way. A very complete and detailed modeling of the movement of cesium through the food chain to man has been developed by Whicker and Kirchner (1987).

Alkaline Earth Elements: Strontium, Barium, and Radium

The radionuclides of beryllium, magnesium, and calcium have been widely studied for the insights they give into metabolic processes, but they are of no health and safety interest.

The distribution of all the alkaline earth elements, with the exception of beryllium, is predominantly to bone, acting much like the native element calcium. Like calcium, strontium, barium, and radium are incorporated in newly formed or remodeling bone.

^{140}Ba has a physical half-life of 128 days, but it gives rise to its daughter ^{140}La. The pair, ^{140}Ba–^{140}La make up a significant fraction of fission product activity from nuclear weapon detonation or power reactor operation. These isotopes will be a major contributor of dose to bone for people nearby in the case of major accidental releases from a nuclear power accident; but because of the relatively short half-life, they constitute a limited environmental hazard except for the "close-in" case just described.

Strontium has been known for many years to be a significant contributor to bone dose and, concomitantly, bone marrow dose. It is a major constituent of fission products and was released in large amounts by the nuclear weapon atmospheric testing that occurred in the United States and abroad during the 1950s and 1960s. Two isotopes of strontium, ^{89}Sr and ^{90}Sr, are important contributors. ^{90}Sr, with a half-life of 28 years, is the more important hazardous radionuclide of these two. Soluble strontium compounds are rapidly absorbed across the gastrointestinal tract to the extent of about 50% of the administered dose. The movement of strontium has also been evaluated and modeled in the food chain by Whicker and Kirchner (1987).

Since strontium is such an efficient tracer of calcium metabolism, it appears in many parts of the food cycle where calcium-rich products are consumed by people. The most important of these are milk and milk products. During the years of intensive atmospheric weapons testing, children received very significant doses from consumed milk and milk products. Because of its importance, a great deal of research has been done on the metabolism and dosimetry of strontium in human beings and animals. Much of this research is incorporated in a very comprehensive model for the retention of the radionuclide in man. The model was developed by the Task Group on Alkaline Earth Metabolism in Adult Man (International Commission on Radiation Protection, 1973).

Myelogenous leukemia, lymphosarcoma, and reticulum cell sarcoma have been reported in dogs and swine given chronic exposure to ^{90}Sr administered by the oral route.

Radium.   The radionuclides of this element are all products of the decay of naturally occurring precursor elements in the earth's crust. The *uranium series*, which starts with ^{238}U, gives rise, after a number of successive decay products, to ^{226}Ra. The radium decays via ^{222}radon to radioactive isotopes of polonium, bismuth, and lead. These, in turn, ultimately decay to the stable isotope of lead, ^{206}Pb. Two other naturally

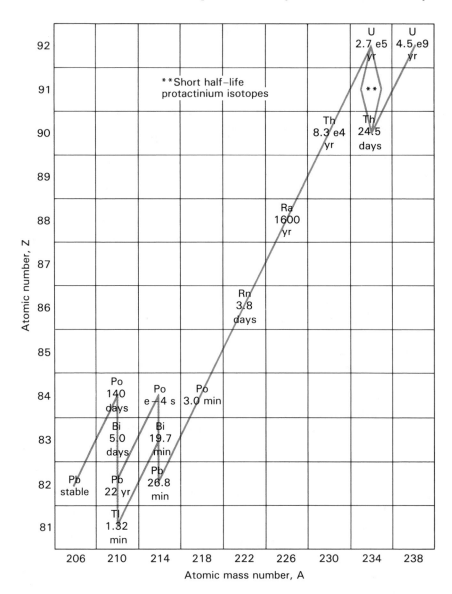

Figure 15.4 The naturally occurring decay chain commencing with ^{238}U and ending with stable ^{206}Pb. For the purposes of this chapter, the radioactive characteristics of radium are the most relevant. The decay chains are extremely complex in detail, and the figure is a simplification for purposes of illustration. The half-life of the nuclide is given in exponential notation, when necessary, in its block in the diagram.

occurring decay chains give rise to two other radioactive isotopes of radium. These are the *thorium* and *actinium* series. The former decay series commences with ^{232}Th and produces ^{224}Ra. The latter commences with ^{235}U and produces ^{223}Ra. Figure 15.4 is a simplified decay scheme for the uranium series. In nature the isotope of radium arising from ^{238}U, ^{226}Ra, is of overriding importance.

Comparison of the effects of radium and strontium.   If radium and strontium isotopes are injected intravenously into animals, the retention for the two nuclides is very dissimilar. At approximately 100 days, injected radium is retained to the extent of about 5% in man while injected strontium is retained to the extent of about 20% to 25%. Both elements are retained after deposition for a very long time in man. A very old study (Martland, 1931) showed that some 20 years after deposition in the human body, radium still remained to the extent of 0.6% of the original amount deposited.

The preponderance of cancers occurring in animals after experimental deposition of strontium radionuclides was cancers of myelogenous origin, with tumors or fractures in bone of lesser significance. After deposition of radium, the tumors were preponderantly of bony origin and the tumors were accompanied by frequent fractures. Lyon and others (1979) reported an increased incidence of childhood leukemias in association with exposure to fallout from continental nuclear weapons testing in the United States. If this finding is indeed substantiated by further examination of the data, which is now in progress by Lyon, then the attribution of cause should undoubtedly be to the dose to the bone marrow from deposited strontium.

Radium dial painters.   At the turn of the century and through the 1920s, young women were employed to apply luminous paint containing radium salts to the numbers and markers on watch faces. This work was accomplished with the use of very fine brushes, which were wet with the lips before they were dipped into the paint pot. As a result, the women ingested the radium-containing paint and the radionuclide was deposited in bone. Many years later these people developed extensive sarcomas of the bone and carcinomas of the sinus epithelium.

Radium isotopes were also used as a therapeutic measure in the past. The individuals receiving these treatments also developed bony tumors after a long latent period. Shortly after World War II, approximately 2000 young people were injected with a shorter half-life isotope of radium (^{224}Ra, half-life of 362 days), either for the treatment of tuberculosis or a serious disease of the spinal column, ankylosing spondylitis. At last report, follow-up on 900 of these patients disclosed 53 with bone sarcomata.

Individuals with radium body burdens have been followed for many years at the Massachusetts Institute of Technology and at the Argonne National Laboratory. These studies have given us extensive knowledge of the metabolism of radium as well as the dose required for induction of malignant neoplasms (Rowland and others, 1970). These data have been used to set limits for radium in the body and have been particularly useful to set limits for plutonium in the body, since this latter element behaves metabolically much like radium. A particularly intriguing approach has been used for estimating the body burden of plutonium likely to produce cancers in human beings. From experimental studies in dogs, the dose response for bony cancer induction is well known for both radium and plutonium. The body burden of radium required to produce bony tumors in human beings is also well known from studies on the dial painters and on individuals receiving radium for therapeutic purposes. The ratio of the mean effective doses for tumor induction in dogs for plutonium and radium is computed, and this ratio is multiplied by the radium body burden required to produce human tumors. The result is an estimate of the body burden of plutonium required to produce tumors in human beings.

Halogen Elements: Iodine

Radioisotopes of fluorine, chlorine, and bromine have been widely used in nuclear medical diagnosis and therapy, but the element of overriding importance in the halogen group is iodine. All the radionuclides of iodine are produced in the fission of uranium or plutonium. Of these, ^{131}I is the most important. In nuclear accidents, iodine is released as a gas and it quickly spreads over a large area. It has a short physical half-life of 8.04 days, but it moves rapidly through the food chain through the route of forage for cows and milk to the human being (Whicker and Kirchner, 1987). Because of its short half-life, protective measures can include storage of food supplies or simply refraining from eating the supplies for a few weeks at most.

Iodine has a complicated metabolic path in the body (see Figure 15.3), but ultimately the isotope is mostly concentrated in a single organ, the thyroid gland. Iodine is soluble in water and it can be quickly absorbed via all routes of entry. The major route of entry will be, however, through ingestion of food supplies contaminated with [131]I. Once incorporated in the thyroid gland, iodine is only slowly excreted, and for the short-lived isotopes of iodine most of the removal will be through radioactive decay.

The most recent event at which large releases of radioisotopes of iodine occurred was the nuclear disaster at Chernobyl in the Soviet Ukraine (Goldman, Catlin, and Anspaugh, 1988). Significant concentrations of ^{131}I were seen in most of western Europe, including the Scandinavian countries. Prompt action was taken by concerned governments to boycott the utilization of contaminated food supplies. A smaller but still very significant release of ^{131}I also occurred in western England in the 1950s from the Windscale nuclear processing plant. In this case, intake of the isotope was controlled by banning the drinking of milk from cows foraging in contaminated areas.

An additional prophylactic measure that reduces the incorporation of iodine into the thyroid gland is the oral administration of iodide ion, usually as its potassium salt. The high levels of blood iodide resulting from the administration of this salt effectively compete with the radionuclide for incorporation into the thyroid gland, and the end result is a sharply reduced level of incorporation of the radionuclide into the thyroid hormones.

Transuranic Elements: Uranium and Plutonium

Uranium. This naturally occurring element is the principal precursor for other naturally occuring radionuclides in the earth's crust. The nuclide of greatest importance is ^{238}U. Also to be found in naturally occurring uranium is the radionuclide ^{235}U to the extent of about 0.7% in the natural mixture of uranium isotopes. The latter isotope is extremely important, since it is the isotope of uranium that most efficiently undergoes fission when bombarded with thermal neutrons. For this reason, ^{235}U is the principally used fuel in nuclear reactors and uranium-based fission-type nuclear weapons.

The principal isotope ^{238}U has a half-life of 4.5×10^9 years (see Figure 15.4), which is long even in terms of geological time. As a result, the number of decays seen from a small sample of uranium will be quite small, and it has been shown repeatedly that the principal hazard from the incorporation of uranium in the body is a chemical one. Uranium is an effective nephrotoxic agent, producing severe kidney damage.

Only ^{235}U can be of any significance among the other isotopes of uranium because it is prepared in enriched form for power production or use in nuclear weapons. Again, its half-life of 7×10^8 years leads to the same conclusion as for ^{238}U as to the consequences of its biological incorporation.

Plutonium. Much has been written and said about this man-made radionuclide since its synthesis in the war years by the Manhattan Project. It has been repeatedly stated that this element is "the most toxic substance known to mankind." The basis of such a statement is hard to understand.

There are two important isotopes of plutonium, ^{238}Pu, with a half-life of 86.4 years, and ^{239}Pu, with a half-life of 24,890 years. Plutonium is nearly entirely a manufactured element, but recently naturally occurring traces of the element have been identified. Its commercial artificial production is in nuclear reactors by the irradiation of uranium isotopes, and it is the principal nuclide used in modern nuclear weapons. The isotope ^{238}Pu, because of its shorter half-life and consequently higher specific activity, is used as a source of power for space vehicles obtained by thermoelectric conversion of the heat produced in radioactive decay.

The single largest available source of plutonium in the environment is that which was placed in the atmosphere by aboveground nuclear weapon testing. The unfissioned plutonium so injected returns to earth as fallout in a global distribution pattern. It is estimated that over 300 MCi of $^{239-240}$Pu was globally dispersed during the weapons testing era. ^{240}Pu is another isotope of plutonium, which need not be of serious concern since it is only present in trace amounts. Another 100 MCi of plutonium was locally deposited in the overseas and continental test areas and surrounding countryside. A careful, worldwide, ongoing study of human cadavers has shown that traces of plutonium can be found in nearly all the samples tested. The greatest concentrations of plutonium in human cadavers have been found in the lung, liver, and skeleton.

Plutonium is a member of the actinide series of elements and it is chemically similar, insofar as biological metabolism is concerned, to radium. Its oxides, the form usually encountered in the environment, are in general very insoluble. Because of the insoluble nature of the plutonium found in the environment, the most important route of entry for the radionuclide is by inhalation of aerosols bearing the element. These aerosols, if the particle sizes are correct, will penetrate to the deep parenchyma of the lung and settle out there. In time they are taken up by lung macrophages and may move to nearby lymph nodes. To the extent that the plutonium solubilizes in the lymph node location, the element will be transported and to a great extent deposited in skeletal bone, following the same pattern as radium.

Absorption of plutonium by ingestion is much less effective than by inhalation if the element is in the insoluble oxide form, and because of the high chemical reactivity of the metal, the oxide will be the most likely state in which it is found. The overall discrimination factor for plutonium for the soil–plant–animal–man chain is about 10^{-5}, assuring limited incorporation in man from environmental sources through ingestion.

From animal experiments, it would be expected that appropriately incorporated plutonium (that is, plutonium in the organ of interest) will produce malignant neoplasms in bone, liver, lungs, and lymph nodes. A relatively large cohort of nuclear industry workers who have had significant exposure to plutonium during their working life has been followed for many years to observe plutonium-related pathology. To date, no plutonium-related disease has been observed in these individuals; but because of the long latent period for cancers of the lung in particular, it is

too early to draw sweeping conclusions about the radiotoxicity of plutonium for man.

Much of the public concern about plutonium stems from awareness
of its very long half-life. In fact, the half-life is of very little relevance
other than determining the rate of nuclear disintegrations from a given
mass of the nuclide. For biological purposes the half-life of radium at
1600 years is not different from that of plutonium at 25,000 years, since
both of these periods are very long compared to the human life-span.

Once deposited in the body, and particularly if deposited in the
skeleton, we would predict that plutonium would be about as effective as
a carcinogenic agent as radium, possibly somewhat more so; but up to
now, no data on human beings are available to confirm this observation.
Inhaled, insoluble plutonium in the lungs will presumably also be an
effective carcinogenic agent for the production of lung tumors.

The Radon Problem

Radon has become a focus of widespread public attention in the last
few years. The reader is referred to Chapter 16 for a description of the
biological hazards associated with exposure to radon gas and its daughter
products.

REFERENCES

CRYER, M. A., and BAVERSTOCK, K. F. (1972) Biological half-life of ^{137}Cs in man.
   Health Phys. **23,** 394–395.

FEINENDEGEN, L. E., and CRONKITE, E. F. (1977) Effects of microdistribution of
   radionuclides on recommended limits in radiation protection, a model. *Curr.
   Top. Radiation Res. Quart.* **12,** 83–99.

GALT, J. M., and TOTHILL, P. (1973) The fate and dosimetry of two lung scanning
   agents, ^{131}IMAA and ^{99m}Tc ferrous hydroxide. *British J. Radiol.* **46,** 272–279.

GOLDMAN, M., CATLIN, R. J., and ANSPAUGH, L. (1988) Health and environmental
   consequences of the Chernobyl nuclear power plant accident. In *Population
   Exposure to the Nuclear Fuel Cycle.* E. L. Alpen, R. O. Chester, and D. R. Fisher,
   eds. Gordon and Breach, New York, pp. 335–342.

INTERNATIONAL COMMISSION ON RADIATION UNITS AND MEASUREMENTS (1976) Report 25, *Conceptual Basis for the Determination of Dose Equivalent.* ICRU,
   Washington, D.C.

—— (1979) Report 32, *Methods of Assessment of Absorbed Dose in Clinical Use
   of Radionuclides.* ICRU, Washington, D.C.

INTERNATIONAL COMMISSION ON RADIOLOGICAL PROTECTION (1973) Publication 20,
   Task Group Report on Alkaline Earth Metabolism in Adult Man. Pergamon
   Press, Elmsford, N.Y.

—— (1978) Publication 30, Part 1. *Limits for Intakes of Radionuclides by Workers*. Pergamon Press, Elmsford, N.Y.

LYON, J. L., and others (1979) Childhood leukemias associated with fallout from nuclear test. *N. Eng. J. Med.* **300**, 397–402.

MARTLAND, H.S. (1931) The occurrence of malignancy in radioactive persons. *Am. J. Cancer* **15**, 2435–2516.

ROWLAND, R. E., and others (1970) *Some dose–response relationships for tumor incidence in radium patients*. Argonne National Laboratory Report, ANL-7760, Part II.

WHICKER, F. W., and KIRCHNER, T. (1987) Pathway: a dynamic foodchain model to predict radionuclide ingestion after fallout deposition. *Health Physics* **19**, 493–499.

SUGGESTED ADDITIONAL READING

NATIONAL COUNCIL ON RADIATION PROTECTION AND MEASUREMENTS (1985) *General Concepts for the Dosimetry of Internally Deposited Radionuclides*. NCRP, Bethesda, Md.

NORWOOD, W. D. (1975) *Health Protection of Radiation Workers*. Charles C Thomas, Springfield, Ill.

THOMPSON, R. C. (1960) Vertebrate radiobiology: metabolism of internal emitters. *Ann. Rev. Nuclear Sci.* **10**, 531–559. (This quite old document is still extremely useful.)

PROBLEMS

1. By direct measurement of the activity of the thyroid gland in several patients who had received ^{131}I for diagnostic purposes, it was observed that the biological elimination half-life for the radionuclide was 5.22 days. What is the biological half-life and what is the value of the biological decay constant?

2. Compute for ^{3}H the mean energy per nuclear disintegration Δ. The SI unit for Δ is $kg\,Gy\,Bq^{-1}\,s^{-1}$, identical with $J\,Bq^{-1}\,s^{-1}$. The effective half-life for tritium in body water is 10.5 days. Assume the injection of 50 MBq of tritium labe' :d water. What is the average absorbed dose in water after the elapse of three half-times? Assume equilibrium conditions of energy deposition as defined in this chapter.

3. Compute the mean energy per nuclear disintegration Δ for ^{99m}Tc if 2.0×10^9 Bq of this isotope is injected into a patient. Assuming that one-half of the administered dose is taken up by the thyroid and that the removal kinetics seems to follow first-order exponential form with a biological half-time of 4 h, what is the dose to the thyroid? Assume for purposes of this problem that the specific absorbed fraction is 0.5 and the thyroid weighs 80 g. Express the dose in gray.

16

Radiation Exposure
from Natural Background
and Other Sources

INTRODUCTION

An adequate understanding of the relative risks from ionizing radiation is only achievable through a clear conceptualization of the sources of that radiation. What is controllable and what is not? There are two major sources of such exposure, natural background and the medical uses of ionizing radiation. The latter includes almost exclusively the diagnostic applications of ionizing radiation in medicine. These applications include the use of x-rays and radioisotopes in diagnostic imaging. Both x-rays and radioisotopes are also used for therapeutic purposes, but, even though the individual doses for these applications are high, the population exposed is limited. Since the population exposed is so small, the contribution to the national population dose is trivial. In addition to the two principal sources of environmental and medical radiation exposure, there are numerous technological applications and consumer use applications for radioactive materials and for ionizing radiation. Some of these are not even well known or understood by the general public. These consumer applications, in general, do not contribute significantly to population dose; but since, in some cases at least, they have been the object of worldwide attention, it is appropriate to record their contribution to the national population dose. An example of radiation exposure from such a consumer application that has received a great deal of attention is the emission of x-rays from television receivers.

Before starting an extended discussion of the sources of population exposure for nonradiation workers, it is necessary to remind the reader of the quantities and units that are unique to the radiation health protection profession. Chapter 1 contains specific definitions of some of these special units, but a brief reminder is in order. The radiobiological units of dose used throughout this text are the rad and the gray, and since both units are in transition, there has been an effort to use both. For health protection purposes, a biologically effective dose has been widely used. This unit has been named the *dose equivalent, H*, and it is defined in Chapter 15 and in equation 15-13, as well as in Chapter 1. The quality factor, Q, is a dimensionless multiplier that corrects for LET-dependent differences in the biological effectiveness of the radiation for which H is being calculated. It is a derivative of the RBE, but since the end point and the response level are not specified, it is not the same as the RBE. The quality factor is a consensus value, agreed on among the health professionals as being an adequate, conservative, and representative value for the effectiveness of the radiation being considered. The factor N is a correction factor for nonuniformities in dose distribution, as well as any other multiplicative factors the health protection professionals deem necessary, and N is generally important only for internally deposited sources of radiation. If the field of radiation is mixed, appropriate corrections are necessary for each type of radiation, and the dose equivalent is the sum of each corrected contributor to the dose. The dose equivalent H is expressed mathematically in equation 16-1, where it is shown as the summation for all radiation types, from Rx_1 to Rx_n, of the dose, corrected by the quality factor Q_i and any other necessary factor N_i.

$$H = \sum_{Rx_1}^{Rx_n} D_i Q_i \ (N \text{ for other factors})_i \qquad (16\text{-}1)$$

The unit for H depends on the unit in which D is expressed. The name of the unit for H when D is expressed in rad is *rem*. The name of the unit for H when D is expressed in gray is *sievert*. Because the doses are generally small, the units millirem and milli- or microsievert are commonly used. A modified approach to H called the *effective dose equivalent, H_E*, will be used in a later section of this chapter.

The doses generally received from the two sources mentioned, natural background and medical applications, are highly dependent on location as well as socioeconomic condition. The general range of doses from natural radiation sources in the United States is about 70 to 110 mrem/yr (0.7 to 1.10 mSv/yr), and the general range for medically related exposures is about 100 mrem/yr (1.0 mSv/yr). Another important issue of social concern is that within recent years there has been a sharp increase in public awareness of the importance of the contribution of naturally occurring radon to exposure. Much of the concern about the importance of natural radon as a contributor to natural background exposure has arisen because of the newly agreed on quality factor, Q.

In the following sections, in addition to the discussion of natural background and medical applications dosage, there will be some discussion of the role of consumer products in contributing to population dose, as well as an examination of the technological enhancement of the natural contribution to background exposure. What is meant by technological enhancement as a contributor to background exposure are those activities of people that cause an increase in the population dose from natural sources. A typical example of technological enhancement would be the increased contribution to population dose from mining and milling tailings resulting from the extraction of uranium from the earth. Increased exposure due to technological enhancement and from consumer products contributes less than 10% to the total nonoccupational exposure of the American population, but both are the result of human activity and should be well understood and controlled.

EXPOSURES TO NATURAL BACKGROUND RADIATION AND RADIOACTIVITY

Natural radiation and radioactivity in the environment, along with diagnostic medical exposure, make up the very largest part of the accumulated annual dose to human beings who are not occupationally exposed to ionizing radiation from other sources during their daily work activity. Often comparisons are made between the annual radiation exposure arising from natural radiation and radioactivity and the exposure for people from all technology-generated sources. The purposes of these comparisons are often to place in perspective the magnitude of potential exposure from technology-generated sources. Many reports on exposure to natural radiation are available. Among these, the most comprehensive are the United Nations Scientific Committee on the Effects of Atomic Radiation (UNSCEAR) reports of 1966, 1972, 1977, and 1982, and the most recent is the National Council on Radiation Protection and Measurements Report No. 94 (1987b). Oakley (1972) examined doses from external radiation only.

In subsequent years it has become evident that the inhalation of the short-lived decay products of ^{222}radon is one of the more important sources of natural exposure, and there has been a significant change in the quality factor (Q) for alpha radiation that changes our perception of the importance of this source of radiation exposure. ^{222}Radon is the daughter product of the decay of ^{226}radium.

Figure 16.1 displays in bar chart form the contributions to external population dose for average North American residents. The data are plot-

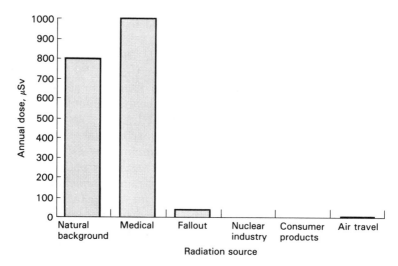

Figure 16.1 Approximate distribution of *external* average annual individual dose from natural and man-made nonoccupational sources. The first block is the average annual individual dose delivered externally from all naturally occurring sources. This value can vary widely depending on location and other factors. The medical dose is the approximate average for the United States and other countries with advanced economies. Exposures from the three sources to the right, nuclear industry, consumer products, and air travel, are all less than 10 μSv/yr. Doses from radioactive material deposited in the body are considered elsewhere in the text.

ted to indicate the relative importance of each of these sources of radiation exposure for individuals who are not occupationally engaged in radiation-related employment. Two important sources of such dose make the other components of exposure seem to be relatively very small: the dose from natural background and the dose from medical diagnostic uses of radiation. Furthermore, it is important to emphasize that the medical diagnostic exposure is very much a characteristic of the state of economic and technological advancement of the nation. For third world countries, this contribution can be as small as 10 mrem/yr (100 μSv/yr).

Naturally Occurring Radionuclides

The sources of natural background radiation can be classified in a two-fold fashion. There is the dose that comes from living in the environment of natural radioactivity at the earth's surface and the dose from direct cosmic radiation arriving at the earth's surface. Of these two, the contribution to dose from naturally occurring radionuclides is much the larger.

Primordial radionuclides.   The primordial radionuclides are principally, but not entirely, the isotopes of three radioactive series. The parent
nuclides for these series are ^{238}U, ^{232}Th, and ^{235}U. If these parent isotopes
are not chemically or physically separated from their daughter products,
the various members of the decay chain attain a state of radioactive
equilibrium in which the apparent rate of decay of each nuclide is the
same as that of the primordial parent. Each of these are examples of
secular equilibrium of radionuclides as described in Chapter 3. An example of the decay scheme for one of the naturally radioactive series, the
uranium series (decay of ^{238}U), is shown in Figure 14.4.

In addition to the primordial nuclides of the three series just described, there are several others that decay directly, in one step, to a
stable nuclide. Of these, the most important are ^{40}K and ^{87}Rb. The radionuclide ^{40}K occurs only to the extent of 0.0118% isotopic abundance
in natural potassium, but since the element is so ubiquitous in living
systems, it contributes as much as one-third of the external terrestrial
and internal dose from natural background. Although the isotopic abundance of ^{87}Rb is higher than that for ^{40}K, its contribution to dose is limited
by the relative scarcity of the element in the earth's crust.

Cosmogenic radionuclides.   The cosmogenic radionuclides are produced by nuclear transformations in elements of the earth's crust or the
atmosphere through interaction with cosmic ray particles reaching the
earth from galactic and extragalactic space. The cosmogenic radionuclides
are a relatively small contributor to dose. They are beta and gamma
emitters with a wide range of half-lives. For exposure of human beings
the four major cosmogenic nuclides are ^{14}C, tritium (^{3}H), ^{7}Be, and ^{22}Na.
Three of these cosmogenic nuclides, ^{14}C, ^{3}H, and ^{22}Na, are of particular
interest since they are important elements in human body composition.

Dose from External Sources

Outdoor exposure.   External exposure for human beings from natural radionuclides is from the radionuclides of the uranium–thorium series. The contribution from ^{40}K in the environment external to the body
is also important, but the ^{40}K in the composition of the human body is a
more significant contributor to dose from internal sources, which will be
considered later. The exposure from external sources is from the heavy
elements in the earth's soil, which are descendants of the uranium and
thorium precursors. The local concentration of these radionuclides and
their decay products varies widely depending on the geologic characteristics of the region. It is a common misconception that igneous rocks are
a more important contributor to this dose source; in fact, sedimentary-
type formations are not a great deal less important. The doses from these

sources external to the body are entirely from gamma rays emitted during the decay of the radionuclides. The quality factor for these radiations is 1, and the doses are expressed in the more familiar units of rad or gray. Any emissions of beta or alpha particles from the decay of natural radionuclides will not contribute to the dose from externally located nuclides because of their low power of penetration. There will be a minor contribution of beta rays to the doses received by the skin.

Beck (1975) has calculated the dose rates in *half-space* (defined as dose in air above an infinite plane surface) from each of the three important contributors to the external dose. The dose is calculated in air at 1 m above the surface for a unit concentration of the source nuclide. He gives the absorbed dose rate in air for ^{40}K as 13 µGy/yr/1 pCi (37 mBq). For ^{238}U and daughters in equilibrium, this value is 139 µGy/yr/1 pCi. For ^{232}Th and daughters in equilibrium, the dose rate is 216 µGy/yr/1 pCi. With the typical values for elemental soil composition, it can be calculated that the ^{232}Th series contributes about 150 to 250 µGy/yr (15 to 25 mrad/yr) to dose from external sources, and ^{40}K is responsible for about the same contribution. The ^{238}U series is a source for about one-half of that value (75 to 125 µGy/yr or 7.5 to 12.5 mrad/yr).

A special circumstance exists for the daughters of the decay of ^{222}Ra. Radon gas that diffuses into the atmosphere decays in a nonequilibrium fashion, and this decay leads to external exposure from principally the lead and bismuth radionuclides that are produced from radon decay. For a typical time-averaged outdoor radon concentration of 7.5 Bq/m^3 (200 pCi/m^3), NCRP Report No. 97 (National Council on Radiation Protection and Measurements, 1988) estimates that the absorbed dose rate in air from these radon daughters would be 232 µGy/yr (2.3 mrad/yr).

The significant external exposure from the radionuclides in soil and from atmospheric radon daughters is from gamma radiation. As just mentioned alpha and beta emitting radionuclides do not contribute significantly to whole body external dose.

The dose to human beings from terrestrial radioactivity varies greatly as a function of the geological composition and history of the area. Typical terrestrial absorbed dose rates from naturally occurring radionuclides in the earth's crust in the United States are as follows: for Atlantic and gulf coastal plains, 150 to 350 µGy/yr (average 23 mrad/yr, 230 µGy/yr); middle America plains, 350 to 750 µGy/yr (average 46 mrad/yr, 460 µGy/yr); and for the Colorado plateau, 750 to 1400 µGy/yr (average 90 mrad/yr, 900 µGy/yr).

Indoor exposure from external sources. The dose from the gamma rays from crustal radionuclides to individuals who are indoors is modified by the materials of construction and by the position of the individual within the structure. Wood, plastic, metal, and glass used in construction contribute very little to modification of exposure through shielding of the external dose, while brick, concrete, masonry, and stone building materials offer significant shielding. UNSCEAR (1982) has used a factor of 1.2 for an indoor/outdoor dose ratio (that is, higher indoors than outdoors), but these data are based mostly on European-style construction methods. The UNSCEAR values presume that the building materials themselves, which contain some amount of crustal radionuclides, contribute more to dose than they remove by shielding.

A summary of North American measurements is shown in Table 16.1 (after data from NCRP Report No. 94, 1987b). These data are presented to make several points. The first is that there exists a widespread belief that the interior of masonry buildings will have very much higher external dose rates than will wood structures, due to the contribution to dose from the crustal radionuclides in the building material. Clearly, this is only marginally so, and the variation would be much less than the regional variation in outdoor exposure rate. The shielding effect of the structure is, again, only marginal, and the 1.2 ratio for wood and masonry structures used by UNSCEAR seems to be excessive. All in all, the indoor to outdoor variation is a trivial concern, even though the data in the table indicate that there is a real but relatively small contribution to dose from radionuclides incorporated into brick and concrete building materials.

Finally, for external exposure from naturally radioactive materials, there is only a small contribution from other than the uranium–thorium series, although ^{40}K in the soil contributes a measurable amount.

TABLE 16.1   Comparison of Indoor and Outdoor Dose Rates in Air as Determined by the Construction Material of the Residence

Reference	Outer Wall Construction	Indoor Dose Rate (% of Outdoor)
Solon and others (1960)	Frame, brick, stone apartments and houses	80 to 100
Lowder and Condon (1965)	Mostly wood frame	70
Yeates, Goldin, and Moeller (1970)	Frame structures	82
	Brick	96
	Steel and concrete	97–106
Lindeken, Jones, and McMillen (1971)	Mostly wood frame	75

DOSE FROM INHALED RADIONUCLIDES

The importance of the short-lived daughters of radon as a source of exposure of the human lung has only been given the significance and attention it deserves in the last decade or so. An earlier NCRP report on background radiation (1975) considered only the contribution from outdoor radon to the dose from external sources. That dose was considered in the compilation of the preceding section. Recently, the contribution to potential lung exposure from the inhalation of radon accumulating indoors has been carefully studied. The report of Nero and Nazaroff (1984) is a landmark reference for this subject. Two NCRP Reports, No. 78 (1984) and No. 97 (1988), explore in depth the issues related to the measurement of radon and its daughters and the dose from this source. In short, the indoor concentration of radon depends on the rate of effusion of radon from the soil as a gaseous decay product of radium, followed by its diffusion from the point of origin into the interior of the house. The pathways are complex and depend on the tightness of the house structure and the air turnover rate in the structure. It has been shown, again contrary to common belief, that the material of which the house is constructed is not an important factor. Release rates of radon from structural materials play only a minimal role in the final dose rate in the structure (NCRP, Report No. 78, 1984).

Radon 222 as a Source of Internal Exposure

The concentration of ^{222}Rn has been measured in many indoor locations in the United States and Canada, and it has been shown to vary widely, depending to a great extent on the exhalation rate of radon from the surrounding earth surfaces. The outdoor concentrations have been widely measured to be in the range of 0.1 to 0.5 pCi/ℓ, while the indoor concentrations have been measured and shown to vary equally widely, between, for example, 0.3 and 10 pCi/ℓ. The higher concentrations of radon seen indoors compared with outdoors are the result of accumulation of the gas in the enclosed spaces of the buildings. Radon is a noble gas that enters into no chemical reactions of importance. It is inhaled and exhaled with very little dose to the lung contributed by the gas itself. The significant dose is from the decay products of radon (see Figure 16.2). The exposure from these radon daughters is received from the alpha rays emitted on the decay of the daughters. Since the daughters are solid elements, not gases, as they are formed from the decay of radon they are adsorbed on aerosol particles, either in the lung or in the external environment. These particles, as they are respired, are deposited on the surfaces of the trachiobronchial tree, where they undergo alpha decay. These short-lived daughters are ^{218}Po (RaA), ^{214}Pb (RaB), ^{214}Bi (RaC),

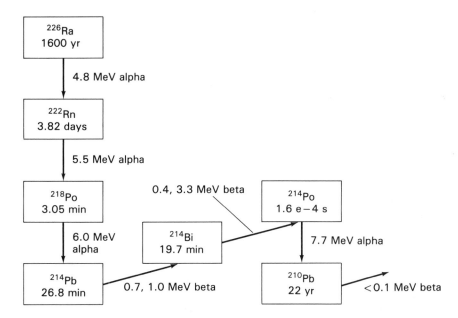

Figure 16.2 Decay scheme for ^{226}Ra giving rise to ^{222}Rn. The energy and type
of radioactive transition are shown alongside the arrows. Note that there are
two important alpha decays occurring after the transition from ^{222}Rn to ^{218}Po.
These are the transitions from ^{218}Po to ^{214}Pb and from ^{214}Po to ^{210}Pb. The
transition from ^{210}Pb gives rise to two additional radionuclides, ^{210}Bi to ^{210}Po,
and ultimately to stable ^{206}Pb.

and ^{214}Po (RaC′). The parenthetical labels are the archaic names for these
radionuclides. They are no longer appropriate, but they are given here
because the radionuclides are still often referred to by these names. Of
the daughter nuclides listed, ^{214}Po and ^{218}Po are the most significant
because of their half-lives and their decay schemes.

Since these short-lived products are rarely in radioactive equilibrium
because of the dynamic processes associated with their formation from a
short-lived gaseous predecessor and their adsorption on airborne aerosols,
the usual units of radioactivity are not appropriate; thus it has become
the practice to talk of exposure to mixtures of these nuclides in terms of
a *working level* (WL) and to state cumulative exposures in *working level
months* (WLM). A working level is defined as any combination of short-
lived daughters that results in the emission of 1.3×10^5 MeV of potential
alpha energy. The readers are referred to NCRP Report 78 (1984) for
extensive detail. That reference gives the conversion factors given in Table
16.2 for dose to bronchial epithelium from working level months. Note
that even though the dose is predominantly from alpha particles of very
high LET the dose is given in rad (gray), not rem or sievert. The discussion
of Q will be taken up shortly. The differences among men, women

TABLE 16.2   Conversion Factors from Working Level
Months to Dose for Various Exposure
Conditions

Underground miners	0.5 rad (0.5 cGy)/WLM
Adult men	0.7 rad (0.7 cGy)/WLM
Adult women	0.6 rad (0.6 cGy)/WLM
Children	1.2 rad (1.2 cGy)/WLM
Infants	0.6 rad (0.6 cGy)/WLM

and children for the conversion factor from WLM to dose are due to the differences in respiratory physiology of each. The principal factor is the respiratory volume per unit time compared to the overall lung volume.

George and Breslin (1980) in a careful analysis of the New York–New Jersey area arrived at an average outdoor concentration of ^{222}Rn of 180 pCi/m^3 and an average indoor concentration of 830 pCi/m^3. NCRP report 78 (1984) converts these values to average annual dose to the lungs, *unadjusted for quality factor*, as follows:

Adult male	190 mrad (1900 μGy)/yr
Adult female	170 mrad (1700 μGy)/yr
Child	300 mrad (3000 μGy)/yr

These authors have not converted the dose to either mrem or mSv per year, since the choice of the quality factor, Q, at that time would have been uncertain because scientific opinion was divided between the quality factor of 10, which had been in use for some years, and a newly proposed quality factor of 20. Scientific evidence in support of the increased quality factor has been accumulating in the meantime, and the factor of 20 has now been widely adopted. Using the new value of 20 for Q for alpha radiation and other high LET radiation, the adult male dose would be 3.8 rem/yr (38 mSv/yr), while the child would receive a dose of 6 rem/yr (60 mSv/yr).

Clearly, these environmental exposures overshadow all other single sources of natural radiation exposure and even exceed by a wide margin the total nonoccupational dose from all sources. For comparison, the annual allowed maximum dose for occupational workers in the radiation industries is 5 rem (50 mSv)/yr. The dose from radon daughters is roughly equal to this occupational limit, which is only rarely reached in occupational practice.

Thorium Series and Its Decay Products

Under normal conditions, only ^{222}Rn and its daughters need be seriously considered in the evaluation of dose from internally deposited nuclides of the uranium–thorium series; however, there are small contributions from some other nuclides. Thoron (^{220}Rn) has a decay chain similar to that of ^{222}Rn, but the half-life of ^{220}Rn is only 55 s; so, in spite of the roughly equal radioactivities of the chain precursors for the uranium and thorium series, the steady-state activity of ^{220}Rn will always be very much less than that for ^{222}Rn. Furthermore, because of its short half-life, losses during diffusion from the soil, where its radium precursor resides, are very great. The only decay product of thoron with a long enough half-life to be of biological consequence is ^{212}Pb (half-life 10.6 h). With its daughter products, ^{212}Bi and ^{212}Po, this nuclide contributes a dose to lung that is about one-sixth of that from the ^{222}Rn daughter products. An interesting observation can be made: the dose from ^{220}Rn daughter products, although low compared to the dose to the lung from ^{222}Rn, is still large compared to other contributors to natural background exposure.

Nonseries Radionuclides

In addition to the uranium–thorium series, which contributes the major portion of the exposure to man from natural radionuclides, there are several "nonseries" primordial radionuclides that exist in the earth's environment. Only two of these make any contribution of significance to the dose received by the human population. These are ^{40}K and ^{87}Rb. It has been suggested earlier that ^{40}K makes a contribution to the dose from external sources. Both nuclides enter into the natural metabolic pathways of the body, potassium as a natural constituent of intracellular electrolytes and rubidium because it acts chemically somewhat like potassium. Since rubidium is a relatively rare element in the earth's crust, it will make only a small contribution to dose after internal deposition. The doses to soft tissue from these radionuclides are as follows (NCRP Report No. 94, 1987b). The quality factor may be taken as 1.0 for these nuclides.

^{40}Potassium: 18 mrem/yr (180 μSv/yr) to gonads
6 mrem/yr (60 μSv/yr) to bone
27 mrem/yr (270 μSv/yr) to bone marrow

^{87}Rubidium: 1 mrem/yr (10 μSv/yr) to gonads
<1 mrem/yr (<10 μSv/yr) to bone
1 mrem/yr (10 μSv/yr) to bone marrow

EXPOSURE FROM COSMIC RAYS AND COSMOGENIC RADIONUCLIDES

Cosmic Rays

The charged particles, primarily protons, from extraterrestrial sources that are incident on the earth's surface have sufficiently high energies to generate a significant flux of secondary particles that penetrate to ground level. The primary galactic flux is attenuated in the first few hundred grams per square centimeter (g/cm^2) of atmosphere. However, because of a multiplicity of nuclear interactions occurring in this region of the atmosphere, there is a net buildup of total particle flux density in the outer first 100 g/cm^2 of the atmosphere. The first-generation products are predominantly neutrons, protons, and pions. The decay of the pions produces further electrons, photons, and muons. As further attenuation proceeds, the particle flux density decreases, and at atmospheric depths equal to greater than 800 g/cm^2 (about 6-km altitude) the highly penetrating muons and the associated electrons are the dominant components of the cosmic ray flux. These particles and to a much lesser extent the gamma rays produced from pion decay are the source of radiation exposure at sea level.

There are significant variations related to latitude and, to a much lesser degree, longitude in the total flux at the earth's surface and at altitude. A long-term average dose from cosmic radiation is shown in Figure 16.3. The data are given as dose in rad on the scale to the right. However, the total dose has a significant fraction contributed from neutrons. In tissue equivalent material at sea level, the neutron dose will be about 5% of the total. At an altitude of 3 km this contribution will be as much as 25% of the total dose from cosmic radiation. It should be noted that with the recent recommendations of regulatory bodies that the quality factor should be as great as 20, as is the case for radon daughter alpha radiation, dose equivalent rate will be increased at altitude. The total dose rates from all cosmic ray sources corrected for the quality factor are shown relative to the left ordinate in Figure 16.3. The doses calculated at sea level and at 3 km are, respectively, 40 mrem/yr (0.4 mSv/yr) and 120 mrem/yr (1.2 mSv/yr). These doses will be of particular significance for high-altitude flying.

Cosmogenic Radionuclides

Cosmic rays that reach the earth's surface produce a variety of spallation and neutron activation reactions in the constituents of the lower atmosphere and in the constituents of the earth. The entire geosphere contains these products in equilibrium concentrations. The most important of these from the point of view of contribution to dose are ^{14}C, ^{3}H,

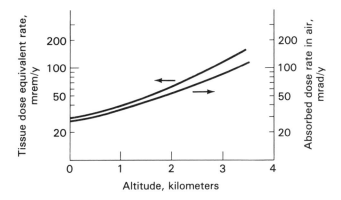

Figure 16.3 Long-term average outdoor absorbed dose in air (mrad/yr) from cosmic radiation as a function of altitude. The absorbed dose rate in air or tissue is shown in the lower curve, relating to the ordinate at right. The upper curve, relating to the ordinate at the left, is the total dose equivalent rate (charged particles plus neutrons) at 5-cm depth in a 30-cm-thick slab of tissue. The dose rates may be converted to mSv/yr and mGy/yr by dividing by 100. (From National Council for Radiation Protection and Measurements Report No. 94, 1987b, by permission)

^{22}Na, and ^{7}Be. Of these ^{14}C and ^{3}H (tritium) are the most important because they are relatively long lived isotopes of elements that are major body constituents. The concentrations of these two radionuclides in the geosphere have been significantly affected by nuclear weapons testing and, to a small extent in the case of ^{3}H, by the operation of the nuclear fuel cycle. Very little of the total natural background dose to humans is contributed by cosmogenic radionuclides. The *total* effective annual dose equivalent from ^{7}Be and ^{22}Na through inhalation and ingestion (drinking water, milk, leafy vegetables, and meat) is estimated to be, respectively, 3×10^{-3} mrem (30 nSv) and 4×10^{-4} mrem (4 nSv). These doses are inconsequential and can be disregarded. The annual dose equivalent for tritium, which is always in equilibrium with the total body water, is 1.2×10^{-3} mrem/yr (12 nSv/yr). For ^{14}C the annual dose equivalent is 1 mrem/yr (10 μSv/yr) to the soft tissues and approximately 3 mrem/yr (30 μSv/yr) for the bone marrow. The nuclides that are produced by cosmogenic action in the upper layers of the earth's crust are inconsequential, since they are produced in small amounts and generally have short half-lives.

SUMMARY OF EXPOSURE FROM NATURAL SOURCES

Figure 16.4 summarizes the data given in the preceding sections of this chapter. When the data are displayed in this fashion, the overwhelming importance of the contribution of radon daughters to the lung dose is

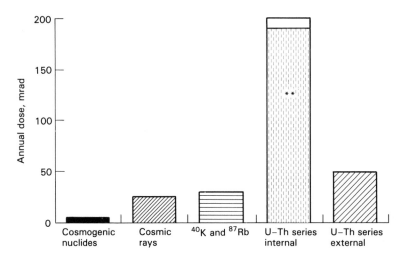

Figure 16.4 The contribution of each of the components of the natural background exposure is shown, with the dose expressed in millirad per year. The vertical column for the U–Th series internal dose is further subdivided to show the overwhelming contribution of the ^{222}Rn daughters, marked **, to the total internal dose from the U–Th series.

seen. These data are not corrected for the quality factor, which would make the importance of the radon daughter exposure even more evident.

Effective Dose Equivalent (H_E)

The *effective dose equivalent* was defined by the ICRP in 1977 as a means to deal with the risks to human beings exposed in highly non-uniform fashion to ionizing radiation (International Commission on Radiation Protection, Report 26, 1977). A particular example of the need for the concept of effective dose equivalent is seen in Figure 16.4. For computation of the effective dose equivalent, a weighting factor is assigned to each important organ by which the dose to that organ is adjusted. The development of the weighting factor for H_E was based on the assignment of a lifetime radiation-related risk for serious genetic defects and for mortality from cancer in various organs. The ICRP assigned a value for the lifetime risk per rem, per million exposed individuals, for each of a number of critical organs for the end points included in the risk estimate. These risk estimates are shown in Table 16.3. The effective dose equivalent is a means of recognizing that when the dose is given to a small portion of the body, that organ is the organ at risk for the future development of radiogenic cancer or, in the case of the gonads, a significant possibility of radiogenically induced mutations. The ICRP used defects in the first two generations only to establish the genetic risk estimate.

TABLE 16.3   Lifetime Radiation Risk Estimates from ICRP (1977) on Which the Weighting Factors for H_E, the Effective Dose Equivalent, Are Based

Organ (end point)	Risk/Rem/ Million	w_t
Gonads (genetic)	40	0.25
Marrow (leukemia)	20	0.12
Bone (cancer)	5	0.03
Lung (cancer)	20	0.12
Thyroid (cancer)	5	0.03
Breast (cancer)	25	0.15
All other organs (cancer)	50	0.30
Total risk	165	1.00

The development of the weighting factors for Table 16.3 (column 3) is straightforward. The weighting factor for the organ is developed as the ratio of the risk for that organ to the total risk. For example, the weighting factor for bone marrow, using leukemia as the risk, would be the quotient of 20/165 = 0.12.

The product desired from application of the dose equivalent model is a risk-weighted dose estimate for a mixture of types of radiation or for radiation of parts of the body. In practice, it is usual to convert doses to single organs from internally deposited radionuclides into effective dose equivalent with the appropriate weighting factor and to sum these with the whole body dose equivalent.

The effective dose equivalent, H_E, is a convenient and useful concept for estimation of a risk-based dose for complex radiation environments or for situations where only small portions of the body are irradiated. It clearly depends on two difficult premises. The first is that there should be well-known and well-understood risk coefficients, such as those shown in column 2 of Table 16.3. The second is that there must be an agreed on quality factor applied to the dose in rad or gray to produce the dose equivalent, H. Fortunately, it will always be possible to return to a set of dose calculations to recompute the effective dose equivalent whenever new agreed on quality factors and risk coefficients are developed.

Table 16.4 is developed from data in NCRP Report 94, in which data similar to that shown in Figure 16.4 are converted into appropriate effective dose equivalents. The weighting factors, w_t, used in Report 94 are slightly different from the ones given in Table 16.3, since NCRP has removed the thyroid risk from their calculations. In a sense this demonstrates the flexibility of the concept of the effective dose equivalent.

TABLE 16.4 Estimated Effective Dose Equivalent per Year for a Member of the Population in the United States and Canada from Natural Background Radiation Sources (NCRP Report 94, 1987b)

Source		Lung	Gonads	Bone	Marrow	Other	Total
		Total Effective Dose Equivalent Rate: mSv/yr (1 mSv = 100 mrem)					
	w_t	0.12	0.25	0.03	0.12	0.48	1.00
Cosmic radiation		0.03	0.07	0.008	0.03	0.13	0.27
Cosmogenic nuclides		0.001	0.002	—	0.004	0.003	0.01
Terrestrial external		0.03	0.07	0.008	0.03	0.14	0.28
Inhaled		2.00	—	—	—	—	2.00
In the body		0.04	0.09	0.03	0.06	0.17	0.40
Totals		2.1	0.23	0.05	0.12	0.44	**3.0**

Table 16.4 still shows the predominant role of radon and its daughters in contributions to the total effective dose equivalent annually. The total effective dose equivalent rate from all natural background sources is 3 mSv/yr (300 mrem/yr); of this, on a risk-based apportionment of dose, the lung receives 2 mSv/yr (200 mrem/yr), almost entirely from radon daughters. Little doubt remains that the most significant risk to human beings from natural background radiation sources is for the lung from radon daughters and the subsequent development of cancer.

EXPOSURE FROM MEDICAL APPLICATIONS

The use of radiation in medicine is clearly the largest single source of *external* exposure of the population in developed countries. There has been continuing effort in the United States over the last two decades to reduce the exposure from this source through improved practices and improved equipment, without the sacrifice of diagnostic efficiency and precision (Food and Drug Administration, 1984). Much of the data in this section on population exposure to medical radiation in the United States have been derived from this publication and its predecessors. The goal of this national dose reduction program has certainly been achieved; but at the same time that the average dose per procedure has been significantly reduced, there is no question that the frequency of application of diagnostic procedures in medicine and dentistry has increased. This is partly the result of new and better technologies, such as computerized tomography, but the frequency of standard examinations such as chest and extremity x-rays has also increased. It remains to be seen what effects on average medical radiation exposure result from some of the newer technologies that do not use ionizing radiation. Nuclear magnetic resonance imaging and ultrasound imaging are already important elements of the medical diagnostic array, and other technologies, such as microwave imaging, are in developmental stages.

Diagnostic X-ray Examinations

Because of the nature of disease distribution in the population, there is a substantial weighting of radiation exposure from diagnostic x-rays toward the older age groups, with over half of the individuals being subjected to x-ray examination being over 45 years and at least a quarter over 64. This result is not unexpected, since the older population will be seen more frequently by their physicians as the aging population group experiences a higher disease rate, as well as a spectrum of diseases in late life that require diagnostic x-ray examination. The skewing of the age distribution has a distinct effect on the computation of the effective dose equivalent.

Since a large proportion of the population is beyond the child-bearing age, there should be a sharp downward reduction in the weighting factor for the gonads and some changes in the risk estimates for cancer because of two factors. The incidence of cancer depends strongly on age and sex, and, to the extent that the radiation induced additional risk is relative and not absolute, the risk estimates would need adjustment. The second factor is the long latent time for the development of radiogenic cancer; for example, the latency for breast and lung cancer can be as long as 25 years. At present, data are not available to make the necessary adjustments to the weighting factors for the effective dose equivalent, and as this parameter is presented in the following sections the standard ICRP weighting factors are used, unless other approaches are indicated.

Estimates have been recently made for the genetically significant dose (dose to the gonads for the child-bearing age group) and the dose to the active bone marrow arising from the diagnostic use of x-rays for the U.S. population as of 1980 (FDA, 1984). These population estimates have been made for each decade since 1960 as a function of the U.S. Food and Drug Administration's Center for Medical Devices and Radiological Health. Depending on the source of the data for dose associated with each procedure, the mean adult active bone marrow dose in 1980 is between 114 mrad/yr (1.14 mGy/yr) and 75 mrad/yr (0.75 mGy/yr), and the genetically significant dose due to these same procedures is approximately 30 mrad/yr (0.3 mGy/yr). These values represent an increase of about 13% from the value calculated in 1970 and an increase of 38% since 1964. Since, at the same time, dose per procedure has generally been sharply reduced for most diagnostic techniques, all the increase in average individual dose from diagnostic medical usage of radiation must be the result of increased frequency.

Recalculation of the diagnostic x-ray doses from the usual values stated in rad or gray, as just given, to the effective dose equivalent, H_E, provides a slightly different view of the 1980 exposure patterns. The per capita annual effective dose equivalent for the 1980 values given is 36 mrem/yr (0.36 mSv/yr). As suggested, there should probably be further weighting of these annual doses to recognize the latency question and the difference in susceptibility to radiogenic cancer as a function of age at exposure and the sex of the exposed person. If such additional weighting factors for age and sex-specific adjustments are applied, the effective dose equivalent drops from the 36 mrem/yr (0.36 mSv/yr) to 23 mrem/yr (0.23 mSv/yr), an appreciable decrement from the approximately 100 mrem mentioned earlier as an unadjusted dose for annual exposure to diagnostic x-rays for medical purposes. Further adjustment of the effective dose equivalent for the genetically significant dose is not necessary for most procedures, since dose to the gonads is trivial in well-performed diagnostic examinations.

Social factors play a very significant role in the determination of the annual population dose from medical x-rays. The extreme point can be made that medicine is practiced in third world countries that does not approach the level of sophistication seen in America and Europe. For diagnostic x-ray procedures generally, poorer and older equipment will be used, and these machines will not incorporate all the new dose reduction technology of the most modern x-ray diagnostic suites. The individual will probably receive a higher dose per procedure, but since fewer individuals enter the medical service stream, the population average will be lower. From this extreme case of third world medicine, there is a broad spectrum of practices affecting the average annual dose to an individual. As a result, a great deal of variation in annual medical x-ray dose is seen throughout the world.

Therapeutic Radiology

High doses of ionizing radiation are routinely used in the control of cancer. Typically, the course of treatment for local control of cancer in a tumor may be 60 Gy (6000 rad). There is a limited population under treatment, and one that, in general, does not have an optimistic after-survival outlook. With these considerations in mind, applications of ionizing radiation for therapy make only a very small contribution to annual average per capita dose.

Nuclear Medicine Procedures

The specialty of nuclear medicine has developed the use of radioactive isotopes for the diagnosis of disease and dysfunction. The radionuclides used are numerous indeed and may be either manufactured or naturally available. The former predominate in modern nuclear medical practice. Nuclear medical procedures are not nearly as widely used as are diagnostic x-rays, but there has been a very significant growth in the volume of nuclear medicine procedures in the last decade or so. The principal growth has been in the use of radionuclides for cardiac function tests. The popularity of radioisotope brain scans has dropped, but the frequency of cardiac function tests with radionuclides has increased dramatically. The population undergoing nuclear medical examinations is even older and generally more ill than that receiving diagnostic x-rays.

There is just as much emphasis in the nuclear medical community as in diagnostic radiology on the principle of dose reduction without compromising diagnostic efficiency. Some important and useful steps have already been introduced. For example, replacement of ^{131}I by ^{123}I will bring about significant reduction in doses to the thyroid gland in future

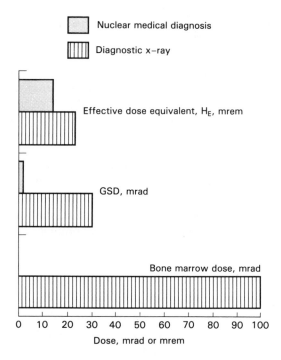

Figure 16.5 Comparison of the doses from diagnostic x-ray procedures and nuclear medicine procedures. The two upper bars compare the average annual effective dose equivalent (H_E) from nuclear medicine and diagnostic x-rays. The middle two bars compare the genetically significant dose (GSD) for the same sources. The bottom bar is the average annual bone marrow dose, in millirad, from diagnostic x-rays. The dose in mSv/yr and mGy/yr can be obtained by dividing by 100.

procedures. Much that was said for diagnostic radiology is also applicable for nuclear medicine. While dose reduction strategies continue to be stressed, the development of new technologies, as well as more widespread use of nuclear medical procedures, will increase the average annual dose from this source. The rapid growth of positron emission tomography, which, in general, leads to a very significant contribution to individual dose, will also increase the average annual per capita dose from nuclear medical procedures.

For nuclear medicine diagnostic procedures, with their highly localized dose and difficult dose distribution calculations, it is the practice to state the annual per capita organ dose only in terms of the effective dose equivalent and to make no attempt at the assessment of a dose in gray, or even dose equivalent. In the United States in 1982, the annual average effective dose equivalent from nuclear medicine was 14 mrem/yr (140 μSv/yr). This value is quite a bit higher than that found for the

United Kingdom, 1.7 mrem/yr (17 μSv/yr), and that for the Soviet Union, 2 to 4 mrem/yr (20 to 40 μSv/yr). Again, as for diagnostic x-ray procedures, social norms and practices play a very large role in determining the average annual per capita dose from the practice of nuclear medicine.

As is the case for the medical use of x-rays, there are also therapeutic applications for the use of radioisotopes; but again, as for x-rays, the number of individuals receiving nuclear medical procedures with therapeutic intent is very small. Even though the dose to the individual patients may be very high, the contribution to per capita average annual dose is insignificant.

Figure 16.5 gives a comparison of the average annual per capita dose in the United States from diagnostic x-ray procedures and from nuclear medicine procedures. At first examination, the diagnostic x-ray contribution seems to be very much the important contributor to annual dose from medical procedures; but if the x-ray dose is compared to the nuclear medical dose in terms of effective dose equivalent (H_E), it is clear that the x-ray dose is only about half again larger than the nuclear medicine dose. This is not surprising, since probably three to five times as many people experience an x-ray diagnostic procedure.

Because of the highly localized nature of the dose from radioisotopes, there is essentially no genetically significant dose from radioisotope procedures.

POPULATION EXPOSURE FROM CIVILIAN NUCLEAR POWER OPERATIONS

Public perception of the sources of public exposure resulting from the production of electric power by nuclear means tends to identify only the actual operations of the reactor and the storing of generated radioactive wastes as the principal contributors to population dose from nuclear power operations. It is important when discussing this issue that the totality of possible exposure sources be identified. There are two valuable and up-to-date sources for data on population exposure as the result of operations of the nuclear fuel cycle, NCRP Report No. 92 (1987a) and Alpen, Chester, and Fisher (1988).

Nuclear Fuel Cycle

The cycle of movement of the uranium fuel through the nuclear power production system serves to identify the sources of radiation exposure associated with each step. Some of these are significantly more important than others in their contribution to public radiation dose. The stages of

the cycle are discussed next. For more detail, the reader is referred to
NCRP Report No. 92 (1987a).

Mining.   Uranium mining in the United States is concentrated in
the Rocky Mountain states. Uranium extraction is about equally divided
between open pit and underground mines. Radon released into the atmosphere is the principal source of public exposure from this element of
the fuel cycle.

Extraction, milling, and refining.   The feedstock for extraction of uranium is ore containing, on average, 0.01% to 0.3% U_3O_8. The feedstock
is crushed, ground, and then leached by either an acid or alkaline process.
The material is then dried and packaged as a product called *yellowcake*.
The product is a semirefined uranium compound, U_3O_8 or $Na_2U_2O_7$, for
the acid and alkaline processes, respectively. The tailings from these
processes contain residual uranium and its radioactive daughter products.
This is the source of public exposure.

Producing uranium hexafluoride.   Uranium hexafluoride can be converted into a vapor form for feedstock to the diffusion enrichment process.
The yellowcake is converted to the hexafluoride by treatment with fluorine
and hydrofluoric acid. Small quantities of oxides and fluorides of uranium
are released to the environment by this process.

Enrichment.   Only ^{235}U is effectively fissionable by thermal neutrons. Natural uranium contains 0.72% of the 235 isotope. Enrichment
is carried out by a diffusion process with uranium hexafluoride vapor.
The final product is enriched to 2% to 4% in ^{235}U, which is satisfactory
as a fuel for light water reactors, but not of a high enough level of enrichment for nuclear weapon manufacture. Small quantities of hexafluoride escape from enrichment plants in air and waste streams.

Fuel fabrication.   Enriched uranium hexafluoride is converted to
uranium dioxide, UO_2, pressed and sintered to form ceramic pellets. The
pellets are assembled in long, small-diameter tubes that are sealed and
tested. The smaller tubes are assembled in fuel rod bundles appropriate
to the design of the reactor in which they will be used. Conversion to the
oxide and subsequent fabrication into fuel rod assemblies will be accompanied by the release of a small amount of airborne uranium, most of
which is trapped in efficient air handling systems in the fabrication fa-
cility. There is the possibility of very small releases as liquid waste.

Power generation.   The fuel rod assemblies are arranged in the core
of the operating power reactor, and during power production sustained

fission of ^{235}U occurs. Approximately one-third to one-fourth of the fuel assemblies are replaced yearly, depending on the level of power production. Within the plant, components become radioactive as the result of nuclear reactions with leakage neutrons from the core. As small leaks occur in the fuel rods, there is the possibility of the transfer of radionuclides to the secondary cooling water loop. There are also deliberate releases of radionuclides in small amounts during plant operation.

Fuel reprocessing.　Expended fuel assemblies removed from the reactor core still contain significant quantities of ^{235}U, which are economically valuable enough to be commercially recovered. As a national policy at the present time, no nuclear fuel is reprocessed in the United States. ^{239}Pu is also recovered as a by-product for use as a nuclear fuel.

At the time of reprocessing, the fuel cladding is removed and the accumulated fission products are released at the same time that the uranium and plutonium are put into solution. The material is dissolved in strong acid and subjected to chemical separation. The purified uranium and plutonium are recycled for further fuel use, and the fission products are concentrated for storage. Public radiation exposure is principally from the fission products. A modern reprocessing plant has no aqueous waste stream, and only gaseous releases are expected under normal plant operations.

Low-level waste disposal.　Any portion of the fuel cycle may give rise to low-level wastes contaminated with fission products, uranium compounds, or transuranic elements. Present practice is to bury these wastes in engineered, shallow-ground burial sites. Public exposure will be from gases released from the sites, such as tritium, ^{14}C, and ^{222}Rn.

Spent fuel storage.　If fuel is not reprocessed or if it is held for reprocessing, it will be stored in sites especially engineered for the necessary cooling of the fuel rods and containment of released radioactivity. Releases will be airborne volatile fission products that escape through cladding defects on the spent fuel rods in storage. Any releases into the cooling water are contained by the plant removal systems.

High-level waste disposal.　The residues from fuel reprocessing will be highly radioactive, and these wastes, as well as spent fuel that is not to be reprocessed, must be stored in safe, long-lived storage sites. Present U.S. policy is to dispose of these wastes in deep geologically secure sites after the waste is prepared in a form designed to exclude the release of radioactivity. Federal regulations have set a standard that public exposure from such storage procedures shall not exceed 0.25 mSv/yr (25 mrem/

yr). It appears possible to achieve this standard by presently available means.

Transportation. Fuel and wastes must be transported from fabrication, use, and disposal sites to other destinations. Public exposure would include low-level direct exposure to radiation from the containers and possible release during road or rail accidents.

Estimation of Population Dose

Population dose from the nuclear fuel cycle will be fundamentally different from that received from the natural background and medical exposures discussed earlier. The latter sources are generally applicable to all the population, and population averages are meaningful. Releases from the nuclear fuel cycle are regional in nature, and some population smaller than the national population will be potentially exposed and at risk. Furthermore, brief exposures to long-lived radionuclides may lead to corporeal deposition of the nuclides and concomitant continuing dose accumulation after the cessation of exposure. For these reasons, two dose concepts are applied in considering population exposure to the nuclear fuel cycle.

Committed dose equivalent. After deposition of a radionuclide in the body from a brief exposure, the nuclide continues to decay in the body and continues to contribute to the organ dose. The nuclide will not only undergo radioactive decay; it will also be excreted from the body by biological processes. The committed dose equivalent is simply the total dose deposited in the organ throughout the residence time of the radionuclide in the body.

Collective dose commitment. For stochastic radiogenic effects, such as cancer and genetic effects, there is a probability associated with any dose that the effect will occur, and this probability can be dealt with collectively for population groups. It is particularly useful to use the collective dose commitment for the case of groups of individuals in which the individuals have widely varying doses. The concept of collective dose was discussed in Chapter 12, but will be briefly reviewed here. If 1 million individuals received 1 cGy each, there would be a collective dose of $1 \times 10^6 \times 0.01$ Gy, or 1×10^4 person-gray. In the same population, if only 100,000 were irradiated to a dose of 10 cGy, the collective dose for the 1 million people would be $1 \times 10^5 \times 0.10$ Gy, or 1×10^4 person-gray. The important conclusion, based on the stochastic nature of the radiogenic effects considered and assuming a linear relationship with dose, is that the same number of radiogenic lesions would be discovered on average for each of these irradiated populations. The collective dose commitment is the same thinking extended to the commited dose, and it will almost always be presented in terms of the effective dose equivalent.

Maximally exposed individual. Although the probabilistic outcome just described is entirely correct, the individual outlook for the 10^5 individuals who received 10 cGy is very different from that for the 10^6 people who received 1 cGy. To deal with this issue, the concept of the maximally exposed individual has been developed. This person, sometimes called the "fence-post person," is declared to be the person who lived closest to the source of external radiation and had dietary and respiratory patterns that maximize the expected dose. Such a person has no reality, but the concept provides a conservative upper limit on potential exposure.

These concepts are discussed at length in NCRP Report No. 92 (1987a), and interested persons are urged to consult this source. It is particularly appropriate to point out that this report does not compute collective dose for the entire U.S. population, since this would make average individual doses trivial; rather it computes collective doses for the number of people in the affected region for each part of the fuel cycle.

Estimated Population Exposures

From NCRP Report No. 92, the data in Tables 16.5 and 16.6 have been extracted and summarized for the exposure of the "fence-post person," the maximally exposed individual, and for the collective dose equivalents to *regional* populations as the result of operations of existing U.S. nuclear fuel cycle facilities.

The highest doses are associated, not surprisingly, with the mining processes. The ^{222}Rn decay chain is almost exclusively responsible for this radiation exposure, and again, as for radon in the natural environment, this exposure is almost entirely from the short-lived daughters of this nuclide. Compared to the 30 mSv/yr (3000 mrem/yr) to basal cells of the bronchial epithelium from the natural background of the short-lived daughters of ^{222}Rn, the doses listed are relatively small.

The careful studies of NCRP Report No. 92 have shown that the airborne releases of radioactivity are overwhelmingly responsible for public exposure from nuclear-power-related activity. Doses from liquid effluents are very small. The high value associated with mining for exposure to external gamma radiation is important, but must be understood in context. At the place where measurements and calculations were made, people were not in residence, and actual exposures from this category are, on average, much below the value given. The value does highlight the national importance given recently to controlling the tailings disposal from uranium mines.

TABLE 16.5   Summary of Radiation Doses to the Maximally Exposed Individual from Existing Nuclear Fuel Cycle Facilities in the United States

Fuel Cycle Stage	Effective Dose Equivalent (mrem/yr), External Gamma	Effective Dose Equivalent (mrem/yr), from Effluents
Mining[a]	200–4000	20–60
Milling	7	0.4–260
Conversion	Negligible	Approx. 3
Enrichment	Negligible	<1
Fabrication	Negligible	<1
Reactor operations	10–15	<5
Low-level waste storage	Negligible	<1
Transportation	20	<1

[a] At locations of waste piles and storage areas of abandoned mines.

TABLE 16.6   Collective Effective Dose Equivalent to Regional Populations Due to Radioactive Effluents from Fuel Cycle Facilities

Fuel Cycle Stage	Collective Effective Dose Equivalent (person-rem/yr)
Mining	
Open pit (air)	1.0
Open pit (water)	0.2
Underground (air)	10
Underground (water)	21
Milling (air)	62
Conversion (mostly air)	
Wet process	0.4
Dry process	2.9
Enrichment (air)	0.002–0.4[a]
Fabrication (air)	0.01–0.7[a]
Reactor operations (air)	0.003–13[a]
(water)	0–40[a]
Low-level waste storage	<4

[a] The ranges for these processes arise as the result of different affected populations at different process sites.

Reprocessing has not been included in the tabulations since none is being undertaken in the United States. It is being done in other countries, however. UNSCEAR (1982) reports that reprocessing is a major contributor to the total collective dose from the fuel cycle. Again, airborne effluents are principally responsible for the dose. The main contributors to the dose are ^{3}H, ^{14}C, ^{85}Kr, and ^{129}I, all of which are long-lived and can diffuse to some distances from the source.

RADIATION EXPOSURE FROM CONSUMER PRODUCTS

There is a variety of consumer products and miscellaneous sources of ionizing radiation that results in exposure of the U.S. population. For the most part these contributions to total nonoccupational dose are trivial or nearly so. However, there is often significant public concern about some of them. One of these sources of public, nonoccupational radiation exposure, cigarette smoking, continues to be of importance. The readers are referred for extensive detail to NCRP Report No. 95 (1987c). Although some of the sources of radiation dose are naturally occurring, they are considered here either because of a special consumer use, such as tobacco, or because technological activities have increased the potential for exposure from the source. The latter has come to be called *technological enhancement*.

Natural Radioactive Products

Tobacco products. Both ^{210}Pb and ^{210}Po have been observed in tobacco and in cigarette smoke, and there is evidence of the presence of these radionuclides in the lungs of smokers (Little and others, 1965). In spite of the 20 odd years that have passed since these data were published and confirmed, there has been little or no public awareness of the importance of tobacco smoking as a contributor to the radiation dose to human lung.

The presence of ^{210}Pb and ^{210}Po in tobacco seems to be primarily the result of absorption from airborne particulates deposited on the foliage of the tobacco plant. Intake via the root system appears to be insignificant, so the soil concentrations of the precursor nuclides seem to be of little relevance. The inhaled volatile compounds of ^{210}Po are probably cleared rapidly from the lung, but the insoluble aerosol-borne ^{210}Pb may accumulate in the lung, and the following ingrowth of ^{210}Po from ^{210}Pb decay can lead to a very significant local dose.

NCRP Report No. 95 (1987c) has reviewed the calculations made by Little and others (1965) and other researchers on the dose to the lung for adult smokers in the United States. It was assumed that 35% of adult males and 30% of adult females smoked an average of 45 cigarettes per day. These figures are from the Public Health Service statistics on cigarette smoking in the United States. The dose to each of 50 to 55 million adult smokers in the United States is estimated to be about 0.8 to 1.0

rad/yr (8 to 10 mGy/yr). If the quality factor of 20 used earlier for alpha emitting isotopes is applied for these alpha emitting isotopes, the annual dose equivalent is 16 rem (160 mSv). The NCRP Report No. 95 (1987c) has concluded that it is not possible, with presently available data, to compute the effective dose equivalent, H_E. Clearly, for smokers, this may be a very significant portion of the total dose to the lung, and it competes with indoor radon as a source of lung exposure. It is clear from the input data used that the doses apply to relatively heavy smokers.

Other contributors.   Domestic water supplies, building materials, fertilizers, and combustible fuels may all contribute to the dose, both external and internal, from uranium–thorium chain nuclides, but these contributions are generally in the range of a few to tens of μSv/yr. For example, coal and natural gas combustion contribute on the order of 1 to 3 μSv/yr to the annual effective dose equivalent for adults in the United States. The annual dose equivalent from building materials to occupants of masonry buildings is on the order of 130 μSv/yr; however, only a fraction of the U.S. population will approach this upper limit.

Glass and ceramic industries have routinely used uranium compounds to produce fluorescence in glassware and for glazes on tiles. Uranium has another unusual application as an additive for artificial teeth for fluorescence and color improvement. Federal regulations allow up to 10% by weight of nonfood glass product to be composed of uranium and thorium. In large volume glass applications, such as glass brick, this limit is 0.05% by weight. All these applications in ceramic and glass manufacturing contribute very little to individual and collective dose.

Television, Video Display Terminals, and Other Electronic Products

These devices have been the object of public and federal regulatory concern for many years, but extensive measurements, as well as the federal regulatory programs, assure that dose contributions from these sources are negligible. The advent in the 1960s of new color receivers with significantly higher accelerating voltages in the color video display tube led to significantly higher dose rates at the front surface of the television picture tube. Dose rates were measured in excess of tens of mR/h at a few centimeters from the front surface of the picture tube. Subsequent regulation and emission standards have reduced that value to a very low number. New emission standards set in the late 1960s established a level of 0.5 mR/h at 5 cm from any portion of the set. These emission standards have been implemented by television equipment manufacturers to the extent that television receivers now emit virtually no ionizing radiation. Good shielding design, including increased wall thickness of the glass envelope of the picture tube, led to the decreased exposure potential. Insufficient data are available, but certainly the present designs of TV receivers are such that the average per capita population dose is very much less than 10 μSv/yr, probably closer to 2 to 3 μSv/yr at the surface of the picture tube.

Video display terminals (VDTs) are the visual display units for modern computer terminals, personal computers, word processors, and other information display devices. The display units have much in common with TV receiver visual display tubes. The ionizing radiation exposure level for VDTs will be at or below the level quoted for TV tubes.

Results of surveys of airport inspection systems, which are also federally regulated, demonstrate that the population dose from this source is less than 2 μrem/yr (0.003 μSv), and these devices can be neglected in a summary of total environmental exposure.

Other Consumer Products

A number of other sources of environmental exposure exist, but their contribution is usually very much less than a small fraction of a microsievert per year. They are smoke, gas, and aerosol detectors, transport of radioactive materials (mostly for medical applications), spark gap irradiators and electron tubes, thorium products, such as fluorescent lamp starters and gas lamp mantles, highway and road construction materials, and others of lesser importance.

Summary: Consumer Products

Among all the potential and actual sources for radiation exposure discussed in this section, only cigarette smoking appears to be of any consequence compared to natural environmental exposure. To what extent the radiation dose to the bronchoepithelial basal cells of the lung is responsible for the excess cancer risk in smokers is not known. Probably the radiogenic component of lung cancer in smokers remains small compared with the overall risk. It is certainly well known that cancers in other organs than lung result from chronic tobacco smoking. It is unlikely that radiation exposure can be responsible for these nonlung cancers.

REFERENCES

ALPEN, E. L., CHESTER, R. O., and FISHER, D. R., eds. (1988) *Population Exposure to the Nuclear Fuel Cycle*. Gordon and Breach, New York.

BECK, H. L. (1975) The physics of environmental radiation fields. In *The Natural Radiation Environment, II*, J. A. S. Adams, W. M. Lowder, and T. F. Gesell, eds. Technical Information Center, U.S. Department of Energy, Washington, D.C.

FOOD AND DRUG ADMINISTRATION (1984) *Nationwide Evaluation of X-Ray Trends (NEXT). Eight Years of Data, 1974–1981.* HHS Publication (FDA) 84–8229. U.S. Government Printing Office, Washington, D.C.

GEORGE, A. C., and BRESLIN, A. J. (1980) The distribution of ambient radon and radon daughters in residential buildings in the New Jersey–New York area. In *The Natural Radiation Environment, III*, T. F. Gesell and W. M. Lowder, eds. Technical Information Center, U.S. Department of Energy, Washington, D.C.

INTERNATIONAL COMMISSION ON RADIOLOGICAL PROTECTION (1977) *Recommendations of the International Commission on Radiological Protection.* Publication 26. Pergamon Press, Elmsford, N.Y.

LINDEKEN, C. L., JONES, D. E., and MCMILLEN, R. E. (1971) *Natural Terrestrial Background Variations between Residences.* USAEC Report UCRL-72964. University of California Livermore Radiation Laboratory, Livermore, Calif.

LITTLE, J. B., and others (1965) Distribution of polonium in pulmonary tissues of cigarette smokers. *New Eng. J. Med.* **273,** 1343.

LOWDER, W. M., and CONDON, W. J. (1965) Measurement of the exposure of human populations to environmental radiation. *Nature* **206,** 658.

NATIONAL COUNCIL ON RADIATION PROTECTION AND MEASUREMENTS (1984) *Evaluation of Occupational and Environmental Exposure to Radon and Radon Daughters in the United States.* Report number 78. NCRP, Bethesda, Md.

——— (1987a) *Public Radiation Exposure from Nuclear Power Generation in the United States.* Report number 92. NCRP, Bethesda, Md.

——— (1987b) *Exposure of the Population of the United States and Canada from Natural Background Radiation.* Report number 94. NCRP, Bethesda, Md.

——— (1987c) *Radiation Exposure of the U.S. Population from Consumer Products and Miscellaneous Sources.* Report number 95. NCRP, Bethesda, Md.

——— (1988) *Measurement of Radon and Radon Daughters in Air.* Report number 97. NCRP, Bethesda, Md.

NERO, A. V., and NAZAROFF, W. W. (1984) Characterizing the source of radon indoors. *Rad. Prot. Dosimetry* **7,** 23.

OAKLEY, D. T. (1972) *Natural Radiation Exposure in the United States.* US-EPA Report ORP/SID-72-1. United States Environmental Protection Agency, Washington, D.C.

SOLON, L. R., and others (1960) Investigation of natural environmental radiation. *Science* **131,** 903.

UNITED NATIONS SCIENTIFIC COMMITTEE ON THE EFFECTS OF ATOMIC RADIATION (1966, 1972, 1977, 1982) *Ionizing Radiation: Sources and Biological Effects.* United Nations, New York.

YEATES, D. B., GOLDIN, A. S., and MOELLER, D. W. (1970) *Radiation from Natural Sources in the Urban Environment.* Harvard School of Public Health, Department of Environmental Health Sciences Report No. HSPH-EHS-70-2. Harvard University, Cambridge, Mass.

SUGGESTED ADDITIONAL READING

ALPEN, E. L., CHESTER, R. O., and FISHER, D. R., eds. (1988) *Population Exposure to the Nuclear Fuel Cycle*. Gordon and Breach, New York.

NATIONAL COUNCIL ON RADIATION PROTECTION AND MEASUREMENTS (1984) *Evaluation of Occupational and Environmental Exposure to Radon and Radon Daughters in the United States*. Report number 78. NCRP, Bethesda, Md.

———— (1987a) *Public Radiation Exposure from Nuclear Power Generation in the United States*. Report number 92. NCRP, Bethesda, Md.

———— (1987b) *Exposure of the Population of the United States and Canada from Natural Background Radiation*. Report number 94. NCRP, Bethesda, Md.

Appendix:
Useful Physical Constants
and Conversion Factors

Physical Constants		
Speed of light in vacuo	c	2.998×10^8 m s^{-1}
Gravitational constant	G	6.67×10^{-11} n m^2 kg^{-2}
Universal gas constant	R	8.31 J mol^{-1} K^{-1}
Avogadro's number	N_0	6.02217×10^{-23} mol^{-1}
Boltzmann's constant	k	1.38×10^{23} molecule^{-1} K^{-1}
Planck's constant	h	6.625×10^{-34} J s or
		4.14×10^{-15} eV s
Elementary charge	e	1.602×10^{-19} C
Atomic mass unit	amu	1.660×10^{-27} kg
Electron rest mass	m_e	9.11×10^{-31} kg or
		0.000549 amu
Proton rest mass	m_p	1.6720×10^{-27} kg or
		1.007276 amu
Neutron rest mass	m_n	1.6744×10^{-27} kg or
		1.008665 amu

Conversion Factors		
Mass energy conversion	c^2	931 MeV amu^{-1} or
		8.99×10^{16} J kg^{-1}
Joule	J	10^7 erg
Electron volt	eV	1.60×10^{-19} J or
		1.60×10^{-12} erg

Author Index

The author index includes the names of multiple authors as they appear in the chapter References. Senior authorship is indicated by **bold** page numbers.

Subject Index

A

Absolute risk vs. relative risk, 276–78
Absorbed dose, 84–86
  kerma, relationship to, 84–86
Alpha particle emission, 41 (*see also* Radioactive decay)
Atomic structure, 22–25
  Auger electron, 46–47
  Bohr atom, 22–25
  hydrogen atom, 213–25
  quantized orbitals, 23
  Rutherford alpha scattering experiment, 22
  Rutherford atomic model, 22
Attenuation coefficients, 51–53
  atomic coefficient, 51–52
  electronic coefficient, 51–52
  energy absorbed, 54
  energy transferred, 54
  interrelationship of coefficients, 52–53
  mass coefficient, 51–52
  mixtures, 77–78

B

Background (*see* Natural background, radiation dose from)
Beta (negative electron) decay, 42–43 (*see also* Radioactive decay)
Binding energy, 39
  mass decrement, 40–41
  mass defect, 39
Bragg peak of ionization, 303–5
  Bethe–Bloch equation, 304
  continuous slowing down approximation, 303
Bremsstrahlung, 53, 71
  fast electrons, 84
  Heitler formulation, 71
Bromodeoxyuridine sensitization, 185–86

C

Cell age and radiosensitivity, 157–61